UNITEXT – La Matematica per il 3+2

Volume 76

For further volumes:
http://www.springer.com/series/5418

Ernesto Salinelli · Franco Tomarelli

Discrete Dynamical Models

 Springer

Ernesto Salinelli
Università del Piemonte Orientale
Dipartimento di Studi per l'Economia e l'Impresa
Novara, Italy

Franco Tomarelli
Politecnico di Milano
Dipartimento di Matematica
Milano, Italy

UNITEXT – La Matematica per il 3+2
ISSN 2038-5722   ISSN 2038-5757 (electronic)

ISBN 978-3-319-02290-1   ISBN 978-3-319-02291-8 (eBook)
DOI 10.1007/978-3-319-02291-8
Springer Cham Heidelberg New York Dordrecht London

Library of Congress Control Number: 2013945789

Translated from the original Italian edition:
Ernesto Salinelli, Franco Tomarelli: Modelli Dinamici Discreti, 3a edizione
© Springer-Verlag Italia 2014

Cover Design: Beatrice B., Milano
Cover figure: "Tempo di fioritura", Gianni Dova, 1964.
Il Castello Arte Moderna e Contemporanea, Milano.
Typesetting with LaTeX: PTP-Berlin, Protago TeX-Production GmbH, Germany (www.ptp-berlin.eu)
Printing and Binding: GECA Industrie Grafiche, San Giuliano Milanese (MI), Italy

Springer is part of Springer Science+Business Media (www.springer.com)

Preface

Mathematical modeling plays a relevant role in many different fields: for instance meteorology, population dynamics, demography, control of industrial plants, financial markets, retirement funds. The few examples just mentioned show how, not only the expert in the field but also the common man, in spite of himself, has to take part in the debate concerning models and their implications, or even he is summoned to polls for the choice between alternative solutions that may achieve some reasonable control of the complex systems described by these models. So it is desirable that some basic knowledge of mathematical modeling and related analysis is spread among citizens of the "global village".

When models refer to time dependent phenomena, the modern terminology specifies the **mathematical model** with the notion of **dynamical system**.

The word **discrete**, which appears in the title, points out the class of models we deal with: starting from a finite number of snapshots, the model allows to compute recursively the state of the system at a discrete set of future times, e.g. the positive integer multiples of a fixed unit of time (the **time step**), without the ambition of knowing the phenomenon under study at every subsequent real value of time, as in the the case of **continuous-time models**.

Sometimes also the physical quantities under measure turn out to be discrete (in this cases they are called "quantized") due to experimental measurement limits or to the aim of reducing the volume of data to be stored or transmitted (signal compression). Since we focus our attention on models where the quantization has no relevant effects, we will neglect it.

In recent years a special attention has been paid to **nonlinear dynamical systems**. As a consequence, many new ideas and paradigms stemmed from this field of investigation and allowed a deeper understanding of many theoretical and applied problems. In order to understand the term **nonlinear**, it is useful recalling the mathematical meaning of **linear problem**: a problem is called linear if there is a direct proportionality between the initial conditions and the resulting effects; in formulae, summing two or more initial data leads to a response of the system which is equal to the sum of the related effects.

In general the theory provides a complete description (or at least a satisfactory numerical approximation) of solutions for linear problems, but the relevant problems in engineering, physics, chemistry, biology and finance are more adequately described by **nonlinear models**.

Unfortunately, much less is known about solutions of nonlinear problems. Often nonlinearity leads to qualitatively complex dynamics, instability and sensitive dependence on initial conditions, to such an extent that some extreme behavior is described by the term **chaos.**

This spellbinding and evocative but somehow misleading expression is used to express the possible extreme complexity in the whole picture of the solutions. This may happen even for the simplest nonlinear deterministic model, as the iteration of a quadratic polynomial. So we are lead to an apparent paradox: in such cases the sensitive dependence on initial conditions (whose knowledge is always affected by experimental error) leads to very weak reliability of any long-term anticipation based on deterministic nonlinear models. The phenomenon is well exemplified by weather forecast; nevertheless short-term forecasting has become more and more affordable now, as everybody knows.

These remarks do not call into question the importance and the effectiveness of nonlinear models analysis, but they simply advise not to extrapolate in the long term the solutions computed by means of nonlinear models, because of their intrinsic instability.

The present volume aims to provide a self-contained introduction to discrete mathematical modeling together with a description of basic tools for the analysis of discrete dynamical systems.

The techniques for studying discrete dynamical models are scattered in the literature of many disciplines: mathematics, engineering, biology, demography and finance. Here, starting by examples and motivation and then facing the study of the models, we provide a unitary approach by merging the modeling viewpoint with the perspective of mathematical analysis, system theory, linear algebra, probability and numerical analysis.

Several qualitative techniques to deal with recursive phenomenon are shown and the notion of explicit solution is given (Chap. 1); the general solutions of multi-step linear difference equations are deduced (Chap. 2); global and local methods for the analysis of nonlinear systems are presented, by focussing upon the issue of stability (Chap. 3). The logistic dynamics is studied in detail (Chap. 4).

The more technical vector-valued case is studied in the last chapters, by restricting the analysis to linear problems.

The theory is introduced step-by-step together with recommended exercises of increasing difficulty. Additional summary exercises are grouped in specific sections. The worked solutions are available for most of them in a closing chapter. Hints and algorithms for numerical simulations are proposed.

We emphasize that discrete dynamical systems essentially consist in the iteration of maps, while computers are very efficient in the implementation of iterative computations. The reader is invited to do the exercises of the text, either by looking for a closed solution (if possible), or by finding qualitative properties of the solution via graphical method and exploitation of the theoretic results provided by the text, or by numerical simulations.

Presently, many software packages oriented to symbolic computation and computer graphics are easily available; this allows to produce on the computer screen those entangling and fascinating images which are called fractals. It can be achieved by simple iterations of polynomials, even without knowing the underlying technicalities of Geometric Measure Theory, the discipline which describes and categorizes these non-elementary geometrical objects.

At any rate some basic notions for the study of these objects are given in Chap. 4 and Appendix F: fractal dimension and fractal measure.

For every proposed exercise, the reader is asked to ponder not only on the formal algebraic properties of the wanted solution, but also on the modeled phenomenon and its meaning, on the **parameter domain** where the model is meaningful, on the **reliability** of model predictions and on the numerical computations starting from initial data which are unavoidably affected by **measurement errors**.

In this perspective Mathematics can play as a feedback loop between the study and the formulation of models. If an appropriate abstraction level and a rigorous formulation of the model are achieved for one single problem, then innovative ideas may develop and be applied to different problems, thanks to the comprehension of the underlying general paradigms in the dynamics of solutions. In short: understanding the whole by using the essential.

We thank Maurizio Grasselli, Stefano Mortola and Antonio Cianci for their useful comments and remarks on the text and Irene Sabadini for the careful editing of many figures. We wish to express our sincere acknowledgements to Francesca Bonadei for her technical support, to Alberto Perversi for his helpful graphic advises and to the students of the program Ingegneria Matematica for their enthusiasm and valued comments. We wish to thank Debora Sesana for her revising work on the figures. Last but not least, we would like to thank Simon Chiossi for his linguistic version of the translation of the Third Italian edition of this book.

Milano   Ernesto Salinelli
October 2013   Franco Tomarelli

Reader's Guide

This book consists of eight chapters, the first seven of which unfold theoretical aspects of the subject, show many examples of applications and offer a lot of exercises. The last chapter collects the worked solutions of most exercises proposed earlier. The text ends with a collection of appendices containing main prerequisites and several summarizing tables.

Chapters 1 and 2 are self-contained: the only prerequisite is the familiarity with elementary polynomial algebra.

Chapter 3 requires the knowledge of the notions of continuity and derivative for functions of one variable.

Chapter 4 relies on some basic topological notions.

Chapsters 5, 6 and 7 study vector-valued dynamical systems by deeply exploiting Linear Algebra techniques.

All prerequisites, even when elementary, are briefly recalled in the text and can be found in the appendices.

Proofs, always printed in a smaller typographical font, are useful for a full comprehension, but can be omitted at first reading.

Chapters 1, 2 and 3 may be taken as syllabus for an elementary teaching modulus (half-semester course): in this perspective, the Sects. 2.5, 2.6, 2.7, 2.8, 3.3, 3.4 and 3.9 can be omitted.

Chapters 4, 5, 6 and 7 may constitute another advanced teaching modulus.

All chapters, except the first one, are rather independent from each other, so to meet the reader's diversified interests. This allows for a big choice in which path to follow when reading, reflecting the many applications of the theory presented. A few possible "road maps" through the various chapters are shown below:

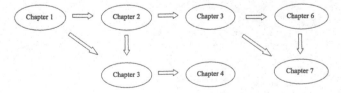

Contents

Preface v

Reader's Guide ix

Contents xi

Symbols and notations xv

1 Recursive phenomena and difference equations 1
    1.1 Definitions and notation 1
    1.2 Examples 3
    1.3 Graphical method 18
    1.4 Summary exercises 20

2 Linear difference equations 25
    2.1 One-step linear equations with constant coefficients 25
    2.2 Multi-step linear equations with constant coefficients 31
    2.3 Stability of the equilibria of multi-step linear equations with
         constant coefficients 40
    2.4 Finding particular solutions when the right-hand side is an
         elementary function 47
    2.5 The $\mathcal{Z}$-transform 50
    2.6 Linear equations with non-constant coefficients 61
    2.7 Examples of one-step nonlinear equations which can be
         turned into linear ones 65
    2.8 The Discrete Fourier Transform 71
    2.9 Fast Fourier Transform (FFT) Algorithm 76
    2.10 Summary exercises 81

3 Discrete dynamical systems:
   one-step scalar equations 85
    3.1 Preliminary definitions 85

3.2 Back to graphical analysis 93
3.3 Asymptotic analysis under monotonicity assumptions 95
3.4 Contraction mapping Theorem 99
3.5 The concept of stability 100
3.6 Stability conditions based on derivatives.................. 105
3.7 Fishing strategies....................................... 110
3.8 Qualitative analysis and stability of periodic orbits 114
3.9 Closed-form solutions of some nonlinear DDS 117
3.10 Summary exercises 122

4 **Complex behavior of nonlinear dynamical systems:**
   bifurcations and chaos 125
   4.1 Logistic growth dynamics............................... 125
   4.2 Sharkovskii's Theorem 129
   4.3 Bifurcations of a one-parameter family of DDS 134
   4.4 Chaos and fractal sets................................. 146
   4.5 Topological conjugacy of discrete dynamical systems 156
   4.6 Newton's method..................................... 160
   4.7 Discrete dynamical systems in the complex plane 163
   4.8 Summary exercises 177

5 **Discrete dynamical systems:**
   one-step vector equations 179
   5.1 Definitions and notation 179
   5.2 Applications to genetics 181
   5.3 Stability of linear vector discrete dynamical systems 191
   5.4 Strictly positive matrices and the Perron–Frobenius Theorem . 199
   5.5 Applications to demography 205
   5.6 Affine vector-valued equations.......................... 208
   5.7 Nonlinear vector discrete dynamical systems 212
   5.8 Numerical schemes for solving linear problems 215
   5.9 Summary exercises 225

6 **Markov chains** 227
   6.1 Examples, definitions and notations...................... 227
   6.2 Asymptotic analysis of models described by absorbing
        Markov chains 240
   6.3 Random walks, duels and tennis matches 244
   6.4 More on asymptotic analysis 250
   6.5 Summary exercises 254

7 **Positive matrices and graphs** 257
   7.1 Irreducible matrices................................... 257
   7.2 Graphs and matrices.................................. 266
   7.3 More on Markov Chains 273

7.4 Algorithm PageRank: why a good search engine seems to read the minds of those who questioned it 278
7.5 Summary exercises 283

8 **Solutions of the exercises** 285
8.1 Solutions of the exercises in Chapter 1 285
8.2 Solutions of the exercises in Chapter 2 292
8.3 Solutions of the exercises in Chapter 3 308
8.4 Solutions of the exercises in Chapter 4 323
8.5 Solutions of the exercises in Chapter 5 333
8.6 Solutions of the exercises in Chapter 6 336
8.7 Solutions of the exercises in Chapter 7 339

Appendix A. Sums and series 343

Appendix B. Complex numbers 347

Appendix C. Basic probability 351

Appendix D. Linear Algebra 353

Appendix E. Topology 363

Appendix F. Fractal dimension 365

Appendix G. Tables of $\mathcal{Z}$-transforms 371

Appendix H. Algorithms and hints for numerical experiments . 375

References 381

Index 383

Symbols and notations

DDS	discrete dynamical system;
$\{I, f\}$	DDS $X_{k+1} = f(X_k)$, $k \in \mathbb{N}$, associated to the map $f : I \mapsto I$, $I \subset \mathbb{R}$;
$\mathbb{N}$	set of natural numbers together with 0;
$\mathbb{Z}$	set of integer numbers;
$\mathbb{R}$	set of real numbers;
$\mathbb{Q}$	set of rational numbers;
$\mathbb{C}$	set of complex numbers;
$\mathbb{R}^n$	n–dimensional Euclidean space;
$\{\mathbf{e}_1, \mathbf{e}_2, \cdots, \mathbf{e}_n\}$	canonical basis of $\mathbb{R}^n$: $\mathbf{e}_j = (0, 0, \ldots, 1, \ldots, 0)$;
$\mathbf{x} = (x_1, \ldots, x_n)$	vector $\mathbf{x} \in \mathbb{R}^n$ with components $x_j \in \mathbb{R}$ with respect to the canonical basis of $\mathbb{R}^n$;
$\|\mathbf{x}\|$	Euclidean norm of $\mathbf{x} \in \mathbb{R}^n$;
$B_R(\mathbf{x})$	n dimensional open ball with radius R centered at $\mathbf{x}$;
∂E	boundary of the set $E \subset \mathbb{R}^n$;
$\overline{E}$	closure of the set $E \subset \mathbb{R}^n$;
ω_n	volume of the unit ball $B_1(\mathbf{0}) \subset \mathbb{R}^n$;
σ_{n-1}	surface area of the unit sphere $S^{n-1} = \partial B_1(\mathbf{0}) \subset \mathbb{R}^n$;
$X = \{X_k\}_{k \in \mathbb{N}}$	sequence of numbers X_k, or trajectory of a scalar DDS;
$\mathbf{X} = \{X_k\}_{k \in \mathbb{N}}$	sequence of vectors $\mathbf{X}_k = \{X_k^h\}_{h=1,\ldots,n}$, or trajectory of a vector-valued DDS;
$m_{i,j}$	$\{i, j\}$-entry of the matrix $\mathbb{M}$;
$\lambda_\mathbb{M}$	dominant eigenvalue (or Perron-Frobenius eigenvalue) of the matrix $\mathbb{M}$;
$\mathbf{V}^\mathbb{M}$	dominant eigenvector (or Perron-Frobenius eigenvector) of the matrix $\mathbb{M}$;

x^+	positive part of the real number x;
x^-	negative part of the real number x;
$o(1)$	infinitesimal;
$o(x^a)$	"little o" of x^a: infinitesimal term, faster than x^a as $x \to 0_+$; $a > 0$;
$O(|x|^a)$	"big O" of $|x|^a$: a term whose modulus is estimated by $C|x|^a$;
$\asymp$	of the same order;
$\sim$	asymptotic to;
$\operatorname{Re} z$	real part of z;
$\operatorname{Im} z$	imaginary part of z;
$|z|$	modulus of $z \in \mathbb{C}$: $|z| = \left((\operatorname{Re} z)^2 + (\operatorname{Im} z)^2\right)^{1/2}$;
$\arg(z)$	argument (or phase) of $z \in \mathbb{C} \setminus \{0\}$: oriented angle between $\mathbf{e}_1$ and z;
$\operatorname{res}(f(z), z_0)$	residue of the function f at the point $z = z_0 \in \mathbb{C}$;
$\operatorname{Ind}(\gamma, z_0)$	winding number of the curve γ with respect to the point $z_0 \in \mathbb{C}$;
$v: \Omega \mapsto E$	map v with domain $\Omega \subset \mathbb{R}^n$ and range in $E \subset \mathbb{R}^m$;
$v(x_+)$	right-limit at the point x of the map v depending on one real variable;
$v(x_-)$	left-limit at the point x of the map v, depending on one real variable;
DFT	Discrete Fourier Transform: $\mathbf{Y} = \operatorname{DFT} \mathbf{y} = n^{-1}\mathbb{F}\mathbf{y}$, $\mathbf{Y}, \mathbf{y} \in \mathbb{R}^n$, $\mathbb{F}_{hk} = \omega^{hk}$, $\omega = e^{2\pi i/n}$;
$(\operatorname{DFT})^{-1}$	inverse Discrete Fourier Transform: $\mathbf{y} = (\operatorname{DFT})^{-1}\mathbf{Y} = \mathbb{F}\mathbf{Y}$;
FFT	Fast Fourier Transform;
$\mathcal{Z}$	Zeta Transform: $x(z) = \mathcal{Z}\{X\}$; X scalar sequence.

1

Recursive phenomena and difference equations

In this first chapter we set the notations used in the text and present the main background motivations to study discrete mathematical models using examples. The systematic analysis of the latter is postponed to subsequent chapters.

The abstract definitions of the first paragraph provide a unified framework for the study of a wide class of problems, some examples of which are presented in the second paragraph.

1.1 Definitions and notation

In the sequel we will consider **sequences** of complex numbers. A sequence is a function $F : \mathbb{N} \to \mathbb{C}$, where [1] $\mathbb{N} = \{0, 1, 2, 3, \dots\}$.

The reader not familiar with complex numbers can, at a first reading, replace $\mathbb{C}$ with $\mathbb{R}$. For convenience the basic properties of complex numbers are recalled in Appendix B.

A generic sequence will be indicated with a capital letter, for example F, while its k-th term will be denoted with a capital letter with a subscript, F_k. In other words $F_k = F(k)$.

Following a common practice, we will identify a sequence F with the ordered set of values it takes: $F = \{F_k\}_{k \in \mathbb{N}}$ or, briefly, $F = \{F_k\}$.

Although the student interested in the description of variables in a succession of steps or temporal instants may find it restrictive to deal only with functions defined on the set of natural numbers, this is not the case. In fact, if $t_0 \in \mathbb{R}$ is any initial instant and $h > 0$ is an increment, the sequence $\{t_0, t_0 + h, t_0 + 2h, t_0 + 3h, \dots\}$ is mapped one-to-one to the sequence $\{0, 1, 2, 3, \dots\}$ by the

---

[1] Note that in other contexts $\mathbb{N} = \{1, 2, 3, \dots\}$. The purely conventional choice is motivated by the problems considered here. In all cases we will study it is useful to have an initial data corresponding to the index 0 as a reference. All results, however, can be rephrased with sequences starting from an index k_0.

E. Salinelli, F. Tomarelli: *Discrete Dynamical Models.*
UNITEXT – La Matematica per il 3+2 76
DOI 10.1007/978-3-319-02291-8_1, © Springer International Publishing Switzerland 2014

transformation

$$k = \frac{t - t_0}{h}.$$

Since the object of our study are quantities that depend on natural numbers, it is not surprising that the **induction principle**, recalled below, turns out to be very useful for the analysis:

Let $P(k)$ be a proposition about the natural number k. If:

  i) there exists $\widetilde{k} \in \mathbb{N}$ such that $P\left(\widetilde{k}\right)$ is true;

  ii) for all $k \geq \widetilde{k}$, if $P(k)$ is true, then $P(k+1)$ is true;

then the proposition $P(k)$ is true for all $k \geq \widetilde{k}$.

For any subset I of $\mathbb{C}$ (in most cases I will be an interval in $\mathbb{R}$), we set:

$$I^n = \underbrace{I \times I \times \cdots \times I}_{n \text{ times}},$$

i.e. I^n denotes the set of the n-tuples $(x_1, x_2, \ldots, x_n)$ such that $x_j \in I$ for all $j = 1, 2, \ldots, n$.

Definition 1.1. *We call **difference equation**[2] **of order** n, the set of equations*

$$g(k, Y_k, Y_{k+1}, \ldots, Y_{k+n}) = 0 \qquad \forall k \in \mathbb{N} \tag{1.1}$$

where n is a positive integer and g is a given scalar function in $n+2$ variables, defined on $\mathbb{N} \times I^{n+1}$

$$g : \mathbb{N} \times I^{n+1} \to \mathbb{R}.$$

*A difference equation of order n is said to be **in normal form** if it is expressed in the form*

$$Y_{k+n} = \phi(k, Y_k, Y_{k+1}, \ldots, Y_{k+n-1}) \qquad \forall k \in \mathbb{N} \tag{1.2}$$

where ϕ is a given function

$$\phi : \mathbb{N} \times I^n \to I. \tag{1.3}$$

Formula (1.2) is a **recursive relationship** that starting from the knowledge of the first k consecutive values of the sequence Y allows, "with a lot of patience", to evaluate step by step all values of the sequence. In many cases, this

---

[2] The name originates from the situation in which

$$g(k, Y_k, Y_{k+1}, \ldots, Y_{k+n}) = \psi(k, Y_{k+1} - Y_k, Y_{k+2} - Y_{k+1}, \ldots, Y_{k+n} - Y_{k+n-1})$$

for a suitable function ψ.

calculation can be prohibitive even with automatic procedures. Hence, it is useful to know an *explicit formula* (say, a non-recursive one) for Y.

Definition 1.2. *A **solution** of (1.1) is a sequence X explicitly defined via*

$$X_k = f(k) \qquad \forall k \in \mathbb{N} \tag{1.4}$$

where $f : \mathbb{N} \to I$ is a given function that renders (1.1) an identity when $f(k)$ replaces Y_k, for any k.
*The set of all solutions of (1.1) is called **general solution** of the equation.*

Theorem 1.3 (existence and uniqueness). *If ϕ is a function as in (1.3), the difference equation in normal form (1.2) has solutions.*
For each choice of the n-tuple $(\alpha_0, \alpha_1, \ldots, \alpha_{n-1}) \in I^n$, the problem with initial data associated with the difference equation in normal form (1.2):

$$\begin{cases} Y_{k+n} = \phi(k, Y_k, Y_{k+1}, \ldots, Y_{k+n-1}) \\ Y_0 = \alpha_0, \quad Y_1 = \alpha_1, \quad \ldots, \quad Y_{n-1} = \alpha_{n-1} \end{cases}$$

has a unique solution.

Proof. By substituting the initial conditions $\alpha_0, \alpha_1, \ldots, \alpha_{n-1}$ in (1.2) with $k = 0$ we obtain exactly one value for Y_n, say α_n; moreover, by (1.3), $(\alpha_1, \alpha_2, \ldots, \alpha_{n-1}, \alpha_n)$ belongs to I^n: substituting the conditions

$$Y_1 = \alpha_1, \quad Y_2 = \alpha_2, \quad \ldots, \quad Y_n = \alpha_n$$

in equation (1.2) we obtain Y_{n+1}. Repeating this procedure defines a sequence Y uniquely, which is a solution of (1.1) in the sense of (1.4). $\qquad\square$

When it is not possible to find an explicit expression for the solution X, it is still useful to show whether the sequence X exhibits a particular property or not: periodicity, asymptotic behavior to a certain value, or even a more complex trend.

1.2 Examples

In this section we give motivations for the study of discrete mathematical models through the discussion of some examples. Their systematic analysis is deferred to later chapters.
The first examples are basic. They are presented in an increasing order of difficulty, and they are grouped by subject. The last examples, not elementary and quite technical at first sight, are intended to illustrate the variety of situations in which difference equations play an important role in applied mathematics.

Example 1.4. A mobile phone call costs α as connection charge and β for each minute of conversation. The total cost of call C_{k+1} after $k+1$ minutes of conversation is the solution of

$$\left\{ \begin{array}{l} C_{k+1} = C_k + \beta \\ C_0 = \alpha \end{array} \right\}$$

which is an example of a problem with initial data for a difference equation of the first order.   □

Example 1.5 (Simple interest). Suppose that the amount Y_{k+1} of a deposit of money calculated at time $k+1$ (the date for the computation of the interests) is given by the amount Y_k of the preceding time plus the interest calculated on the basis of a constant rate r on the initial deposit D_0. The difference model describing the behavior of the deposit corresponds to the problem:

$$\left\{ \begin{array}{l} Y_{k+1} = Y_k + rY_0 \\ Y_0 = D_0 \end{array} \right.$$

whose solution is (the reader can prove it by using the induction principle):

$$Y_k = (1 + rk)\, D_0. \tag{1.5}$$

This model describes the trend of a capital invested in a corporate or government bond with a fixed yield, for a number of years corresponding to the life of the bond.   □

Example 1.6 (Compound interest). Let us modify the previous example as follows: the interest is calculated always at a constant rate r on the sum deposited in the previous period, and then added to the deposit. The recursive model that gives the amount X_{k+1} of the deposit at time $k+1$ becomes:

$$\left\{ \begin{array}{l} X_{k+1} = X_k + rX_k = (1+r)\, X_k \\ X_0 = D_0 \end{array} \right.$$

whose solution reads (also this formula can be proved by induction):

$$X_k = (1+r)^k\, D_0. \tag{1.6}$$

This model describes the dynamics of a bank deposit whose interest rate does not change.

Note that if $r > 0$ the capital growth with compound interest is much faster than with simple interest; in fact, even by neglecting terms in the expansion of Newton's binomial, we obtain

$$(1+r)^k > 1 + kr + \frac{k\,(k-1)}{2} r^2, \qquad \forall k > 2.$$

Then by substituting this inequality in (1.5) and (1.6), we get a quantitative comparison between the two types of *financial regimes*:

$$X_k - Y_k > \frac{k\,(k-1)}{2} r^2 D_0.$$ □

Example 1.7. A **fixed-reverse bond** provides coupons that vary over time but according to a rule which is established at the issue date. A typical example is given by a title of the duration of 15 years giving coupons of 7% for the first 5 years and then an annual coupon of $2,50\%$ to maturity. By denoting Z_k the sum of the invested capital and the coupons up to time k and assuming one does not reinvest the coupons, we find the model is

$$Z_{k+1} = Z_k + r(k)\,Z_0$$

with $r(k) = \begin{cases} 0,07 & \text{if } k = 0, 1, \ldots, 5 \\ 0,025 & \text{if } k = 5, 6, \ldots, 14 \\ 0 & \text{if } k \geq 15 \end{cases}$.

Observe that this model does not take into account changes of the bond price on the secondary market. □

Example 1.8. A borrower secures a **mortgage** loan of an amount S_0 at time $k = 0$. This sum is returned in constant payments of amount R, starting at time $k = 1$. Let r be the constant interest rate by which the interest on the remaining debt S_k is calculated. The latter at the time $k + 1$, $k \geq 1$, is given by

$$S_{k+1} = S_k + rS_k - R = (1+r)\,S_k - R\,.$$ □

Example 1.9. At time $k = 0$, a sum is invested in a security that pays a fixed amount coupon C at integer times $k \geq 0$. If r denotes the compound interest rate at which one reinvests the coupons received, then the amount M_{k+1} of the capital at time $k + 1$ is the sum of the amount M_k of the capital at time k, of the interests rM_k matured on such principal and of the coupon C paid at time $k + 1$ (with $M_0 = C$):

$$M_{k+1} = M_k + rM_k + C = (1+r)\,M_k + C.$$ □

Example 1.10 (Cobweb model). Suppose that the quantities Q^d_{k+1} and Q^o_{k+1} of certain goods, respectively *demanded* and *supplied* on a market at time $k + 1$, are functions of their price P_k:

$$Q^d_{k+1} = a - bP_{k+1} \qquad Q^o_{k+1} = -c + dP_k$$

where the positive constants $a, b, c, d > 0$ are known. This means that if the price of the goods increases, the purchased quantity decreases while the supplied quantity increases. Notice that in order to have a meaningful model from the economical viewpoint we should consider only prices in the interval $[P_{min}, P_{max}] := [c/d, a/b]$, with parameters such that $c/d < a/b$.

The *equilibrium condition of the market* $Q_{k+1}^d = Q_{k+1}^o$ leads to a problem in the unknown price, expressed as a difference equation with initial condition p^*:

$$\begin{cases} P_{k+1} = -\dfrac{d}{b} P_k + \dfrac{a+c}{b} \\ P_0 = p^* \end{cases}.$$

The word "cobweb" in the model's name will be clarified in the following. The reader might capture the meaning naively by solving the system "graphically".

Example 1.11 (Model of radioactive decay). Consider a population of n particles that decay according to the following law: their number decreases by 20% per unit of time.[3]

If P_k denotes the number of particles after k time intervals, we have

$$\begin{cases} P_{k+1} = 0.8 P_k \\ P_0 = n \end{cases}.$$

In analogy to Example 1.6, we obtain

$$P_k = (0.8)^k P_0 = e^{-k \ln(5/4)} n.$$

For this reason we speak of *exponential decay*.

Observe that the presence of non-integer values of P_k leads to considering real values (the correct description in suitable macroscopic units of measure, if n is very large).

The model remains discrete in time despite the continuous description of the population's magnitude.

We ask ourselves the following questions:

(1) what is the **mean lifetime** m of these particles?

(2) what is the **half-life** d of the population?

(3) which one of these two quantities is larger?

We emphasize the fact that the first two questions are different.

In the first case we want to determine the mean lifetime, i.e. the number

$$m = \frac{1}{n} \sum_{k=1}^{+\infty} k \left(P_k - P_{k+1} \right). \tag{1.7}$$

---

[3] The time interval can be very long: it is determined by the nature of the particles.

By substituting the value of P_k in (1.7), we obtain:

$$m = \frac{1}{n} \sum_{k=1}^{+\infty} nk \, (0.8)^k (1 - 0.8) = 0.2 \sum_{k=1}^{+\infty} k \, (0.8)^k = 0.16 \sum_{k=1}^{+\infty} k \, (0.8)^{k-1} = 4 \, .$$

The second question asks to determine the greatest integer k such that

$$P_k \geq \frac{1}{2} P_0 \, ,$$

which is equivalent to finding

$$\max \left\{ k \in \mathbb{N} : \ (0.8)^k \geq \frac{1}{2} \right\} .$$

Then the identities $\ln \left(2 \cdot (0.8)^k \right) = \ln 2 + k \, (\ln 4 - \ln 5)$ and $\ln 1 = 0$ entail

$$d = \max \left\{ k \in \mathbb{N} : \ k \leq \frac{\ln 2}{\ln 5 - \ln 4} \right\} = \text{integer part of} \left(\frac{\ln 2}{\ln 5 - \ln 4} \right) =$$
$$= \text{integer part of } 3.1063 = 3 \, .$$

We observe that neither quantity (mean lifetime and half-life) depends on the initial population: they both depend only on the percentage (20%) of particles decaying in the unit of time; note, moreover that $m > d$. This inequality does not hold in general: depending on the type of population and the corresponding laws of decay, one can have different possibilities: $m > d$, $m = d$, $m < d$. For example, consider a population of 3 particles (or individuals) $\mathcal{A}$, $\mathcal{B}$ and $\mathcal{C}$: $\mathcal{A}$ lives 10 years, $\mathcal{C}$ lives 20 years (see Fig. 1.1). Then:

- if $\mathcal{B}$ lives 15 years, then $d = m = 15$ years;
- if $\mathcal{B}$ lives b years, with $10 \leq b < 15$, then $d = b < m$;
- if $\mathcal{B}$ lives b years, with $15 < b < 20$, then $m < b = d$.

Finally, we observe that the mean lifetime may not even be defined if the population levels are decaying at a slower rate than the exponential decay.

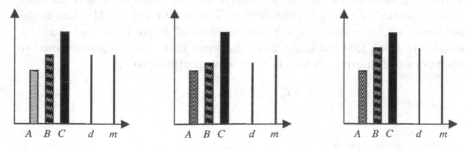

Fig. 1.1 Mean lifetime and half-life in three examined cases

For instance, if $P_k = \dfrac{1}{k}P_0$, the mean lifetime is infinite:

$$\frac{1}{P_0}\sum_{k=1}^{+\infty}k\,(P_k - P_{k+1}) = \sum_{k=1}^{+\infty}k\left(\frac{1}{k} - \frac{1}{k+1}\right) = \sum_{k=1}^{+\infty}\frac{1}{k+1} = +\infty.$$

On the other hand, to ensure that the half-life is well defined, it is sufficient to assume that $\lim_k P_k = 0$, which means that the population tends to extinction.

For instance, if $P_k = \dfrac{1}{k}P_0$ (as above), then $d = 2$.   □

Example 1.12 (Malthus model[4]). This simple and by-now-classical model describes a population (biological or not), where the number Y_k of individuals at time k is given by a constant τ times the number Y_{k-1} of individuals at the previous time $k - 1$. The quantity $\tau - 1$ is called **intrinsic growth rate**:

$$\boxed{Y_{k+1} = \tau Y_k \qquad \tau > 0.}$$

The intrinsic growth rate $\tau - 1$ represents the difference between the birth and mortality rates. Obviously, the population grows if $\tau > 1$ and decreases if $0 < \tau < 1$.

We observe that, according to the selected value of τ, the model describes a behavior quantitatively identical to the case of the compound interest (if we put $\tau = 1 + r$) or the radioactive decay if $0 < \tau < 1$. Such models are referred as *exponential growth* because of the analytical expression of their solutions.

Example 1.13 (Logistic growth). In Chap. 2 we will see in detail that, if $\tau > 1$, the Malthus model exhibits solutions with exponential growth. Several authors have criticized this model since it implies an unlimited and too fast growth. In case of large crowds, even if one ignores the competition with other species and the habitat has unlimited resources, one should at least take into account factors of intraspecific competition. The easiest way to take into account such social phenomena consists in correcting the Malthus model. The Verhulst model describes a more aggressive attitude and a decreased tendency to reproduction in cases of overcrowding. The equation of the Malthus model is "corrected" by introducing a *nonlinear term* of degree two, since it is reasonable to think that the competition between individuals is proportional to the number of meetings, which in turn is proportional to Y_k^2:

$$Y_{k+1} = \tau Y_k - \omega Y_k^2 \qquad \tau, \omega > 0 .$$

---

[4] Thomas Robert Malthus, 1766-1834.

By setting $H = \tau/\omega$ we can write

$$Y_{k+1} = \tau Y_k \left(1 - \frac{Y_k}{H}\right) \qquad \tau, H > 0 . \tag{1.8}$$

This model is called **logistic growth.**
The quantity H is called **sustainable total**: in fact, if one starts with a population Y_0 such that $0 < Y_0 < H$, and with a parameter τ that verifies $0 < \tau < 4$, then all subsequent values Y_k stay between 0 and H. Moreover, it is not possible to start from values $Y_0 > H$ as this would imply $Y_k < 0$ for all k, and the model would be meaningless from the biological standpoint.
Scaling by $X_k = Y_k/H$ we can rewrite the model as

$$X_{k+1} = \tau X_k (1 - X_k) \qquad \tau > 0$$

where the population is expressed as a fraction of total sustainable.
Correspondingly, any meaningful initial value X_0 must satisfy the inequality $0 < X_0 < 1$.
Let us consider the contribution to the variation $X_{k+1} - X_k$ in a single time step: for small populations the linear term $(\tau - 1) X_k$ prevails, while for large populations the quadratic term with negative coefficient $-\tau X_k^2$ is more significant. In this context "small" means close to zero, while "large" means the opposite and hence makes sense also for values less than 1.

Example 1.14 (Lotka–Volterra model). A region is populated by two species: *preys*, denoted by P, feed on plants and *predators*, denoted by C, which are carnivorous and feed on preys. Preys, in absence of predators, grow in "Malthusian way" at a constant rate τ; the presence of predation reduces the growth rate of preys by a term proportional to the size of the populations of preys and predators; in absence of preys, predators become extinct. Analytically, the model is expressed by a system of two coupled equations that describe the evolution of the two populations:

$$\begin{cases} P_{k+1} = (1 + a) P_k - b\, C_k P_k \\ C_{k+1} = (1 - c) C_k + d\, C_k P_k \end{cases} \qquad a, b, c, d > 0, \qquad \forall k \in \mathbb{N}.$$

Note that the effect of predation is proportional to the number of encounters between preys and predators; this model assumes that this number is proportional to the size of the two populations, i.e. to the product $C_k P_k$. This product takes the name of *impact factor*.

Example 1.15 (Fibonacci numbers[5]). A rabbit breeder wonders how many male-female couples of rabbits may be obtained in a year, from a single

---

[5] Fibonacci is the familiar nickname of Leonardo Pisano (1175-1240).

couple, assuming that every month each couple, which is at least two months' old, gives birth to a new couple.
For simplicity, we can assume that:

- no couple dies;
- the first couple is made of newborn rabbits;
- the gestation time is one month;
- sexual maturity is reached after the first month.

If C_k is the number of pairs at time k (expressed in months), from the assumptions we deduce the recursive model:

$$\begin{cases} C_{k+1} = C_k + C_{k-1}, & k \in \mathbb{N} \setminus \{0\} \\ C_0 = 0, \quad C_1 = 1 \end{cases}.$$

Unlike the previous examples, this is a **two-step model**: in order to obtain the value C_{k+1} it is necessary to know the values C_k and C_{k-1} corresponding to the two previous months. By substitution, one obtains a sequence of numbers called **Fibonacci numbers**; we list the first 13 numbers of the sequence:

$$C = \{0, 1, 1, 2, 3, 5, 8, 13, 21, 34, 55, 89, 144, 233, \ldots\}.$$

Coming back to the question asked by the breeder, we report in a table the number of months and the corresponding number of pairs of rabbits:

months k	0	1	2	3	4	5	6	7	8	9	10	11	12	13
couples C_k	0	1	1	2	3	5	8	13	21	34	55	89	144	233

We notice that the value of C_k grows very fast with k. For instance $C_{25} = 75025$: a huge farming in two years only.
We emphasize that in order to calculate the values of C_k with $k > 30$, even with a computer, one should use some care to avoid very long computation time or memory problems. These difficulties are typical of recursive operations, even when they consist in very simple operations. The problem is due to the fact that the computer repeats (unnecessarily) a large number of times the same operations if the computed values are not retained. This is the reason why one seeks explicit formulas for the calculation that do not require the knowledge of all previous values (see in this respect Appendix H).
Moreover, not only C_k but also the number N_k of operations necessary to the calculation of C_k grows very quickly (as a function of k), if the values of C_k are not tabulated, after their evaluation.
In the Fibonacci model of rabbit breeding not only the time steps are discrete, but also the quantity under study (the number of rabbits) is discrete (they are integers numbers). However, a complete description of the solutions requires the use of real numbers, namely the enlargement of the numerical set where the model has been introduced (as discussed in Chap. 2).

Example 1.16 (Leslie model [14]). The population models presented above do not take into account any possible structure of the population based on age, sex or other categorization.

To take into account at least the different ages of the individuals belonging to a biological population, the population can be divided into n disjoint classes of age, each of equal span. For example, the first class may be constituted by individuals aged less than or equal to 20, the second by individuals between 20 and 40 year old, and so on. In this way, 5 age classes would be sufficient to describe in a satisfactory manner a human population in a given region.

In general, given a population and the number n of age classes, the unit of time is the conventional maximum age divided by n. So the number of individuals that constitute the total population at time k is denoted by $\mathbf{Y}_k$. For each k, $\mathbf{Y}_k$ is a vector with n components Y_k^j, $j = 1, \ldots, n$, each of which indicates the number of individuals in the j-th class, i.e. those aged between $j - 1$ and j, in the fixed unit of temporal measurement.

The presence of two indices should not scare: the first one refers to time, the second to the age group. For example, if we consider age classes of 5 years for the human population of a fixed region, then Y_5^1 is the number of people belonging to the youngest class (age less than 5 years) evaluated at the time of the fifth census (surveys generally take place every five years).

The number Y_{k+1}^1 of newborns at time $k + 1$ is the sum of births in each age group Y_k^j (say Y_k^j times the birth rate φ_j):

$$Y_{k+1}^1 = \sum_{j=1}^{n} \varphi_j Y_k^j.$$

The number of individuals Y_{k+1}^j in older classes ($j = 2, 3, \ldots, n$) at time $k + 1$ is given by the "survivors" inside Y_k^{j-1} k:

$$Y_{k+1}^j = \sigma_{j-1} Y_k^{j-1} \qquad j = 2, \ldots, n, \quad 0 \le \sigma_{j-1} \le 1.$$

The values φ_j and σ_j are characteristic of the various populations, and are statistical indicators of the expected birth and survival. They are all non-negative and typically φ_j is small for extreme values of j ($j = 0$ and $j = n$). By putting $\mathbf{Y}_k = \begin{bmatrix} Y_k^1 & Y_k^2 & \ldots & Y_k^n \end{bmatrix}^T$, the model is described in a concise vector equation

$$\mathbf{Y}_{k+1} = \mathbb{L} \mathbf{Y}_k$$

i.e., for each component j,

$$Y_{k+1}^j = \sum_{h=1}^{n} \mathbb{L}_{j,h} Y_k^h$$

where

$$
\mathbb{L} = \begin{bmatrix} \varphi_1 & \varphi_2 & \varphi_3 & \cdots & \varphi_{n-1} & \varphi_n \\ \sigma_1 & 0 & 0 & \cdots & 0 & 0 \\ 0 & \sigma_2 & 0 & \cdots & 0 & 0 \\ \vdots & \vdots & \vdots & \ddots & \vdots & \vdots \\ 0 & 0 & 0 & \cdots & \sigma_{n-1} & 0 \end{bmatrix}.
$$

Notice that at each time step the total population is obtained by summing the number of individuals that compose the various age classes: $\sum_{j=1}^{n} Y_k^j$.

Example 1.17 (Gambler's ruin). Two players A and B play heads or tails with an unfair coin (showing heads in 40% of the cases). At every toss they bet one Euro. A always bets on heads, B always bets on tails. At the beginning, A has a Euros, B has b Euros, with $a > b > 0$. The game continues until one of the players takes all of the other's money.

One wonders what is the probability that A will win all the money owned by B, and the probability of the opposite outcome.

We refer to Chap. 2 (Exercise 2.36) and to Chap. 6 (Exercise 6.3) for two different ways to model and solve the problem.

Example 1.18. During a tennis match between two players the score is 40 to 15. Assuming we know the player in the lead has a given probability $p \in (0,1)$ to win each ball, we would like to determine how likely it is he will win the match.

For an analysis of this situation, see Example 6.30.

Example 1.19 (Solving an ordinary linear differential equation whose coefficients are expressed by elementary functions).

Consider the following differential equation, known as the **Airy equation**,

$$
y'' = xy \qquad\qquad x \in \mathbb{R}. \tag{1.9}
$$

Note that the equation, although linear, has non-constant coefficients.

We try to determine solutions of (1.9) of the form[6]

$$
y(x) = \sum_{n=0}^{+\infty} A_n x^n.
$$

Differentiating termwise, we obtain

$$
y'(x) = \sum_{n=1}^{+\infty} n\, A_n\, x^{n-1} \qquad\qquad y''(x) = \sum_{n=2}^{+\infty} n\,(n-1)\, A_n x^{n-2},
$$

---

[6] To successfully apply the method, it is not necessary that the coefficients be elementary functions: it is sufficient to have analytic functions, i.e. functions which can be expressed locally as the sum of a convergent power series of the independent variable.

and substituting into (1.9), we get the identity

$$\sum_{n=2}^{+\infty} n(n-1) A_n x^{n-2} = \sum_{n=0}^{+\infty} A_n x^{n+1}. \qquad (1.10)$$

Furthermore

$$\sum_{n=2}^{+\infty} n(n-1) A_n x^{n-2} = 2A_2 + \sum_{n=3}^{+\infty} n(n-1) A_n x^{n-2} = \boxed{\text{by setting } m=n-3}$$

$$= 2A_2 + \sum_{m=0}^{+\infty} (m+2)(m+3) A_{m+3} x^{m+1}$$

and since the index is just a symbol, one can set $m = n$ in the series on the right-hand side of (1.10) and get the equivalent identity:

$$2A_2 + \sum_{m=0}^{+\infty} (m+2)(m+3) A_{m+3} x^{m+1} = \sum_{m=0}^{+\infty} A_m x^{m+1}.$$

The equality must hold true for each $x \in \mathbb{R}$, hence the coefficients of each power must be the same on both sides:

$$A_2 = 0 \qquad\qquad (m+2)(m+3) A_{m+3} = A_m.$$

We obtain the recursive relation

$$A_{m+3} = \frac{A_m}{(m+2)(m+3)} \qquad m \in \mathbb{N}. \qquad (1.11)$$

Given the values $A_0 = y(0)$ and $A_1 = y'(0)$ and taking into account $A_2 = 0$, the recursive expression (1.11) allows to calculate all coefficients. Notice that $A_{3k+2} = 0$ for each k, and the first terms of the expansion of the solution are:

$$y(x) = A_0 \left(1 + \frac{x^3}{2 \cdot 3} + \frac{x^6}{2 \cdot 3 \cdot 5 \cdot 6} + \cdots \right) + A_1 \left(x + \frac{x^4}{3 \cdot 4} + \frac{x^7}{3 \cdot 4 \cdot 6 \cdot 7} + \cdots \right).$$

Example 1.20 (Discretization of an ordinary differential equation by the finite difference method).
In the study of the **Cauchy problem**[7]

$$\begin{cases} u'(x) = f(x, u(x)) \\ u(x_0) = u_0 \end{cases} \qquad (1.12)$$

---

[7] That is, a problem which consists in finding the solution of a differential equation which is constrained by a suitable initial condition.

where f is a known function, x varies in a suitable interval centered at x_0 and u is an unknown function, one can obtain approximate solutions by substituting the derivative of the unknown function by appropriate difference quotients. For instance, the **backward Euler method** consists in fixing a step $h > 0$ for the increment of the variable x which separates the points $x_k = x_0 + hk$, $k \in \{0, 1, \ldots, n-1\}$, then at each point x_k substituting the *forward difference quotient:*

$$\frac{u(x_k + h) - u(x_k)}{h} = \frac{u(x_0 + h(k+1)) - u(x_0 + hk)}{h}$$

to $u'(x)$ in (1.12) and evaluating the function $f(x, u(x))$ at x_{k+1}.
If we consider a function u^h which approximates u and set $U_k = u^h(x_0 + hk)$, then:

$$\boxed{\frac{U_{k+1} - U_k}{h} = f(x_{k+1}, U_{k+1}).}$$

In other words

$$U_{k+1} = U_k + hf(x_0 + (k+1)h, U_{k+1}) \qquad k = 0, 1, \ldots, n-1.$$

For each given k, the relationship above is an implicit equation in U_{k+1} if U_k is known: the values U_{k+1} are obtained by solving iteratively, starting from $U_0 = u(x_0) = u_0$.
The approximation U_k of $u(x_k)$ should get better as h approaches zero.

Alternatively, the **Euler method** consists in calculating the approximation of f in x_k and leads to

$$\boxed{\frac{U_{k+1} - U_k}{h} = f(x_k, U_k).}$$

Given U_k, the above relationship is an explicit equation in the unknown U_{k+1}. However, the advantage of solving an explicit equation is counterbalanced by a worse numerical stability; for this reason the implicit method is preferable, provided one can actually solve the implicit equation for U_{k+1}. $\qquad \square$

Example 1.21 (Discretization of the heat equation). Consider the following problem with initial and boundary conditions for the heat equation in a bar

$$\begin{cases} \dfrac{\partial v}{\partial t} = \dfrac{\partial^2 v}{\partial x^2} & x \in (0,a), \ t > 0 \\[2mm] v(x,0) = f(x) & x \in [0,a] \\[2mm] v(0,t) = v(a,t) = 0 & t > 0 . \end{cases} \qquad (1.13)$$

This problem describes the evolution of the temperature $v = v(t,x)$ at time t and at a given point x in a bar, under the assumptions that the temperature obeys Fourier's law, has the value f at time 0 and is controlled by thermostats at the bar's extremities. Assume moreover $f(0) = f(a) = 0$.

Let us fix the step $s = a/(N+1)$ of the spatial variable, $N \in \mathbb{N} \setminus \{0\}$, and $h > 0$ as time step; then define $t_k = hk$ with $k \in \mathbb{N}$. In this way we obtain a grid of points in space-time (see Fig. 1.2)

$$(x_j, t_k) = (js, hk) \qquad\qquad j = 0,1,\ldots,N+1, \quad k \in \mathbb{N}.$$

If the solution v of (1.13) has continuous derivatives up to the second order, then at every point (x_j, t_k) of the grid Taylor's formula gives:

$$\frac{\partial v}{\partial t}(x_j, t_{k+1}) = \frac{v(x_j, t_{k+1}) - v(x_j, t_k)}{h} + O(h)$$

$$\frac{\partial^2 v}{\partial x^2}(x_j, t_{k+1}) = \frac{v(x_{j+1}, t_{k+1}) - 2v(x_j, t_{k+1}) + v(x_{j-1}, t_{k+1})}{s^2} + O(s).$$

These relationships suggest to approximate the heat equation with the difference equation:

$$\frac{v(js, h(k+1)) - v(js, hk)}{h} =$$

$$= \frac{v((j+1)s, h(k+1)) - 2v(js, h(k+1)) + v((j-1)s, h(k+1))}{s^2}$$

which is a discretization based on the **backward Euler** method in time, where the second spatial derivative is approximated by a three-point formula. By denoting $V_{j,k}$ an approximation of $v(js, hk)$ and reordering the terms, we obtain the system:

$$\begin{cases} V_{j,k} = \left(1 + \dfrac{2h}{s^2}\right) V_{j,k+1} - \dfrac{h}{s^2}\left[V_{j-1,k+1} + V_{j+1,k+1}\right] & j = 1,\ldots,N, \quad k \in \mathbb{N} \\[2mm] V_{0,k+1} = V_{N+1,k+1} = 0 & k \in \mathbb{N}. \end{cases}$$
$$(1.14)$$

It is reasonable to set: $V_{j,0} = f(js)$ for $j = 0, \ldots, N+1$.

Problem (1.14) can be solved recursively (with the aid of automatic computations): the values $V_{j,0}$ are known, while the $V_{j,1}, j = 1, \ldots, N$, can be deduced from the values $V_{j,0}, j = 1, \ldots, N$.

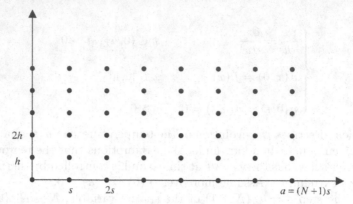

Fig. 1.2 Discretization grid for the heat equation in a bar

Introducing the $N \times N$ matrix

$$
\mathbb{A} =
\begin{bmatrix}
1 + \frac{2h}{s^2} & -\frac{h}{s^2} & 0 & \cdots & 0 \\
-\frac{h}{s^2} & 1 + \frac{2h}{s^2} & -\frac{h}{s^2} & \cdots & 0 \\
0 & -\frac{h}{s^2} & 1 + \frac{2h}{s^2} & \cdots & \vdots \\
\vdots & \vdots & \vdots & \ddots & -\frac{h}{s^2} \\
0 & \cdots & 0 & -\frac{h}{s^2} & 1 + \frac{2h}{s^2}
\end{bmatrix}
\qquad
\mathbf{V}_k =
\begin{bmatrix}
V_{1,k} \\
V_{2,k} \\
\vdots \\
V_{N,k}
\end{bmatrix}
$$

we can rewrite the previous system as a **vector-valued difference equation of the first order, with constant coefficients**

$$
\mathbb{A}\mathbf{V}_{k+1} = \mathbf{V}_k \tag{1.15}
$$

with initial condition $\mathbf{V}_0 = \begin{bmatrix} f(s) \ f(2s) \ \cdots \ f(Ns) \end{bmatrix}^T$.

The **Euler method** leads to the following vector-valued difference equation

$$
\mathbf{W}_{k+1} = \mathbb{B}\mathbf{W}_k \tag{1.16}
$$

with the initial condition $\mathbf{W}_0 = \begin{bmatrix} f(s) \ f(2s) \ \cdots \ f(Ns) \end{bmatrix}^T$, where $W_{j,k}$ approximates the value of $v(js, hk)$ and $\mathbb{B}$ is the $N \times N$ matrix defined by

$$
\mathbb{B} =
\begin{bmatrix}
1 - \frac{2h}{s^2} & \frac{h}{s^2} & 0 & \cdots & 0 \\
\frac{h}{s^2} & 1 - \frac{2h}{s^2} & \frac{h}{s^2} & \cdots & 0 \\
0 & \frac{h}{s^2} & 1 - \frac{2h}{s^2} & \cdots & \vdots \\
\vdots & \vdots & \vdots & \ddots & \frac{h}{s^2} \\
0 & \cdots & 0 & \frac{h}{s^2} & 1 - \frac{2h}{s^2}
\end{bmatrix}.
$$

In both schemes, $\mathbb{A}$ and $\mathbb{B}$ have a band structure and are positive definite matrices (the second one only if $4h < s^2$).

We observe that the solution of the Euler scheme amounts to iteratively determine $\mathbf{W}_{k+1}$ starting from $\mathbf{W}_k$: the numerical solution of this problem consists, at each step, in the multiplication of a matrix of order N by an N-dimensional vector.

The solution of the backward Euler scheme consists in evaluating $\mathbf{V}_{k+1}$ starting from $\mathbf{V}_k$: in this case the numerical solution consists at each step in the solution of a system of N algebraic equations (instead of inverting $\mathbb{A}$, which in practice can have high dimension, one takes advantage of its special structure: see Appendix D and Sect. 5.8). □

Example 1.22 (Three-moments equation for a supported beam).
Consider a heavy and homogeneous elastic bar (see Fig. 1.3), of length L, supported at $N - 1$ equally spaced points (at a distance $\delta > 0$: $N\delta = L$).
Suppose that there are no forces acting on the bar with the exception of its weight and of the constraint reaction at the supporting points. We want to calculate the *bending moments* M_k, $k = 1,, \ldots, N - 1$, at the supporting points. Let M_k be the bending moment at the k-th support. Consider three consecutive supports (labeled by $k-1, k, k+1$), place the origin at the middle point and denote by x the oriented distance from that point.
The bending moment at the point x is given by:

$$M(x) = \begin{cases} M_k + (M_k - M_{k-1})\dfrac{x}{\delta} & -\delta \le x \le 0 \\[2mm] M_k + (M_{k+1} - M_k)\dfrac{x}{\delta} & 0 \le x \le \delta \end{cases} \tag{1.17}$$

From the theory of elastic bars we know that the vertical displacement of the beam is described by the graph of $y = y(x)$ solution of

$$YIy'' = M(x), \tag{1.18}$$

where YI is the bending rigidity of the bar, which we assume independent of x (Y is Young's modulus, I the moment of inertia with respect to an axis through the barycenter).

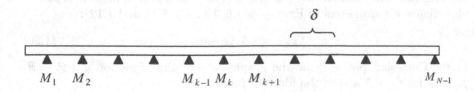

Fig. 1.3 Bar supported at equally spaced points

By integrating the differential relation (1.18) and using (1.17), we obtain

$$YIy' = \begin{cases} M_k x + (M_k - M_{k-1})\dfrac{x^2}{2\delta} + c_1 & -\delta \le x \le 0 \\[3mm] M_k x + (M_{k+1} - M_k)\dfrac{x^2}{2\delta} + c_2 & 0 \le x \le \delta \end{cases} \tag{1.19}$$

Since y' must be continuous at $x = 0$, we obtain $c_1 = c_2 = c$.
By integrating (1.19):

$$YIy = \begin{cases} \dfrac{1}{2}M_k x^2 + (M_k - M_{k-1})\dfrac{x^3}{6\delta} + cx & -\delta \le x \le 0 \\[3mm] \dfrac{1}{2}M_k x^2 + (M_{k+1} - M_k)\dfrac{x^3}{6\delta} + cx & 0 \le x \le \delta \end{cases}$$

Since the vertical displacement y vanishes at support points $(x = 0, \pm\delta)$, we obtain two equations

$$\frac{1}{2}M_k\delta^2 - (M_k - M_{k-1})\frac{\delta^2}{6} - c\delta = 0$$

$$\frac{1}{2}M_k\delta^2 + (M_{k+1} - M_k)\frac{\delta^2}{6} + c\delta = 0.$$

By adding them together we get the **three-moments equation**:

$$M_{k-1} + 4M_k + M_{k+1} = 0.$$

1.3 Graphical method

In order to understand the rough behavior of the solutions of a first order difference equation

$$X_{k+1} = f(X_k) \qquad \forall k \in \mathbb{N}$$

with $f : I \to \mathbb{R}$, I interval in $\mathbb{R}$, it may be convenient to use a graphical representation that identifies the trend, although for a few initial values of k, of the sequence X_k.

We illustrate this method in a particular case (which corresponds to a particular choice of parameters in Examples 1.5, 1.8, 1.9, 1.11 and 1.12):

$$X_{k+1} = 1.5\,X_k - 0.5. \tag{1.20}$$

In the Cartesian plane, draw the graphs of the affine function $f : \mathbb{R} \to \mathbb{R}$, $f(x) = 1.5\,x - 0.5$ and of the identity function $id(x) = x$.

Now choose an initial value X_0 on the horizontal axis, for example $X_0 = 2$. Then, the value $X_1 = f(X_0) = 2.5$ can be obtained graphically by drawing

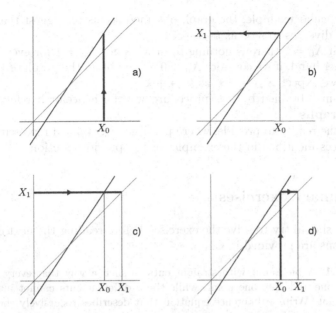

Fig. 1.4 Implementation of the graphical method

a vertical segment from $(X_0, 0)$ to the line $y = 1.5\,x - 0.5$ (Fig. 1.4a), and a horizontal segment to the vertical axis (Fig. 1.4b).

To determine X_2 from these it is necessary to score the value X_1 on the horizontal axis. To do so, move X_1 horizontally until it hits the line $y = x$, and then project vertically to the horizontal axis (Fig. 1.4c). We draw (Fig. 1.4d) all the paths segments in a single graph, deleting the horizontal segments which are swept back and forth. We iterate the procedure by drawing a vertical segment from the point (X_1, X_1) to the graph of the function f (Fig. 1.5a).

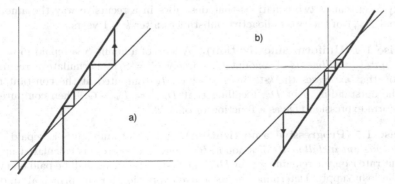

Fig. 1.5 Graphical method: $f(x) = 1.5x - 0.5$.   a) $X_0 = 2$; b) $X_0 = 0.5$

In this particular example, the graph obtained seems to suggest that the sequence X_k diverges to $+\infty$ as $k \to +\infty$.

Note that if $X_0 = 1$ there is nothing to draw, since $X_k = 1$ for every k.

On the other hand, if we consider $X_0 = 0.5$ we obtain the graph of Fig. 1.5b. This time we expect $X_k \to -\infty$ as $k \to +\infty$.

The diagrams obtained by this procedure will be henceforth referred to as **cobweb graphs**.

We invite the reader to provide, where possible, a graphical representation of the sequences identified in the examples of the previous section.

1.4 Summary exercises

The reader should try to solve the exercises before reading the next chapters. The solutions are provided in Chap. 8.

Exercise 1.1. A pie is cut with straight cuts in such a way that every two cuts intersect in one and only one point, while three different cuts cannot intersect at the same point. Write a difference equation that describes recursively the number N_k of portions after k cuts.

Exploit the result to answer the question: what is the maximum number of slices (possibly different from each other) in which you can divide a round cake by k straight cuts?

Exercise 1.2. Referring to Example 1.8, verify that if

$$R \leq rS_0$$

it is impossible to repay the debt!

Exercise 1.3. Two radioactive (unstable) substances X and Y whose initial amount is, respectively, Y_0 and X_0, decay according to the following scheme after one year:

- a fraction $0 < p_Y < 1$ of Y is transformed into X;
- a fraction $0 < p_X < 1$ of X is transformed into an inert material.

Identify a system of two equations that describes in a recursive way the amounts Y_{k+1} and X_{k+1} of the two radioactive substances after $k + 1$ years.

Exercise 1.4 (Uniform amortization). A loan of amount S is repaid in n *constant capital installments* of amount C, namely $C = S/n$. Formulate a recursive equation that expresses the k-th interest share I_k computed at the constant rate r on the outstanding debt D_k, recalling that $D_{k+1} = D_k - C$. Then compute its closed-form expression (D_k as a function of only k).

Exercise 1.5 (Progressive amortization). A loan of amount S is repaid in n annual *constant installments* of amount R, where the interest is calculated at the constant rate r on the remaining debt D_k. The first installment shall be paid one year after the loan supply. Determine the recursive expression for the principal amount C_k and the interest portion I_k.

Exercise 1.6. A sum S is put today $(k = 0)$ on a bank account. Knowing that the interest is calculated at an annual interest rate r in the regime of compound interest, that the bank charges an annual amount C to cover expenses and assuming no other transactions occur, give the recursive expression of the amount on the account after $k + 1$ years.

Exercise 1.7. An investment of initial capital C_0 after d days is C_d, i.e. it gives a percentage *return* equal to $100 (C_d - C_0) / C_0$. Determine the internal *annual compound rate of return*, i.e. the solution s of the equation (see Example 1.6)

$$C_d = (1 + s/100)^{d/365} C_0 \,.$$

The calculation of s is used to evaluate the investment regarding the purchase of a *zero-coupon bond*, that is, a fixed-rate bond without coupons.

Exercise 1.8. With reference to the case of radioactive decay (Example 1.11), given a fraction r, $0 < r < 1$, of particles which decay in time units, compare the mean lifetime $m(r)$ and the half-life $d(r)$ when r varies.

Exercise 1.9. After building a tower of k blocks, a child wants to put them back in their basket. For this purpose he takes one or two at a time.
Describe with a recursive equation the number of different ways in which the operation can be completed.

Exercise 1.10 (Hanoi Tower problem, Lucas[8], 1883). A number of disks of different radius are stacked on one of three rods, from the largest at the bottom to the smallest at the top, see Fig. 1.6.
If Y_k is the minimum number of moves needed to shift k disks onto the right-most rod, determine the relation between Y_k and Y_{k+1}. A move consists in moving a disk from one rod to another, and no disk can be stacked over one of smaller radius.

Exercise 1.11. In the previous exercise, add a rule that prevents from moving disks from the left rod to the right one, and calculate the minimum number of moves to move all disks in this situation.

Exercise 1.12. The sides of an equilateral triangle have length 1 and are divided into k equal parts; by drawing segments parallel to the sides between the subdivision points, X_k equilateral triangles arise.
Determine a recursive relationship linking X_{k+1} to X_k.

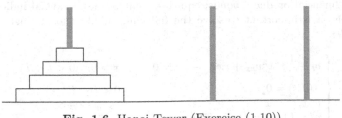

Fig. 1.6 Hanoi Tower (Exercise (1.10))

---

[8] François Edouard Anatole Lucas, 1842-1891.

Exercise 1.13. Study the cobweb model graphically (Example 1.10) under the assumption

$$\frac{d}{b} = 0,9 \qquad \frac{a+c}{b} = 2.$$

It should be noted that the resulting picture accounts for the name of this model.

Exercise 1.14. Determine a recursive formula for the coefficients A_n of the series $\sum_{n=0}^{+\infty} A_n x^n$ of a solution y of the following differential equations of second order ($x \in \mathbb{R}$):

$$\text{Bessel equation} \quad x^2 y'' + xy' + (x^2 - k^2) y = 0$$

$$\text{Hermite equation} \quad y'' - 2xy' + 2ky = 0$$

$$\text{Laguerre equation} \quad xy'' + (1 - x) y' + ky = 0$$

$$\text{Legendre equation} \quad ((1 - x^2) y')' + k (k + 1) y = 0$$

where $k \in \mathbb{N}$ is assigned and y', y'' denote the first and second derivatives of y with respect to x.

Exercise 1.15. Discretize the following differential equations with the backward Euler method:

$$(1) \qquad y' = xy \qquad x_0 = 0;$$
$$(2) \qquad y' = ay(1 - y) \quad x_0 = 0.$$

Exercise 1.16. Discretize the following partial differential equations with finite difference methods:

a) *Laplace equation* $\qquad \dfrac{\partial^2 u}{\partial x^2} + \dfrac{\partial^2 u}{\partial y^2} = 0;$

b) *D'Alembert equation* $\qquad \dfrac{\partial^2 v}{\partial t^2} - \dfrac{\partial^2 v}{\partial x^2} = 0.$

Exercise 1.17 (Black & Scholes equation). The differential equation

$$v_t + \frac{1}{2}\sigma^2 x^2 v_{xx} + rxv_x - rv = 0$$

in the unknown $v = v(x, t)$, where σ (**volatility**) and r (**risk-free interest rate**) are known parameters, is called equation of Black & Scholes. By solving this differential equation one can determine the price v of a derivative financial instrument (e.g., option, forward contract, future) at time t and for a certain price x of the underlying financial product (bonds, equities, commodities, financial indices).

For example, it is important to solve the following problem for a single purchase option (*call*):

$$\begin{cases} v_t + \dfrac{1}{2}\sigma^2 x^2 v_{xx} + rxv_x - rv = 0 & x > 0, \quad 0 < t < T \\ v(0, t) = 0 & 0 < t < T \\ v(x, t) \sim x & \text{as } x \to +\infty \\ v(x, T) = \max(x - E, 0) \end{cases}$$

where E denotes the *strike price* established by the contract.

Note that mathematically this is a *backward problem*: given $v(x,T)$, with $T > 0$, you look for two unknowns: the contract price $v(x,0)$ at time 0 and the quantity $-\dfrac{\partial v}{\partial x}(x,0)$ (denoted by Δ), which has great importance to balance the portfolio of the financial institution (*writer*) issuing the contract with respect to risk.

Prove that with the transformations

$$x = E\,e^y \quad t = T - 2\tau/\sigma^2 \quad v(x,t) = E\,w(y,\tau) \tag{1.21}$$

$$w(y,\tau) = e^{\alpha x + \beta \tau}\,u(y,\tau) \tag{1.22}$$

and the appropriate choices of the parameters α, β, the problem is translated into the "forward problem" for the heat equation (with initial data at $t = 0$, see Example 1.21). Thew latter can be solved using the techniques of Sect. 5.8.

2

Linear difference equations

In this chapter we present some techniques to solve linear difference equations with one or more steps, modeling the dynamics of scalar quantities. Nonlinear models and the vector-valued case will be addressed respectively in Chaps. 3-4 and 5-6.

In the linear scalar case, the particular structure of the set of solutions allows us to characterize them completely. This is true even in the vector-valued case, as we shall see in Chap. 5, while in the nonlinear case this is not possible.

2.1 One-step linear equations with constant coefficients

Definition 2.1. *A linear homogeneous difference equation of the first order with constant coefficients is the set of infinitely many relations*

$$X_{k+1} = aX_k \qquad \forall k \in \mathbb{N}, \tag{2.1}$$

where a is a given real or complex parameter.

The previous equation is called *linear homogeneous* because the right-hand side is a linear function of X_k; *of the first order* because it expresses a one-step relationship between two consecutive values of the sequence X only; with *constant coefficients* since a does not depend on k.

Formula (2.1) controls "growth" phenomena according to a *geometric progression* of common ratio a: see Examples 1.6, 1.11, 1.12 and Exercise 2.13.

By successive substitutions, from (2.1) we obtain the list of relationships:

$$X_1 = aX_0$$
$$X_2 = aX_1 = a^2 X_0$$
$$\ldots = \ldots$$
$$X_{k-1} = aX_{k-2} = a^{k-1} X_0$$
$$X_k = aX_{k-1} = a^k X_0$$

leading to the following statement.

E. Salinelli, F. Tomarelli: *Discrete Dynamical Models.*
UNITEXT – La Matematica per il 3+2 76
DOI 10.1007/978-3-319-02291-8_2, © Springer International Publishing Switzerland 2014

Theorem 2.2. *For each initial condition X_0, problem (2.1) admits the unique solution:*

$$X_k = a^k X_0 \qquad k \in \mathbb{N}. \tag{2.2}$$

Proof. We prove (2.2) by induction.

If $k = 1$, equation (2.2) is trivially satisfied. Assume now that $X_k = a^k X_0$ holds for a fixed index $k \geq 1$. Then:

$$X_{k+1} = aX_k = \boxed{\text{induction hypothesis}}$$

$$= a\left(a^k X_0\right) = \boxed{\text{properties of powers}}$$

$$= a^{k+1} X_0.$$

That is to say, the sequence (2.2) satisfies (2.1) for $k + 1$ too: by the induction principle the claim follows. □

Example 2.3. Consider a population whose evolution is described by the Malthus model $X_{k+1} = \tau X_k$ (see Example 1.12). This population, starting from an initial value X_0, reaches at time k the value

$$X_k = \tau^k X_0. \qquad □$$

A slower growth model (see Examples 1.4 and 1.5) is the one with constant additive increments, typical of the evolution of an *arithmetic progression*:

$$X_{k+1} = X_k + b \qquad \forall k \in \mathbb{N}, b \in \mathbb{R}$$

whose solution is given by $X_k = X_0 + kb$, as one can easily verify.

The two types of evolution just described are special cases of affine equations.

Definition 2.4. *An **affine difference equation, of the first order, with constant coefficients,** is a set of relationships in the unknown X:*

$$\boxed{X_{k+1} = aX_k + b \qquad \forall k \in \mathbb{N}} \tag{2.3}$$

*where a and b are fixed constants with $a \neq 0$. It is natural to call **equilibrium** of equation (2.3) each solution of the algebraic equation*

$$x \in \mathbb{R}: \quad (a - 1)x + b = 0,$$

as the constant sequence assuming such value is a solution of (2.3).

Recall that the difference equation (2.3) corresponds to infinitely many algebraic equations that must be verified at the same time.

We will often use the term **orbit** as a synonym of solution of (2.3), because the values of a sequence have the natural interpretation of positions reached by the system described by the difference equation.

Theorem 2.5. *For each initial value $X_0 \in \mathbb{R}$, the difference equation (2.3) has one and only one solution $X = \{X_k\}$. The explicit expression of such solutions (or orbits) have different formulations, depending on whether $a = 1$ or $a \neq 1$:*

1)

$$\boxed{\quad \text{if } a = 1 \quad \text{then} \quad X_k = X_0 + kb \qquad k \in \mathbb{N} \quad}$$

and in this case

(i)   if $b \neq 0$, there is no equilibrium and all orbits are divergent;
(ii)   if $b = 0$, all values are equilibria and all orbits are constant;

2)

$$\boxed{\quad \text{if } a \neq 1 \quad \text{then} \quad X_k = (X_0 - \alpha)\, a^k + \alpha \qquad k \in \mathbb{N} \quad} \qquad (2.4)$$

where $\alpha = b/(1 - a)$ is the unique equilibrium. Furthermore, in this case:

(i)   if $|a| < 1$, then $\lim\limits_{k} X_k = \alpha$ (in fact $\lim\limits_{k} a^k = 0$, and all the corresponding orbits converge to α);

(ii)   if $|a| > 1$, then $\lim\limits_{k} |X_k| = +\infty$ (in fact $\lim\limits_{k} |a^k| = +\infty$, and all the corresponding orbits are divergent, except the (constant) one corresponding to $X_0 = \alpha$);

(iii) if $a = -1$, then the equilibrium is $\alpha = b/2$, and all the solutions oscillate around it:

$$X_{2k} = \alpha + (X_0 - \alpha) = X_0 \qquad k \in \mathbb{N}$$
$$X_{2k+1} = \alpha - (X_0 - \alpha) = b - X_0 \quad k \in \mathbb{N}.$$

Therefore, when $a = -1$, among the solutions there is a unique constant orbit (for $X_0 = \alpha = b/2$) and infinitely many periodic orbits (for $X_0 \neq b/2$), whose values are $\{X_0, b - X_0, X_0, b - X_0, \dots\}$:

$$\boxed{\begin{array}{l} \text{if } a = -1 \text{ then:} \\[2mm] X_{2k} = X_0 \\[2mm] X_{2k+1} = (-1)^{2k+1} X_0 + \dfrac{1 - (-1)^{2k+1}}{2} b = 2\alpha - X_0 = b - X_0 \end{array}}$$

that is, X is periodic with period 2 and oscillates around $\alpha = b/2$. In this case the oscillations have half-width $|X_0 - \alpha| = |X_0 - b/2|$.

Proof. The only non-trivial fact is the explicit representation (2.4) relative to the case $a \neq 1$. This fact can be proved by induction, by showing that the sequence (2.4) describes a solution of (2.3) with initial value X_0: by uniqueness (for a fixed initial condition), formula (2.4) describes all the solutions as X_0 varies.

It is also interesting, however, to recover formula (2.4) directly, instead of showing its "a posteriori" validity, because the constructive proof allows a better understanding of its meaning. We have to prove that case 2) is qualitatively similar to the homogeneous one ($b = 0$, Theorem 2.1), up to affine changes of coordinates (in $\mathbb{R}$). Under the assumption $a \neq 1$, we study the sequences (for each X_0)

$$Z_k = X_k - \alpha \qquad \forall k \in \mathbb{N} \tag{2.5}$$

where $\alpha = b/(1 - a)$ is the equilibrium of equation (2.3). Taking X_k from (2.5) and substituting it in (2.3), we find

$$Z_{k+1} + \alpha = a(Z_k + \alpha) + b$$

$$Z_{k+1} + \frac{b}{1 - a} = aZ_k + \frac{ab}{1 - a} + b$$

$$Z_{k+1} = aZ_k$$

whose solution is $Z_k = a^k Z_0$, from Theorem 2.2. Replacing this in (2.5) gives

$$X_k = Z_k + \alpha = a^k(X_0 - \alpha) + \alpha = a^k\left(X_0 - \frac{b}{1 - a}\right) + \frac{b}{1 - a} . \qquad \square$$

Figures 2.1 and 2.2 describe the qualitative behavior of the orbits in the cases discussed in Theorem 2.5.

Remark 2.6. When $a \neq 1$, expression (2.4) shows that the distance $|X_k - \alpha|$ between the k-th term of X and the equilibrium is a fixed multiple of the k-th term of the geometric progression of common ratio $|a|$:

$$|X_k - \alpha| = |X_0 - \alpha| \, |a|^k .$$

Therefore, if $|a| < 1$, we obtain an estimate for the high speed of convergence to the equilibrium and of the improvement of the error bound at each step:

$$|X_{k+1} - \alpha| = |X_k - \alpha| \, |a| .$$

Example 2.7. If one supposes $r \neq 0$ in Example 1.8, as the equilibrium is R/r, from (2.4) it follows

$$S_k = \left(S_0 - \frac{R}{r}\right)(1 + r)^k + \frac{R}{r} = (1 + r)^k S_0 - \frac{(1 + r)^k - 1}{r} R . \qquad \square$$

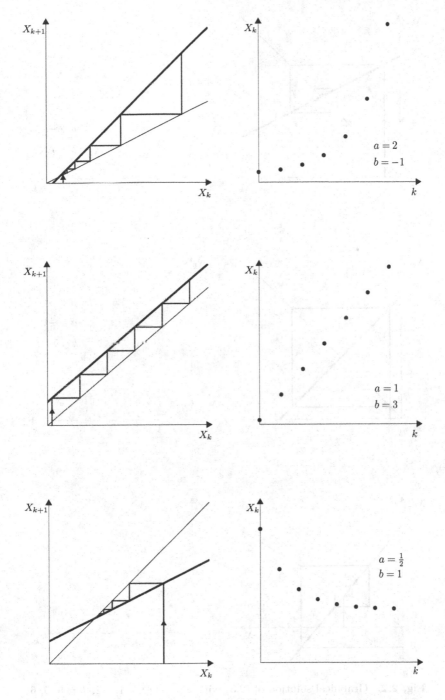

Fig. 2.1. Graphical solution of (2.3) with $a = 2,\ 1,\ 1/2,\ b = -1,\ 3,\ 1$

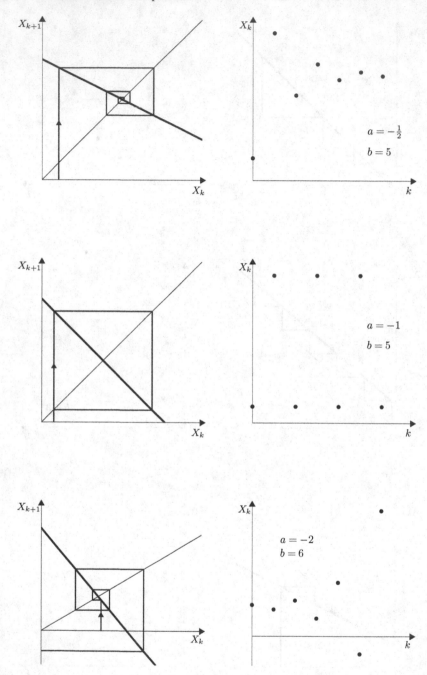

Fig. 2.2. Graphical solution of (2.3) with $a = -1/2$, -1, -2, $b = 5$, 5, 6

Example 2.8 (Cobweb model). Let us go back to the supply and demand model (Example 1.10). If the price level follows the recursive law

$$P_{k+1} = -\frac{d}{b}P_k + \frac{a+c}{b} \qquad a, b, c, d > 0$$

then $-d/b < 0$. Since the equilibrium of the model is

$$\alpha = \frac{a+c}{b+d},$$

given an initial price P_0, by (2.4) the expression in closed form of the equilibrium price at time k is

$$P_k = \left(-\frac{d}{b}\right)^k \left(P_0 - \frac{a+c}{b+d}\right) + \frac{a+c}{b+d} \qquad k \in \mathbb{N}.$$

Notice that the behavior of the price will be oscillating as $-d/b$ is negative, and the convergence (if any) to the equilibrium depends on the ratio between the slopes of the demand and supply functions.

In particular, if $d < b$ then the prices P_k will converge quickly to the equilibrium α when $k \to +\infty$. If $d > b$ the price will not stabilize. $\square$

Exercise 2.1. Given the following linear equations with constant coefficients

(1) $\begin{cases} X_{k+1} = X_k - 1 \\ X_0 = 0 \end{cases}$ (2) $\begin{cases} X_{k+1} = \frac{1}{3}X_k + 2 \\ X_0 = 1 \end{cases}$ (3) $\begin{cases} X_{k+1} = 4X_k - 1 \\ X_0 = \frac{1}{4} \end{cases}$

(4) $\begin{cases} X_{k+1} = -\frac{1}{5}X_k + 3 \\ X_0 = 3 \end{cases}$ (5) $\begin{cases} X_{k+1} = -X_k + 1 \\ X_0 = 2 \end{cases}$ (6) $\begin{cases} X_{k+1} = -3X_k + 5 \\ X_0 = 1 \end{cases}$

i)   determine their closed-form solution;
ii)   study the asymptotic behavior of the solution;
iii) draw the cobweb plot and the evolution of the solution.

Exercise 2.2. Prove Theorem 2.5 by iterating (2.3), starting from the initial value X_0.

2.2 Multi-step linear equations with constant coefficients

In this section we shall provide a complete description of the solutions to linear equations with constant coefficients, also in the case of several steps. For this purpose it is useful to introduce first some definitions.

Definition 2.9. We define **sum** of any two sequences F and G as the sequence H whose element H_k is the sum of the corresponding elements of the two sequences: $H_k = F_k + G_k$. We will write: $H = F + G$.

Definition 2.10. *We define **product of a sequence** F **by a complex number** α, the sequence W whose generic element is obtained by multiplying the corresponding element of F by α: $W_k = \alpha F_k$. We will write $W = \alpha F$.*

Definition 2.11. *Two sequences F and G are said to be **linearly dependent** if there exist two constants α and β, not both zero, such that $\alpha F + \beta G$ is the zero sequence. If this is not the case F and G are said to be **linearly independent**.*
*Similarly, n sequences F^j, $j = 1, 2, \ldots, n$, are said to be **linearly dependent** if there exist n constants $\alpha_1, \alpha_2, \ldots, \alpha_n$ not all zero, such that $\sum_{j=1}^{n} \alpha_j F^j$ is the null sequence ($\alpha_1 F_k^1 + \alpha_2 F_k^2 + \cdots + \alpha_n F_k^n = 0$, $\forall k$).*

The set of complex-valued sequences is an infinite-dimensional **vector space** when endowed with the two operations introduced above. This means that the sum of (a finite number of) sequences or the product of a sequence by a constant are still sequences. Moreover, for any positive integer n, it is possible to find n linearly independent sequences.

Definition 2.12. *Given n constants a_0, a_1, $\ldots$, a_{n-1} with $a_0 \neq 0$, we call **linear homogeneous difference equation, of order** n, **with constant coefficients**, the system of infinitely many relations*

$$X_{k+n} + a_{n-1}X_{k+n-1} + \cdots + a_1 X_{k+1} + a_0 X_k = 0 \quad \forall k \in \mathbb{N} \qquad (2.6)$$

The expression **n-step equation** will be used for n-th order equation.

Theorem 2.13 (Superposition principle). *If X and Y are solutions of (2.6), then also $X + Y$ and cX are solutions for each constant c.*
Therefore, the solution set of (2.6) is a vector space.

Proof. A simple substitution in (2.6) proves the statement.   □

The previous theorem suggests the convenience of looking for "a sufficient number of solutions" forming a basis for the solution space.
By analogy with the case of first order linear equations with constant coefficients, one can look for solutions of the homogeneous equation (2.6) of the form

$$X_k = c\lambda^k, \qquad c \neq 0, \ \lambda \neq 0.$$

If such a sequence solves (2.6), by substitution one obtains:

$$\lambda^k \left(\lambda^n + a_{n-1}\lambda^{n-1} + \cdots + a_1\lambda + a_0 \right) = 0 \qquad \forall k \in \mathbb{N}$$

that is λ must be a solution of

$$\lambda^n + a_{n-1}\lambda^{n-1} + \cdots + a_1\lambda + a_0 = 0.$$

We shall see that all the solutions of (2.6) can be obtained starting from these ones.

These considerations justify the following definitions.

Definition 2.14. *We call **characteristic polynomial**, denoted by $\mathcal{P}(\lambda)$, associated with the homogeneous equation (2.6), the following polynomial of degree n in the variable λ:*

$$\mathcal{P}(\lambda) = \lambda^n + a_{n-1}\lambda^{n-1} + \cdots + a_1\lambda + a_0$$

The equation

$$\lambda \in \mathbb{C}: \quad \mathcal{P}(\lambda) = 0 \tag{2.7}$$

*is called **characteristic equation** associated to the homogeneous equation.*

By the Fundamental Theorem of Algebra, equation (2.7) has exactly n complex solutions: $\lambda_1, \ldots, \lambda_n$ (each one is counted with its multiplicity). One or more roots λ_j can be complex even if the constants $a_0, a_1, \ldots, a_{n-1}$ are all real.

Theorem 2.15. *Equation (2.6) has infinitely many solutions. Among these solutions, n linearly independent ones can be determined; moreover, the set of solutions of (2.6) is an n-dimensional subspace of the space of the sequences. To each simple (i.e. of multiplicity 1) solution λ of the characteristic equation (2.7) is associated a solution of the difference equation of the form*

$$F_k = \lambda^k$$

To each solution λ of multiplicity m of the characteristic equation the following linearly independent solutions are associated:

$$F_k^0 = \lambda^k, \quad F_k^1 = k\lambda^k, \quad F_k^2 = k^2\lambda^k, \quad \ldots \quad F_k^{m-1} = k^{m-1}\lambda^k \tag{2.8}$$

Every solution of the difference equation (2.6) is a linear combination of these solutions.

When the coefficients a_j of equation (2.6) are real, the results of Theorem 2.15 can be refined: in particular, it is interesting to express the solutions of (2.6) in real form, even if some roots of the characteristic polynomial are complex. More precisely: if $a_0, a_1, \ldots, a_{n-1}$ are real and $\lambda = a + ib$ is a solution of (2.7), with a and b real and $b \neq 0$, then $\overline{\lambda} = a - ib$ is a solution too. To this pair of solutions the following two linearly independent real solutions are associated

$$F_k = \rho^k \cos(\theta k) \qquad G_k = \rho^k \sin(\theta k)$$

where ρ is the modulus and θ the argument of $a + ib$:

$$\rho = \sqrt{a^2 + b^2} \qquad \cos\theta = \frac{a}{\sqrt{a^2 + b^2}} \qquad \sin\theta = \frac{b}{\sqrt{a^2 + b^2}}.$$

One proves the previous statement by writing in real form the following expression:

$$c_1 (a + ib)^k + c_2 (a - ib)^k \qquad a, b \in \mathbb{R}.$$

More precisely, if $a + ib = \rho(\cos\theta + i\sin\theta)$, then:

$$(a + ib)^k = \rho^k(\cos\theta + i\sin\theta)^k = \boxed{\text{de Moivre's formula}}$$
$$= \rho^k(\cos(\theta k) + i\sin(\theta k)).$$

By operating in the same way with $(a - ib)^k$, one concludes that, for each choice of c_1 and c_2, there exist two constant $\widetilde{c}_1$ and $\widetilde{c}_2$ which depend on c_1, c_2 and k, such that

$$c_1 (a + ib)^k + c_2 (a - ib)^k = \widetilde{c}_1 \rho^k \cos(\theta k) + \widetilde{c}_2 \rho^k \sin(\theta k).$$

de Moivre's formula implies that the linear combinations with coefficients in $\mathbb{C}$ of the two sequences $\{\rho^k \cos(\theta k)\}$ and $\{\rho^k \sin(\theta k)\}$ span the same two-dimensional space of complex sequences generated by linear combinations of $\{\lambda^k\}$ and $\{\overline{\lambda}^k\}$.

Analogously, if $a_0, a_1 \ldots, a_{n-1}$ are real, then to each complex solution $\lambda = a + ib$ of multiplicity m of (2.7) and to the solution $\overline{\lambda} = a - ib$, which is also of multiplicity m, the following $2m$ linearly independent solutions are associated:

$$F_k^0 = \rho^k \cos(\theta k), \quad F_k^1 = k\rho^k \cos(\theta k), \quad \ldots, \quad F_k^{m-1} = k^{m-1}\rho^k \cos(\theta k)$$

$$G_k^0 = \rho^k \sin(\theta k), \quad G_k^1 = k\rho^k \sin(\theta k), \quad \ldots, \quad G_k^{m-1} = k^{m-1}\rho^k \sin(\theta k)$$

Proof of Theorem 2.15. To each n-tuple of initial values $X_0, X_1, \ldots, X_{n-1}$ corresponds exactly one solution, which can be determined by computing the X_k iteratively.

Each n-tuple of initial values is a linear combination of the n linear independent initial values

$$1, 0, 0, \ldots 0, 0$$
$$0, 1, 0, \ldots 0, 0$$
$$0, 0, 1, \ldots 0, 0$$
$$\cdots$$
$$0, 0, 0, \ldots 0, 1$$

Further, since by the superposition principle linear combinations of solutions of (2.6) are solutions too, we conclude that the set of solutions is an n-dimensional subspace of the space of sequences. Notice that no root of $\mathcal{P}$ is zero, as $a_0 \neq 0$. Therefore the n sequences boxed above are non-trivial. It remains to prove that they are solutions and they are linearly independent.

Let λ be a solution of equation $\mathcal{P}(\lambda) = 0$. Then, by replacing X_k with λ^k on the left of (2.6), one obtains

$$\lambda^{k+n} + a_{n-1}\lambda^{k+n-1} + \cdots + a_0\lambda^k = \lambda^k \mathcal{P}(\lambda) \qquad (2.9)$$

which is zero by assumption. Therefore, $X_k = \lambda^k$ is a solution.

Observe that if λ is a root of multiplicity $m > 1$ of the equation $\mathcal{P}(\mu) = 0$, then[1]

$$\mathcal{P}(\lambda) = \mathcal{P}'(\lambda) = \cdots = \mathcal{P}^{(m-1)}(\lambda) = 0. \qquad (2.10)$$

Let λ be a solution of multiplicity 2 of the characteristic equation. Then, by replacing X_k with $k\lambda^k$ in the left-hand side of (2.6), we obtain

$$(k+n)\lambda^{k+n} + (k+n-1)a_{n-1}\lambda^{k+n-1} + \cdots + a_0 k\lambda^k = k\lambda^k \mathcal{P}(\lambda) + \lambda^{k+1}\mathcal{P}'(\lambda) \quad (2.11)$$

because

$$\mathcal{P}'(\lambda) = n\lambda^{n-1} + (n-1)a_{n-1}\lambda^{n-2} + \cdots + 2a_2\lambda + a_1.$$

Since, by (2.10), $\mathcal{P}(\lambda) = \mathcal{P}'(\lambda) = 0$, from (2.11) it follows that $k\lambda^k$ is a solution of (2.6).

Now, let λ be a root of multiplicity $m \geq 2$ of the characteristic equation. Fixed n, for $1 \leq h \leq m$ we define

$$Q_{h,k}(\mu) = (k+n)^{h-1}\mu^{k+n} + (k+n-1)^{h-1}a_{n-1}\mu^{k+n-1} + \cdots + k^{h-1}a_0\mu^k =$$

$$= \sum_{s=0}^{n} (k+s)^{h-1} a_s \mu^{k+s}.$$

$Q_{h,k}(\mu)$ is the polynomial of degree less than or equal to $k + n$ in the variable μ that is obtained by replacing X_k with $k^{h-1}\mu^k$ in (2.6).

If $Q_{h,k}$ can be expressed as a linear combination of $\mathcal{P}$ and of its derivatives up to order $h - 1$ with coefficients which are polynomial in μ of degree less than or equal

---

[1] In fact, if λ is a root of multiplicity h of the equation $\mathcal{P}(\mu) = 0$, we can write:

$$\mathcal{P}(\mu) = (\mu - \lambda)^h \mathcal{Q}(\mu)$$

where $\mathcal{Q}$ is a polynomial of degree $n - h$ in μ. Differentiating, one obtains (2.10).

to k (notice that (2.11) tells us that this is true for $m = 2$), i.e. if

$$Q_{h,k}(\mu) = \sum_{j=0}^{h-1} c_{j,k}(\mu) \mathcal{P}^{(j)}(\mu) \qquad \text{with deg } c_{j,k} \le k, \qquad (2.12)$$

then from (2.10) it follows $Q_{h,k}(\lambda) = 0$, therefore $k^{h-1}\lambda^k$ is a solution of (2.6) for each $1 \le h \le m$.

Now we prove (2.12), by induction on h.

If $h = 1$, the statement is true because (2.9) writes as

$$Q_{1,k}(\mu) = \mu^k \mathcal{P}(\mu) \qquad \forall k \in \mathbb{N}.$$

Assume now that there exists an integer $1 \le h < m$ such that (2.12) is true. Then, for each k, we obtain:

$$Q_{h+1,k}(\mu) = \sum_{s=0}^{n} (k+s)^h a_s \mu^{k+s} = \sum_{s=0}^{n} (k+s)^{h-1} a_s (k+s) \mu^{k+s} =$$

$$= \sum_{s=0}^{n} (k+s)^{h-1} a_s \mu \frac{d}{d\mu}\left(\mu^{k+s}\right) = \mu \frac{d}{d\mu} \sum_{s=0}^{n} (k+s)^{h-1} a_s \mu^{k+s} =$$

$$= \mu \frac{d}{d\mu} Q_{h,k}(\mu) = \boxed{\text{by induction}}$$

$$= \mu \sum_{j=0}^{h-1} \left(c_{j,k}(\mu) \mathcal{P}^{(j+1)}(\mu) + c'_{j,k}(\mu) \mathcal{P}^{(j)}(\mu) \right) =$$

$$= \sum_{j=0}^{h} \widetilde{c}_{j,k} \mathcal{P}^{(j)}(\mu)$$

having set $\widetilde{c}_{j,k} = \mu \left(c_{j-1,k}(\mu) + c'_{j,k}(\mu) \right)$. Therefore also $Q_{h+1,k}(\mu)$ can be expressed as a linear combination of $\mathcal{P}$ and its derivatives up to order h with coefficients that are polynomials in μ of degree less than or equal to k.

Eventually we prove the linear independence of the solutions that we have found. In the case of n distinct roots $\lambda_1, \ldots, \lambda_n$ it is sufficient to prove the linear independence of the first n values of the sequences, namely

$$\det \begin{bmatrix} 1 & \lambda_1 & \lambda_1^2 & \ldots & \lambda_1^{n-1} \\ 1 & \lambda_2 & \lambda_2^2 & \ldots & \lambda_2^{n-1} \\ \vdots & \vdots & \vdots & \ddots & \vdots \\ 1 & \lambda_n & \lambda_n^2 & \ldots & \lambda_n^{n-1} \end{bmatrix} \ne 0.$$

This inequality is true if the λ_j's are distinct because such determinant (**Vandermonde determinant**[2]) is

$$(\lambda_n - \lambda_{n-1})(\lambda_n - \lambda_{n-2}) \cdots (\lambda_n - \lambda_1)(\lambda_{n-1} - \lambda_{n-2}) \cdots (\lambda_{n-1} - \lambda_1) \cdots (\lambda_2 - \lambda_1).$$

---

[2] Alexandre-Théophile Vandermonde, 1735-1796.

If λ is a root of multiplicity m of the characteristic equation and $c_1, c_2, \ldots, c_m$ are real numbers such that

$$c_1 \lambda^k + c_2 k \lambda^k + c_3 k^2 \lambda^k + \cdots + c_m k^{m-1} \lambda^k = 0 \qquad k \in \mathbb{N}$$

then

$$c_1 + c_2 k + c_3 k^2 + \cdots + c_m k^{m-1} = 0 \qquad k \in \mathbb{N}$$

By the identity principle of polynomials it follows $c_1 = c_2 = \cdots = c_m = 0$.
We omit the details in the case where simple and multiple roots are simultaneously present. $\qquad\square$

Summing up, Theorem 2.15 allows to build n linearly independent solutions: in fact, the roots (computed with their multiplicity) of the characteristic polynomial of a difference equation of order n are exactly n in number, thanks to the Fundamental Theorem of Algebra.
All the solutions of equation (2.6) are linear combinations of the n linearly independent solutions of Theorem 2.15. Any such combination is called **general solution** of (2.6) because we obtain all solutions by varying its coefficients. More precisely, each choice of the initial values $X_0, X_1, \ldots, X_{n-1}$, determines a unique set of coefficients and consequently a unique solution.

Example 2.16. The difference equation

$$X_{k+3} - 4X_{k+2} + 5X_{k+1} - 2X_k = 0$$

has characteristic equation $\lambda^3 - 4\lambda^2 + 5\lambda - 2 = 0$, which is equivalent to

$$(\lambda - 1)^2 (\lambda - 2) = 0.$$

Thus, we have the simple root 2 and the double root 1. The general solution of the difference equation is

$$X_k = c_1 2^k + c_2 + c_3 k \qquad k \in \mathbb{N}$$

where c_1, c_2 and c_3 are arbitrary constants.
If the initial conditions $X_0 = 0$, $X_1 = 1$ and $X_2 = 0$ are imposed to the difference equation, by substitution one obtains the system:

$$\begin{cases} c_1 + c_2 = 0 \\ 2c_1 + c_2 + c_3 = 1 \\ 4c_1 + c_2 + 2c_3 = 0 \end{cases} \quad \text{equivalent to} \quad \begin{cases} c_1 = -2 \\ c_2 = 2 \\ c_3 = 3 \end{cases}$$

hence $X_k = -2^{k+1} + 3k + 2$. $\qquad\square$

Example 2.17. The second order linear difference equation with constant coefficients

$$\begin{cases} C_{k+1} = C_k + C_{k-1} & k \in \mathbb{N} \setminus \{0\} \\ C_0 = 0, \quad C_1 = 1 \end{cases}$$

describes the Fibonacci numbers (see Example 1.10) recursively. Its characteristic polynomial $\lambda^2 - \lambda - 1$ admits the roots:

$$\lambda_1 = \frac{1 + \sqrt{5}}{2} \qquad \lambda_2 = \frac{1 - \sqrt{5}}{2}.$$

Therefore, the general solution is given by $C_k = c_1 (\lambda_1)^k + c_2 (\lambda_2)^k$. Furthermore, from

$$C_0 = c_1 + c_2 = 0, \qquad C_1 = c_1 \frac{1 + \sqrt{5}}{2} + c_2 \frac{1 - \sqrt{5}}{2} = 1$$

it follows $c_1 = +1/\sqrt{5}, \quad c_2 = -1/\sqrt{5}$. Summing up:

$$\boxed{C_k = \frac{1}{\sqrt{5}} \left(\frac{1 + \sqrt{5}}{2} \right)^k - \frac{1}{\sqrt{5}} \left(\frac{1 - \sqrt{5}}{2} \right)^k \qquad k \in \mathbb{N}}$$

Notice that from

$$|\lambda_1| = \left| \frac{1 - \sqrt{5}}{2} \right| < 1$$

we obtain $\lim_k \lambda_1^k = 0$, hence

$$\lim_k \frac{C_{k+1}}{C_k} = \lim_k \frac{\lambda_2^{k+1}}{\lambda_2^k} = \lambda_2 = \frac{1 + \sqrt{5}}{2}.$$

Moreover, recalling that all Fibonacci numbers are integers, from

$$\left| \frac{1}{\sqrt{5}} \lambda_1^k \right| < \frac{1}{2}$$

it follows that C_k is the closest integer to $\frac{1}{\sqrt{5}} \lambda_2^k$ or, if one prefers, it is the integer part of $\frac{1}{\sqrt{5}} \lambda_2^k + \frac{1}{2}$. $\qquad\qquad\qquad\qquad\qquad\qquad\square$

Example 2.18. We solve the equation $X_{k+2} + X_k = 0$ with initial conditions $X_0 = 0$, $X_1 = 1$.
First method: it is a second-order linear homogeneous difference equation, with constant coefficients. Its characteristic equation $\lambda^2 + 1 = 0$ has two complex conjugate roots $\lambda_{1,2} = \pm i$, so the general solution of the equation is

$$X_k = c_1 i^k + c_2 (-i)^k \qquad k \in \mathbb{N}.$$

By imposing the initial conditions, we obtain

$$\begin{cases} 0 = X_0 = c_1 + c_2 \\ 1 = X_1 = ic_1 - ic_2 \end{cases} \quad \text{that is} \quad \begin{cases} c_1 = -i/2 \\ c_2 = i/2 \end{cases}$$

Second method: the (equivalent) real formulation of the general solution is

$$X_k = d_1 \cos(k\pi/2) + d_2 \sin(k\pi/2) \qquad k \in \mathbb{N}$$

obtained from $|i| = |-i| = 1$, $\pi/2 = \arg(i)$. By substituting the initial data, we obtain $X_k = \sin(k\pi/2)$.

Third method: one can observe that the terms with even and odd index are uncoupled. By setting:

$$Y_k = X_{2k} \qquad\qquad Z_k = X_{2k+1}$$

one obtains

$$Y_{k+1} + Y_k = 0 \qquad\qquad Z_{k+1} + Z_k = 0$$
$$Y_k = (-1)^k Y_0 \qquad\qquad Z_k = (-1)^k Z_0$$
$$X_{2k} = (-1)^k X_0 \qquad\qquad X_{2k+1} = (-1)^k X_1.$$

By substituting the initial values $X_{2k} = 0$, $X_{2k+1} = (-1)^k$, $\forall k$.   □

Let us consider the n-**step linear non-homogeneous difference equation**

$$\boxed{X_{k+n} + a_{n-1}X_{k+n-1} + \cdots + a_1 X_{k+1} + a_0 X_k = b \qquad \forall k \in \mathbb{N}}$$

where $b, a_0, a_1, \ldots, a_{n-1}$ are given constants with $a_0 \neq 0$.

The general solution X_k is the sum of the general solution of the homogeneous problem (2.6) and a term of the type ak^m, where m is the multiplicity of the root 1 of the characteristic polynomial ($m = 0$ if 1 is not a root) and a is a constant to be determined by substitution.

Example 2.19. Let us consider the difference equation

$$X_{k+2} - 2X_{k+1} + X_k = 1. \tag{2.13}$$

The homogeneous equation $X_{k+2} - 2X_{k+1} + X_k = 0$, associated to (2.13), has a characteristic equation with double root 1. We look for a solution of the given equation in the form

$$Y_k = ak^2$$

where a is a real constant to be determined. By substituting Y_k in (2.13), we obtain:

$$a(k+2)^2 + -2\left[a(k+1)^2\right] + ak^2 = 1;$$

by matching the coefficients of the powers of k we obtain $a = 1/2$.

In conclusion, the general solution of (2.13) is $X_k = c_1 + c_2 k + \dfrac{1}{2}k^2$.   □

For the analysis of more general cases, see sections 2.4 and 2.5.

Exercise 2.3. Discuss, as a and b vary in $\mathbb{R}$, the asymptotic behavior of the solution of the two-step homogeneous linear equation

$$X_{k+2} + aX_{k+1} + bX_k = 0 \qquad k \in \mathbb{N}, b \neq 0 .$$

Exercise 2.4. Compute the general solution of the following difference equations:

(1) $X_{k+2} - 2X_{k+1} - 3X_k = 0$ (2) $X_{k+2} - \sqrt{3}X_{k+1} + X_k = 0$

(3) $X_{k+3} - X_{k+2} + X_{k+1} - X_k = 0$ (4) $X_{k+3} + 5X_{k+2} + 7X_{k+1} + 3X_k = 0$

(5) $X_{k+2} - 2X_{k+1} - 3X_k = 2$ (6) $X_{k+3} - X_{k+2} + X_{k+1} - X_k = -1$.

2.3 Stability of the equilibria of multi-step linear equations with constant coefficients

We have already observed in the case of the one-step linear difference equation $X_{k+1} = aX_k + b$ that equilibria (a solution of $(1 - a)x = b$) may behave in rather different ways: they can "attract" the solutions corresponding to any given initial value X_0 (when $|a| < 1$), or "repel" them (when $|a| > 1$). If $a = -1$ they neither attract nor repel, as the solutions are periodic; furthermore, if a solution starts near the equilibrium then it remains close to it. If $a = 1$ and $b \neq 0$ there are no equilibria, whereas for $b = 0$ all points are equilibria, hence they do not attract or repel but exhibit a "stability" property.

In this section we pin down the nature of these properties of equilibria in the more general case of multi-step linear equations.

We start with the analysis of a two-step linear problem $(a_0 \neq 0)$

$$X_{k+2} + a_1X_{k+1} + a_0X_k = b \qquad k \in \mathbb{N}. \tag{2.14}$$

An **equilibrium** of a two-step linear difference equation with constant coefficients is a value $\alpha \in \mathbb{R}$ such that the constant sequence $X_k = \alpha$, $\forall k \in \mathbb{N}$, is a solution of (2.14).

An equilibrium α is said to be **attractive** if, for each pair of initial data X_0 and X_1, the corresponding solution tends to the equilibrium:

$$\lim_k X_k = \alpha.$$

An equilibrium α is said to be **stable** if starting from initial conditions close to the equilibrium, one remains close to it, that is, if for each $\varepsilon > 0$ there exists a $\delta > 0$ such that

$$|X_0 - \alpha| + |X_1 - \alpha| < \delta \quad \Rightarrow \quad |X_k - \alpha| < \varepsilon \qquad \forall k \in \mathbb{N}.$$

As α is an equilibrium of (2.14) if and only if it is a solution of the equation

$$x + a_1x + a_0x = b,$$

we can deduce that:

- if $1 + a_1 + a_0 \neq 0$, then there exists a unique equilibrium $\alpha = b/(1 + a_1 + a_0)$ of (2.14);
- if $1 + a_1 + a_0 = 0$ and $b = 0$, then every $\alpha \in \mathbb{R}$ is an equilibrium;
- if $1 + a_1 + a_0 = 0$ and $b \neq 0$, then there are no equilibria.

If λ_1 and λ_2 are the roots of the characteristic polynomial, then by the general solution formula one obtains the following conclusions on the stability of equilibria:

- α is attractive if and only if $|\lambda_1| < 1$ and $|\lambda_2| < 1$;
- α is stable if and only if:
  { both roots λ_1, λ_2 have modulus less than or equal to 1, unit modulus roots (if present) are all simple and, when $b \neq 0$, no root is equal to 1 };
- if α is attractive, then it is stable too.

Let us consider the n-step linear difference equation with constant coefficients

$$X_{k+n} + a_{n-1}X_{k+n-1} + a_{n-2}X_{k+n-2} + \cdots + a_0 X_k = b \qquad k \in \mathbb{N} \quad (2.15)$$

where $a_0, a_1, \ldots, a_{n-1}, b$ are real or complex constants and $a_0 \neq 0$.

Definition 2.20. *An **equilibrium** of equation (2.15) is a number α corresponding to a constant solution $X_k = \alpha$.*

By substituting, for each k, $X_k = \alpha$ in (2.15), one easily gets:

- if $1 + a_{n-1} + a_{n-2} + \cdots + a_0 \neq 0$, then there exists a unique equilibrium:

$$\alpha = b / \left(1 + a_{n-1} + a_{n-2} + \cdots + a_0\right);$$

- if $1 + a_{n-1} + a_{n-2} + \cdots + a_0 = 0$, then: if $b = 0$ each $\alpha \in \mathbb{R}$ is an equilibrium; if $b \neq 0$ there are no equilibria.

Definition 2.21. *An equilibrium α of (2.15) is called **stable equilibrium** if for each $\varepsilon > 0$ there exists a $\delta > 0$ such that*

$$\sum_{j=0}^{n-1} |X_j - \alpha| < \delta \quad \Rightarrow \quad |X_k - \alpha| < \varepsilon \qquad \forall k \in \mathbb{N}.$$

*In other words, initial values close to the equilibrium generate orbits that remain (uniformly with respect to k) close to it. Otherwise it is called **unstable**.*

Definition 2.22. *An equilibrium α of equation (2.15) is said to be an **attractive equilibrium** if for each n-tuple of initial values $X_0, X_1, \ldots, X_{n-1}$, the corresponding solution X_k satisfies*

$$\lim_k X_k = \alpha.$$

Theorem 2.23. *An equilibrium α of the linear equation (2.15) is:*

1) *attractive if and only if each solution λ of the characteristic equation satisfies $|\lambda| < 1$;*
2) *stable if and only if each solution λ of the characteristic equation satisfies $|\lambda| \leq 1$, and possible roots with unit modulus are simple and, if $b \neq 0$, different from 1;*
3) *stable, if attractive.*

Proof. The change of coordinates $Y_k = X_k - \alpha$ reduces us to the homogeneous system with the same characteristic equation in the unknown Y:

$$Y_{k+n} + a_{n-1}Y_{k+n-1} + a_{n-2}Y_{k+n-2} + \cdots + a_0 Y_k = 0.$$

α is stable (respectively attractive) for X if and only if 0 is stable (respectively attractive) for Y. Statements 1) and 2) follow by the explicit general solutions

$$Y_k = \sum_{\lambda_j} \sum_{l=0}^{m-1} c_{jl} k^l \lambda_j^k$$

recalling that:

$$|\lambda| < 1 \quad \Rightarrow \quad \lim_k \left(k^l \lambda^k \right) = 0$$

$$l \geq 1, |\lambda| = 1 \quad \Rightarrow \quad \lim_k \left| k^l \lambda^k \right| = +\infty \, .$$

Statement 3) immediately follows from 1) and 2). $\qquad\square$

Example 2.24. Given the two-step linear equation

$$6X_{k+2} - 5X_{k+1} + X_k = 2$$

we compute the equilibrium and discuss its stability.
The equation $(6 - 5 + 1)\alpha = 2$ forces $\alpha = 1$. The roots of the characteristic equation $6\lambda^2 - 5\lambda + 1 = 0$ are $\lambda_1 = 1/2$ and $\lambda_2 = 1/3$. Hence $\alpha = 1$ is stable and attractive. The general solution is $X_k = c_1 2^{-k} + c_2 3^{-k} + 1$.
Figure 2.3 shows the first values of the solution $X_k = 2^{3-k} - 3^{2-k} + 1$ corresponding to the initial conditions $X_0 = 0$ and $X_1 = 2$.

Example 2.25. Let us consider the equation $X_{k+2} - 2X_{k+1} + 2X_k = 0$. The equilibrium is $\alpha = 0$. The solutions of the characteristic equation $\lambda^2 - 2\lambda + 2\lambda = 0$ are $\lambda_{1,2} = 1 \pm i$. The equilibrium is unstable as $|\lambda_{1,2}| = \sqrt{2}$.
The general solution of the equation is:
$$X_k = c_1 (1 + i)^k + c_2 (1 - i)^k = 2^{k/2} \left(c_1 \cos \left(\frac{\pi}{4}k \right) + c_2 \sin \left(\frac{\pi}{4}k \right) \right) \qquad k \in \mathbb{N}.$$
If $X_0 = \varepsilon > 0$ and $X_1 = 0$, by substitution one obtains $c_1 = \varepsilon$ and $c_2 = -\varepsilon$. According to what has already been said, starting from these initial values $\lim_k |X_k| = +\infty$ for each positive ε.

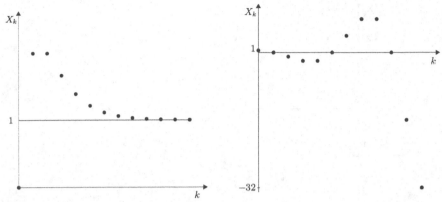

Fig. 2.3. Graph of
$X_k = 2^{3-k} - 3^{2-k} + 1$

Fig. 2.4. Graph of
$X_k = 2^{k/2}(\cos(k\pi/4) - \sin(k\pi/4))$

If we start from the equilibrium $X_0 = X_1 = 0$ then the solution remains constant: $X_k = 0$ for each k. Figure 2.4 shows the behavior of the particular solution corresponding to the initial values $X_0 = 1$, $X_1 = 0$.   □

Example 2.26. Let us consider the equation $X_{k+3} + X_k = 0$. The equilibrium is $\alpha = 0$. The solutions of the associated characteristic equation $\lambda^3 + 1 = 0$ are

$$\lambda_1 = -1 \qquad \lambda_{2,3} = \frac{1 \pm i\sqrt{3}}{2} = \cos\frac{\pi}{3} + i\sin\frac{\pi}{3}.$$

Therefore $|\lambda_j| = 1$ with distinct roots: the equilibrium α is stable but not attractive. The general solution $X_k = c_1 (-1)^k + c_2 \cos(k\pi/3) + c_3 \sin(k\pi/3)$ is the sum of a 2 periodic term and of two 3 periodic terms. So, X_k is periodic with period 6. The behavior of the particular solution corresponding to initial values $X_0 = 1$, $X_1 = 0$ and $X_2 = 2$ is shown in Fig. 2.5.

Example 2.27. Let us consider the equation $5X_{k+2} - 8X_{k+1} + 5X_k = 0$. The equilibrium is $\alpha = 0$. The characteristic equation $5\lambda^2 - 8\lambda + 5 = 0$ has distinct solutions $\lambda_{1,2} = 4/5 \pm i3/5$, both of unit modulus. Consequently, α is stable but not attractive. If $\widetilde{\theta}$ is the unique angle (in radians) in $(0, \pi/2)$ that solves $\cos\theta = 4/5$, then the general solution $X_k = c_1 \cos\left(\widetilde{\theta}k\right) + c_2 \sin\left(\widetilde{\theta}k\right)$ oscillates around the equilibrium but *it is not periodic*. In fact, the angle $\widetilde{\theta} = \arccos(4/5)$ has an irrational quotient with 2π.   □

The gist of the last two examples is that the general solution contains terms of the form

$$Z_k = c_1 \cos(\theta k) + c_2 \sin(\theta k) \qquad (2.16)$$

when the characteristic equation has solutions of the form $\lambda_{1,2} = \cos\theta \pm i\sin\theta$. It is useful to have a clear idea about the behavior of (2.16). This sequence

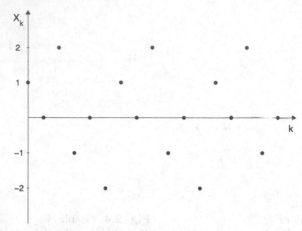

Fig. 2.5 Graph of $X_k = (-1)^k + \frac{2\sqrt{3}}{3}\sin(k\pi/3)$, sequence of period 6

presents a stable behavior: if c_1 and c_2 are small, then Z_k remains small: $|Z_k| \leq \sqrt{c_1^2 + c_2^2}$ for each k. Furthermore:

- if $\theta = 2\pi n/m$ where n, m are positive coprime integers, then Z is a periodic sequence of minimum period m; namely $Z_{k+m} = Z_k$, $\forall k \in \mathbb{N}$, and this last relation is false for values less than m;
- if $\theta \neq 2\pi n/m$ for any pairs of integers n, m with $m \neq 0$, then Z is not periodic.

Summarizing the qualitative analysis of the orbits for the n-step linear homogeneous difference equation (2.6), we can say that the **natural modes** of the solutions (integer powers of the roots of the characteristic polynomial, see (2.8)) have a behavior that depends on the position of the corresponding root λ in the complex plane (in particular $|\lambda| > 1$ implies divergence, $|\lambda| < 1$ implies convergence to zero, $|\lambda| = 1$ implies boundedness).
The different behaviors are illustrated in Fig. 2.7.

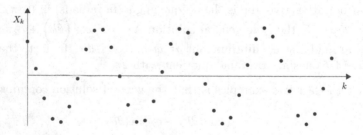

Fig. 2.6 Graph of $X_k = 4\cos\left(\tilde{\theta}k\right) - 4\sin\left(\tilde{\theta}k\right)$: non periodic sequence

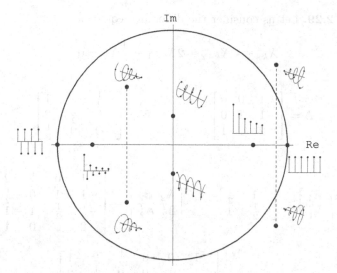

Fig. 2.7 Natural modes of the solutions for a linear homogeneous difference equation with constant coefficients, according to the location of the real and imaginary parts of the eigenvalue λ in the complex plane

The computation of the solutions of the characteristic equation of a linear homogeneous difference equation with constant coefficients of order n involves the solution of an algebraic equation of degree n (see Appendix B, Algebraic equations). In practice this problem for $n \geq 3$ is solved numerically (see e.g. Newton's method in sections 4.7 and 4.8). Some criteria based only on the knowledge of the coefficients of the equation have been developed in order to study the stability of the zero equilibrium of these equations. Here is a famous one.

Theorem 2.28 (Schur). *Let $a_0, a_1, \ldots, a_{n-1}$ be the coefficients of equation (2.6). Given the following triangular matrices of order n*

$$
\mathbb{A} = \begin{bmatrix} 1 & 0 & 0 & \cdots & 0 \\ a_{n-1} & 1 & 0 & \cdots & 0 \\ a_{n-2} & a_{n-1} & 1 & \cdots & 0 \\ \vdots & \vdots & \vdots & \ddots & \vdots \\ a_1 & a_2 & a_3 & \cdots & 1 \end{bmatrix}
\qquad
\mathbb{B} = \begin{bmatrix} a_0 & a_1 & a_2 & \cdots & a_{n-1} \\ 0 & a_0 & a_1 & \cdots & a_{n-2} \\ 0 & 0 & a_0 & \cdots & a_{n-3} \\ \vdots & \vdots & \vdots & \ddots & \vdots \\ 0 & 0 & 0 & \cdots & a_0 \end{bmatrix}
$$

we denote by $\mathbb{A}_1, \mathbb{A}_2, \ldots, \mathbb{A}_n$, and $\mathbb{B}_1, \mathbb{B}_2, \ldots, \mathbb{B}_n$ their NW principal minors. Then the zero equilibrium of equation (2.6) is attractive if and only if the determinants of the following (block) matrices are positive:

$$
\begin{bmatrix} \mathbb{A}_1 & \mathbb{B}_1 \\ \mathbb{B}_1^T & \mathbb{A}_1^T \end{bmatrix}, \quad
\begin{bmatrix} \mathbb{A}_2 & \mathbb{B}_2 \\ \mathbb{B}_2^T & \mathbb{A}_2^T \end{bmatrix}, \quad
\begin{bmatrix} \mathbb{A}_3 & \mathbb{B}_3 \\ \mathbb{B}_3^T & \mathbb{A}_3^T \end{bmatrix}, \quad \ldots, \quad
\begin{bmatrix} \mathbb{A}_n & \mathbb{B}_n \\ \mathbb{B}_n^T & \mathbb{A}_n^T \end{bmatrix}.
$$

Example 2.29. Let us consider the difference equation

$$X_{k+3} - X_{k+2} + 2X_{k+1} - \frac{1}{2}X_k = 0 \, .$$

We have:

$$A = \begin{bmatrix} 1 & 0 & 0 \\ -1 & 1 & 0 \\ 2 & -1 & 1 \end{bmatrix} \qquad B = \begin{bmatrix} -\frac{1}{2} & 2 & -1 \\ 0 & -\frac{1}{2} & 2 \\ 0 & 0 & -\frac{1}{2} \end{bmatrix}$$

and then

$$\begin{bmatrix} A_1 & B_1 \\ B_1^T & A_1^T \end{bmatrix} = \begin{bmatrix} 1 & -\frac{1}{2} \\ -\frac{1}{2} & 1 \end{bmatrix} \qquad \begin{bmatrix} A_2 & B_2 \\ B_2^T & A_2^T \end{bmatrix} = \begin{bmatrix} 1 & 0 & -\frac{1}{2} & 2 \\ -1 & 1 & 0 & -\frac{1}{2} \\ -\frac{1}{2} & -\frac{1}{2} & 1 & -1 \\ 2 & 0 & 0 & 1 \end{bmatrix}$$

$$\begin{bmatrix} A_3 & B_3 \\ B_3^T & A_3^T \end{bmatrix} = \begin{bmatrix} 1 & 0 & 0 & -\frac{1}{2} & 2 & -1 \\ -1 & 1 & 0 & 0 & -\frac{1}{2} & 2 \\ 2 & -1 & 1 & 0 & 0 & -\frac{1}{2} \\ -\frac{1}{2} & 0 & 0 & 1 & -1 & 2 \\ 2 & -\frac{1}{2} & 0 & 0 & 1 & -1 \\ -1 & 2 & -\frac{1}{2} & 0 & 0 & 1 \end{bmatrix} \, .$$

Since

$$\det \begin{bmatrix} A_1 & B_1 \\ B_1^T & A_1^T \end{bmatrix} = \frac{3}{4} > 0 \, ; \ \det \begin{bmatrix} A_2 & B_2 \\ B_2^T & A_2^T \end{bmatrix} = -\frac{9}{4} < 0 \, ; \ \det \begin{bmatrix} A_3 & B_3 \\ B_3^T & A_3^T \end{bmatrix} = \frac{243}{64} > 0$$

we conclude that the zero equilibrium is not attractive.

Exercise 2.5. Calculate the equilibria (if present) of the equation

$$X_{k+2} - \frac{1}{4}X_k = \frac{3}{2}$$

and discuss their stability. Compute also the particular solution corresponding to the initial data $X_0 = 1$ and $X_1 = 0$, and draw their cobweb graphs.

Exercise 2.6. Given the equation $2X_{k+2} + 2X_{k+1} + X_k = 0$: (a) compute the general solution (both in complex and real forms); (b) compute the equilibria (if they exist) and study their nature; (c) compute the particular solution corresponding to the initial data $X_0 = 0$ and $X_1 = -1$.

Exercise 2.7. Given the equation $X_{k+2} - 4X_{k+1} + 5X_k = 6$: (a) compute the general solution; (b) compute the equilibria (if they exist) and study their nature.

Exercise 2.8. Given the equation $2X_{k+2} + 2X_{k+1} + X_k = -10$: (a) compute the general solution; (b) compute the equilibria (if they exist) and study their nature.

Exercise 2.9. Show that the zero equilibrium $\alpha = 0$ of the second order homogeneous equation

$$X_{k+2} + a_1 X_{k+1} + a_0 X_k = 0$$

is stable and attractive if and only if the following relations

$$a_0 < 1 \qquad 1 + a_1 + a_0 > 0 \qquad 1 - a_1 + a_0 > 0 \,,$$

which are equivalent to

$$a_0 < 1 \qquad\qquad |a_1| < 1 + a_0 \,,$$

hold.

2.4 Finding particular solutions when the right-hand side is an elementary function

We now study the non-homogeneous linear case, when the right-hand side is not constant, i.e. it depends on k. In this section we compute the solutions with the basic method called **undetermined coefficients method**. In the next section we discuss a less elementary but more general technique, that is the $\mathcal{Z}$-transform.

Definition 2.30. *Given a non-zero constant a_0 and a sequence F, we call* ***affine difference equation*** *(or* ***non-homogeneous linear***) *of order n with constant coefficients, the following system of infinitely many equations in the unknown X:*

$$\boxed{X_{k+n} + a_{n-1}X_{k+n-1} + \cdots + a_1X_{k+1} + a_0X_k = F_k \qquad \forall k \in \mathbb{N}} \qquad (2.17)$$

Remark 2.31. The non-homogeneous equation (2.17) is often called **complete equation**, and equation (2.6) having the same coefficients as (2.17) and null right-hand side is called **homogeneous equation associated** to (2.17).

Theorem 2.32. *If Y is a particular solution of the complete equation (2.17), then every solutions of (2.17) can be expressed as $Y+X$, where X is a solution of the corresponding homogeneous equation.*

The theorem just stated shows us the way to solve (2.17). Once one has found the general solution of the corresponding homogeneous equation, it is sufficient to compute a particular solution of the complete equation: the general solution of the complete equation is the sum of these two solutions.
Notice that if F_k is constant and 1 is not a root of the characteristic polynomial, then an equilibrium of (2.17) exists and it is also a particular solution.

Example 2.33. Let us compute the general solution of the equation

$$X_{k+2} - 4X_{k+1} + 3X_k = 2^k \qquad k \in \mathbb{N}.$$

The characteristic equation of the homogeneous equation $\lambda^2 - 4\lambda + 3 = 0$ admits $\lambda_1 = 1$ and $\lambda_2 = 3$ as solutions. Therefore, the general solution of the homogeneous equation is

$$X_k = c_1 + c_2 3^k \qquad k \in \mathbb{N}.$$

Let us now test a particular solution of the complete equation of the form $\overline{X}_k = a2^k$, where a is a real constant to be determined. By substituting this expression into the complete equation, we obtain

$$a2^{k+2} - 4a2^{k+1} + 3a2^k = 2^k \qquad k \in \mathbb{N}$$

and dividing by 2^k,

$$4a - 8a + 3a = 1 \quad \Rightarrow \quad a = -1.$$

Eventually, the general solution of the complete equation is

$$X_k = c_1 + c_2 3^k - 2^k \qquad k \in \mathbb{N}. \qquad \qquad \Box$$

Example 2.34. Let us compute the general solution of the equation

$$X_{k+2} - 4X_{k+1} + 3X_k = b^k \qquad k \in \mathbb{N}$$

where b is a real and positive constant different from 1. From the previous example we know the general solution of the homogeneous equation. Let us try to find a particular solution of the complete equation of the form $\overline{X}_k = \beta b^k$. By substituting, we obtain

$$\beta b^{k+2} - 4\beta b^{k+1} + 3\beta b^k = b^k \qquad k \in \mathbb{N}$$

and dividing by $b^k \neq 0$

$$\beta b^2 - 4\beta b + 3\beta = 1 \quad \Rightarrow \quad \beta \left(b^2 - 4b + 3 \right) = 1.$$

The last equation identifies the constant β univocally, provided the quantity in brackets is different from zero, i.e. provided b is not a solution of the characteristic equation related to the homogeneous equation (i.e. $b \neq 1$ and $b \neq 3$). It is easy to see that, if $b = 1$ or $b = 3$, then a particular solution of the complete equation has the form $\overline{X}_k = \beta k b^k$. $\qquad \Box$

The method used in the examples allows us to determine an explicit solution in cases where the right-hand side F is a polynomial, an exponential function or a trigonometric function, as summarized in Table 2.1.

Exercise 2.10 (Sum of an arithmetic progression). Prove that

$$1 + 2 + 3 + \cdots + (k-1) + k = \frac{k(k+1)}{2} \qquad k \in \mathbb{N}.$$

Exercise 2.11. Prove the following formula, which provides the sum of the squares of the first k positive integers

$$1 + 4 + 9 + \cdots + k^2 = \frac{1}{6}k(2k+1)(k+1).$$

Exercise 2.12. Calculate the sum of the cubes of the first k integers.

Exercise 2.13 (Sum of a geometric progression). Given the complex number z, set

$$P_k = 1 + z + z^2 + \cdots + z^k = \sum_{n=0}^{k} z^n.$$

Compute P_k as an explicit function of k.

Exercise 2.14. Compute $f_{h,k} = z^h + z^{h+1} + \cdots + z^{h+k-1} + z^{h+k}$, for given $z \in \mathbb{C}$ and $h, k \in \mathbb{N}$.

Exercise 2.15. Find the values of $z \in \mathbb{C}$ for which the geometric series $\sum_{k=0}^{+\infty} z^k$ is convergent and compute its sum (see Appendix A, 18)-26)).

Exercise 2.16. Prove that the number $0.9999999\ldots$ (where the dots represent an infinite sequence of digits all equal to 9) coincides with 1.

Exercise 2.17. Make the solution of Exercise 1.3 explicit.

Exercise 2.18. Compute the general solution of the following difference equations:

1) $X_{k+2} - 2X_{k+1} - 3X_k = (-3)^k$
2) $X_{k+2} - 2\sqrt{3}X_{k+1} - 4X_k = \left(3 + 2\sqrt{3}\right)k$
3) $X_{k+3} - X_{k+2} + X_{k+1} - X_k = \sin\left(\frac{\pi}{2}k\right)$
4) $X_{k+3} + 5X_{k+2} + 7X_{k+1} + 3X_k = (-1)^k.$

Table 2.1. Methods for finding particular solutions of equation (2.17)

F_k	particular solution
β^k	$c\beta^k$
k^m	$c_m k^m + c_{m-1}k^{m-1} + \cdots + c_1 k + c_0$
$k^m \beta^k$	$\beta^k \left(c_m k^m + c_{m-1}k^{m-1} + \cdots + c_1 k + c_0\right)$
$\sin(\theta k)$ o $\cos(\theta k)$	$c_1 \sin(\theta k) + c_2 \cos(\theta k)$
$\beta^k \sin(\theta k)$ o $\beta^k \cos(\theta k)$	$\beta^k \left(c_1 \sin(\theta k) + c_2 \cos(\theta k)\right)$

The sequence in the second column must **not** be a solution of the homogeneous equation.
If the sequence in the second column is already a solution of the homogeneous equation, then it must be multiplied by a power of k with exponent equal to the multiplicity of the corresponding root.

2.5 The $\mathcal{Z}$-transform

The $\mathcal{Z}$-transform is a mathematical tool of great importance in signal processing. Here we illustrate quickly how it can be used as method for solving a linear difference equation with non constant right-hand side. As a matter of fact, this technique allows to determine particular solutions with assigned initial data even if the right-hand side is not an elementary function of k.
Often, in this context, both terms **source** and **input** are used as synonymous of right-hand side of the difference equation, and the term **output** as synonymous of solution.

Definition 2.35. *The $\mathcal{Z}$-transform of a sequence $X = \{X_k\}$ (sometimes denoted by $\mathcal{Z}\{X\}$) is a function $x(z)$ of the complex variable z defined as the sum of a power series whose coefficients are the terms of the sequence X:*

$$\mathcal{Z}\{X\}(z) = x(z) = \sum_{k=0}^{+\infty} \frac{X_k}{z^k}$$

The function x is defined for all complex numbers z for which the series is convergent. For instance, if X is a bounded sequence, then the series converges for each z such that $|z| > 1$. More generally, if there exists a constant $c \in (0, +\infty)$ such that $|X_k| < c^k$ for each k, then the series converges for each $z \in \mathbb{C}$ such that $|z| > c$.
The reader who is not familiar with complex numbers can consider the restriction to the half-line $\{x \in \mathbb{R} : x > c\}$, where $c > 0$.

It is immediate to check the **linearity of the $\mathcal{Z}$-transform**:

$$\mathcal{Z}\{aX + bY\} = a\mathcal{Z}\{X\} + b\mathcal{Z}\{Y\} \qquad \forall a, b \in \mathbb{C}, \quad \forall X, Y$$

Theorem 2.36. *The $\mathcal{Z}$-transform of a sequence is differentiable in the set of convergence. More precisely, its has derivatives of every order.*

We present below some elementary but important examples of $\mathcal{Z}$-transforms: their names (and the corresponding notation) are borrowed from the terminology used in the theory of discrete signals.

Example 2.37 (Heaviside signal, U). If $U_k = 1$ for each $k \in \mathbb{N}$, then, having set $u(z) = \mathcal{Z}\{U\}$, we have:

$$u(z) = \frac{z}{z - 1}.$$

In fact, recalling the formula of the sum of a convergent geometric series, if $|z| > 1$, then $|1/z| < 1$ and

$$u(z) = \sum_{k=0}^{+\infty} \frac{1}{z^k} = \frac{1}{1 - 1/z} = \frac{z}{z-1}. \qquad \square$$

Example 2.38 (Pulse signal, K^0). If $K_0^0 = 1$ and $K_k^0 = 0$ for each $k \geq 1$, then

$$k^0(z) = 1$$

where $k^0(z) = \mathcal{Z}\{K^0\}$. $\qquad \square$

Example 2.39 (Kronecker signal, K^n). If $K_n^n = 1$ and $K_k^n = 0$ for each $k \neq n$, then

$$k^n(z) = z^{-n}$$

where $k^n(z) = \mathcal{Z}\{K^n\}$. $\qquad \square$

Notice that K^0 and K^n are sequences, whereas K_k^0 and K_k^n are their corresponding k-th terms.

Example 2.40 (Linear signal, L). If $L_k = k$ for each $k \in \mathbb{N}$, then

$$l(z) = \frac{z}{(z-1)^2}.$$

where $l(z) = \mathcal{Z}\{L\}$. In fact $\quad l(z) = \sum_{k=0}^{+\infty} \frac{k}{z^k} = -z \sum_{k=0}^{+\infty} \frac{d}{dz}\left(\frac{1}{z^k}\right) =$

$-z\dfrac{d}{dz}\left(\displaystyle\sum_{k=0}^{+\infty} \frac{1}{z^k}\right) = -z\dfrac{d}{dz}\left(\dfrac{z}{z-1}\right) = \dfrac{z}{(z-1)^2}$ where we have swapped the

derivative operator with the series sum, as it is allowed within the domain of convergence of a power series. $\qquad \square$

Now we describe how "scrolling" the index of a sequence acts on the $\mathcal{Z}$-transform of the sequence itself: the extreme simplicity of this action is of great utility in the study of difference equations and in the treatment of discrete signals with delay.

Definition 2.41. *Given a sequence* $X = \{X_0, X_1, X_2, \dots\}$, *we call* **forward shift** *the transformation* τ_n *defined by*

$$\tau_n X = \{\underbrace{0, 0, \dots, 0}_{n}, X_0, X_1, X_2, \dots\}.$$

Abusing the notation, we will write $\tau_n X = \{X_{k-n}\}_{k \in \mathbb{N}}$.

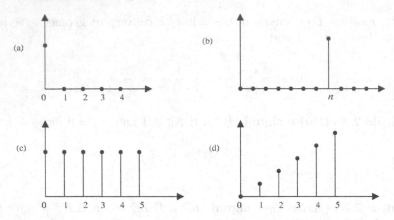

Fig. 2.8 In the presentation of discrete signals it is useful to view graphically the terms of sequences: (a) K^0; (b) K^n; (c) U; (d) L

Definition 2.42. *We call **backward shift** the transformation τ_{-n} that associates the sequence*

$$\tau_{-n}X = \{\, X_n,\, X_{n+1},\, X_{n+2}, \ldots \}$$

to the sequence $X = \{X_0, X_1, X_2, \ldots\}$. Notice that now, in addition to index scrolling, the first n terms of the sequence X are lost.
Abusing the notation, we will write $\tau_{-n}X = \{X_{k+n}\}_{k \in \mathbb{N}}$.
Here we show some examples of sequences (left column) together with their $\mathcal{Z}$-transforms (right column). This list can be used to find also the inverse $\mathcal{Z}$-transform (left column) of a given function (right column).

$$
\begin{array}{ll}
\tau_1 X & z^{-1}x\,(z) \\[4pt]
\tau_n X & z^{-n}x\,(z) \\[4pt]
\tau_{-1} X & zx\,(z) - zX_0 \\[4pt]
\tau_{-2} X & z^2 x\,(z) - z^2 X_0 - zX_1.
\end{array}
$$

The next result shows how the $\mathcal{Z}$-transform acts on shifts in the general case, leaving the proof as exercise.

Theorem 2.43. *Referring to Definitions 2.41 and 2.42 any sequence X fulfils*

$$\mathcal{Z}\{\tau_n X\} = z^{-n}x\,(z)$$
$$\mathcal{Z}\{\tau_{-n}X\} = z^n x\,(z) - z^n X_0 - z^{n-1}X_1 - \cdots - zX_{n-1}\,.$$

From the previous examples it is easy to compile reference tables that allow to $\mathcal{Z}$-transform the sequences and to perform the inverse operation, i.e. to compute the inverse $\mathcal{Z}$-transform of functions defined in the complement of a disk of the complex plane: these results are summarized in the tables of Appendix G.

Let us see first, with an example, how to use the $\mathcal{Z}$-**transform method** to solve difference equations with constant coefficients whose right-hand side is a known function of the index.

Example 2.44. Determine a non-recursive expression of X solving

$$\begin{cases} X_{k+2} - 2X_{k+1} + X_k = 3k & k \in \mathbb{N} \\ X_0 = 0, \quad X_1 = 2 . \end{cases}$$

The solution of this problem (which exists and is unique) could be calculated recursively or with the method of undetermined constants described in the previous section, as $F_k = 3k$ is a polynomial in k. Instead we apply the $\mathcal{Z}$-transform method.

Having set $x(z) = \mathcal{Z}\{X\}$, by Table Z.1 in Appendix G we obtain $\mathcal{Z}\{k\} = z/(z-1)^2$ and, by the linearity of $\mathcal{Z}$, $\mathcal{Z}\{3k\} = 3z/(z-1)^2$. Hence, by transforming the equation and taking into account the shifts, we obtain an algebraic equation in the unknown $x(z)$:

$$\left(z^2 x - z^2 X_0 - z X_1\right) - 2\left(zx - zX_0\right) + x = \frac{3z}{(z-1)^2}$$

equivalent to

$$x(z) = \frac{3z}{(z-1)^4} + \frac{2z}{(z-1)^2} .$$

At this point, again from Table Z.1,

$$\left(\mathcal{Z}^{-1}\left\{\frac{3z}{(z-1)^4}\right\}\right)_k = \frac{1}{2}k(k-1)(k-2)$$

$$\left(\mathcal{Z}^{-1}\left\{\frac{2z}{(z-1)^2}\right\}\right)_k = 2k$$

and we obtain the solution

$$X_k = 2k + \frac{1}{2}k(k-1)(k-2) . \qquad \square$$

Let us re-examine Example 2.44 with general source and initial data:

$$X_{k+2} - 2X_{k+1} + X_k = F_k \qquad \text{with fixed} \quad X_0, X_1 .$$

This problem is equivalent to consider any sequence F rather than the particular case $3k$ and any initial values X_0 and X_1. We can observe that the relation between $f = \mathcal{Z}\{F\}$ and $x = \mathcal{Z}\{X\}$ is very simple:

$$x(z) = \frac{1}{z^2 - 2z + 1} g(z)$$

where g is the sum of f ($\mathcal{Z}$ transform of right-hand side: source or input) and φ, a correction[3] that depends on the initial values and the equation: $\varphi(z) = (z^2 - 2z) X_0 + zX_1$. In this context $h(z) = (z^2 - 2z + 1)^{-1}$ is called **transfer function**[4]: the transfer function is the transform of the solution (or output) H corresponding to having the impulse K^0 concentrated at zero as right-hand side, with null initial data.

In order to fully exploit this fact when dealing with the applications, it is useful to recall the definition of discrete convolution and characterize the $\mathcal{Z}$-transform of the discrete convolution.

Definition 2.45. *We call **discrete convolution** of two sequences X and Y the sequence, denoted by $X * Y$, whose k-th term is*

$$(X * Y)_k = \sum_{h=0}^{k} X_h Y_{k-h}.$$

It is easy to verify that $X * Y = Y * X$ and $(aX) * Y = a(X * Y)$ for each couple of sequences X, Y and for each constant a.

Theorem 2.46 ($\mathcal{Z}$-transform of the discrete convolution). *If X and Y are two sequences whose $\mathcal{Z}$-transform is defined on a non-empty domain, then their discrete convolution $X * Y$ admits a $\mathcal{Z}$-transform (defined on the intersection domain) and*

$$\boxed{\mathcal{Z}\{X * Y\}(z) = x(z) y(z)}$$

So, if H is the solution of a two-step linear difference equation with constant coefficients, initial data $X_1 = X_0 = 0$ and the unit impulse K^0 as right-hand side, the solution of the same equation with $X_1 = X_0 = 0$ and F in place of K^0 is

$$H * F.$$

---

[3] Notice that such a term is the $\mathcal{Z}$-transform of the solution of the homogeneous equation ($F_k \equiv 0$) having the same initial values. Notice also that this solution can be directly computed (in a more elementary way) using the techniques of Sect. 2.2).

[4] More precisely, transfer function of the **solver filter** of the difference equation. Given the homogeneous equation and the initial data, the solver filter corresponds to the linear operation that associate to each right-hand side the solution of the nonhomogeneous equation.

More generally, for each source F and for every choice of initial data X_0, X_1 (not necessarily null), the solution of the same equation is

$$H * G$$

where H is the solution corresponding to the impulse K^0 and null initial datum, while $G = F + D$ and the term D depends on the initial data.

In Example 2.44 one can verify that the solution X is equal to the discrete convolution of $\{3k\}_{k \in \mathbb{N}}$ and $\mathcal{Z}^{-1}\left\{\left(z^2 - 2z + 1\right)^{-1}\right\}$ plus a correction due to the initial data. We recall that the transfer function is the reciprocal of the characteristic polynomial associated with the left-hand side of the difference equation.

Notice that *the transfer function depends only on the left-hand side, it is computed independently of, hence can be used for all transformable right-hand side and any given initial datum.*

Example 2.47. By exploiting the Discrete Convolution Theorem, we prove that the inverse transform of $\left(z^2 - 2z + 1\right)^{-1}$ has general term $X_k = k - 1$ for $k \geq 1$ and $X_0 = 0$.

In fact:

$$\mathcal{Z}^{-1}\left\{\frac{1}{z^2 - 2z + 1}\right\} = \mathcal{Z}^{-1}\left\{\frac{1}{z - 1}\right\} * \mathcal{Z}^{-1}\left\{\frac{1}{z - 1}\right\} = \boxed{\text{Example 2.35}}$$

$$= \mathcal{Z}^{-1}\left\{\frac{1}{z}u\left(z\right)\right\} * \mathcal{Z}^{-1}\left\{\frac{1}{z}u\left(z\right)\right\} = \boxed{\text{Theorem 2.41}}$$

$$= \mathcal{Z}^{-1}\left\{\mathcal{Z}\left\{\tau_1 U\right\}\right\} * \mathcal{Z}^{-1}\left\{\mathcal{Z}\left\{\tau_1 U\right\}\right\} =$$

$$= \tau_1 U * \tau_1 U,$$

whose k-th term (as $\tau_1 U = \{0, 1, 1, 1, \dots\}$) is 0 when $k = 0$ and

$$\sum_{h=0}^{k} \left(\tau_1 U\right)_h \left(\tau_1 U\right)_{k-h} = k - 1 \qquad \text{for } k \geq 1. \qquad \square$$

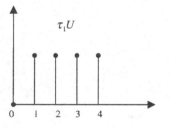
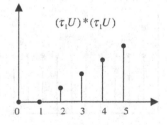

Fig. 2.9 $\mathcal{Z}^{-1}\left\{\left(z^2 - 2z + 1\right)^{-1}\right\}$

Summarizing the previous results, the following theorem holds.

Theorem 2.48. *Let F be a sequence admitting $\mathcal{Z}$-transform (that is, there exists $a > 0$ such that $|F_k| \le a^k$ for each $k \in \mathbb{N}$). Then the unique solution of*

$$\begin{cases} X_{k+n} + a_{n-1}X_{k+n-1} + \cdots + a_1 X_{k+1} + a_0 X_k = F_k \\ X_0 = X_1 = \cdots = X_{n-1} = 0 \end{cases}$$

is given by $X = T * F$ where $T = \mathcal{Z}^{-1}\left\{ \left(z^n + a_{n-1}z^{n-1} + \cdots + a_1 z + a_0\right)^{-1} \right\}$.

We recall below the definition of **moving averages**, often used in the statistical treatment of data, and we observe that they have an easy interpretation in terms of discrete convolution.

Definition 2.49. *The **moving average** (or **rolling mean**) of amplitude n of a sequence X is the sequence of the arithmetic means of n consecutive terms of the sequence X:*

$$\left(M^n \left[X\right]\right)_k = \frac{X_{k-n+1} + X_{k-n+2} + \cdots + X_k}{n} .$$

Thus:

$$\left(M^n \left[X\right]\right)_k = \frac{1}{n}\sum_{j=0}^{n-1} X_{k-j} = \left(X * \frac{1}{n}\left(U - \tau_n U\right)\right)_k .$$

It may be useful to set conventionally the values of X_k for $k < 0$, for instance all equal to zero: in this way $\left(M^n \left[X\right]\right)_k$ is defined also for $0 \le k < n$.

A moving average is used to filter out occasional fluctuations that are insignificant in detecting any long-term trends.

In the analysis of economic and financial phenomena, the moving average is particularly indicated when one knows the size of the business cycle of the considered quantity: for instance, it is used to "de-seasonalize" recorded data (an operation which is essential to correctly estimate inflation or purchases of raw materials).

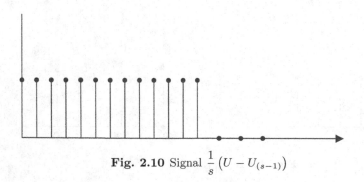

Fig. 2.10 Signal $\dfrac{1}{s}\left(U - U_{(s-1)}\right)$

Example 2.50. Let us consider the quotations X_k of an equity over a period of 160 days (we omit for simplicity the closing days of the stock exchange and assume that the equity is quoted in each of 160 consecutive days). In Fig. 2.11 we have reported the values of X_k (Fig. 2.11 a), the behavior of $M^{10}[X]$ (Fig. 2.11 b) and $M^{20}[X]$ (Fig. 2.11 c).

Moving averages are one of the tools of the "technical analysis" of equities, often used as attempts to extrapolate future values or trends in the short term. It is worth noting that these estimates should be treated with great caution. In fact, even in the short term, the error estimates of an extrapolation require some regularity[5], while the available time series are extremely irregular: in this sense the example, in addition to being a real case, also has a typical behavior. This is why technical analysis should be combined with the analysis of the economic fundamentals that affect the equity. The graphs of the example display the regularizing effect of the convolution. □

As shown in the examples, tables are very useful to compute $\mathcal{Z}$-transforms and its inverse. If tables are not available or do not contain the functions for which one has to compute the inverse transform, it is possible to use a standard technique of complex analysis that we recall here without proof.

Theorem 2.51 (Inversion formula of the $\mathcal{Z}$-transform). *Let f be a $\mathcal{Z}$-transform of a sequence F, with f complex-differentiable on $\mathbb{C}$ except for a finite number of points $z_1, z_2, \ldots, z_N$, such that* [6]

$$\text{for } j = 1, \ldots, N \quad \exists \text{ a positive integer } p_j \colon \exists \lim_{z \to z_j} (z - z_j)^{p_j} f(z) \quad \text{finite.}$$

Then the general term F_k of F is given by

$$F_k = \sum_{j=1}^{N} \text{res}\left(z^{k-1} f(z), z_j\right)$$

*where the **residue of f in** z_j, denoted by $res(f, z_j)$, is a number which can be calculated in the following way:*

(a) if $p_j = 1$ then $\text{res}(f, z_j) = \lim\limits_{z \to z_j} (z - z_j) f(z)$;

(b) if $p_j > 1$ then $\text{res}(f, z_j) = \lim\limits_{z \to z_j} \dfrac{1}{(p_j - 1)!} \left(\dfrac{d}{dz}\right)^{p_j - 1} \left((z - z_j)^{p_j} f(z)\right).$

---

[5] Here regularity means existence and continuity of a fixed number of derivatives: for instance, $X_k = f(k)$ with $f \in C^N(\mathbb{R})$. The extrapolation is precise only for quantities characterized by analytic functions f, a situation very far from that of the values listed in stock markets.

[6] The assumptions on f, although seemingly restrictive, occur in most practical problems. They can also be weakened further: it is enough to require that f has a finite number of isolated singularities.

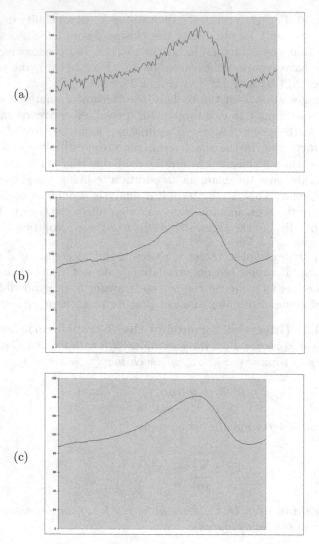

Fig. 2.11 a) Graph of the piecewise linear prolongation; b) graph of the 10 days mean average; c) graph of the 20 days mean average

In the sequel we present some examples of applications of difference equations to **electrical networks**.

Example 2.52. Let us consider the electrical circuit in Fig. 2.12 that has $2N$ resistors of resistance R and a DC electrostatic generator of voltage V (the source or input). We want to find the current's intensity I_k at the k-th horizontal resistor.

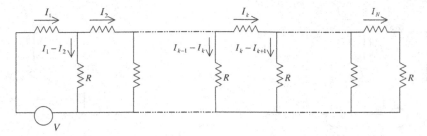

Fig. 2.12 The electrical circuit of Example 2.52

This problem (and similar ones that include various combinations of resistors, capacitors and inductors) can be formulated using *Kirchhoff's circuit laws*:

- at each node the algebraic sum of the current intensities is zero;
- at each mesh the algebraic sum of the potential differences is zero.

We recall that if I denotes the intensity of the current, then the potential difference across a resistor of resistance R is equal to the product IR. Applying the second law, to the first mesh, to the k-th mesh, with $2 \leq k \leq N-1$, and to the N-th mesh, one obtains:

$$I_1 R + (I_1 - I_2)R - V = 0$$
$$I_k R + (I_k - I_{k+1})R - (I_{k-1} - I_k)R = 0$$
$$I_N R + I_N R + (I_N - I_{N-1})R = 0.$$

From the first equation $I_2 = 2I_1 - V/R$. The second one

$$I_{k+2} - 3I_{k+1} + I_k = 0 \qquad 1 \leq k \leq N-2,$$

is solved by

$$I_k = A \left(\frac{3 + \sqrt{5}}{2} \right)^k + B \left(\frac{3 - \sqrt{5}}{2} \right)^k . \qquad (2.18)$$

Taking into account the initial data I_1 and I_2, one obtains

$$A = -\frac{1}{10R} \left[\left(\sqrt{5} - 5 \right) I_1 R + \left(3\sqrt{5} - 5 \right) V \right]$$

$$B = \frac{1}{10R} \left[\left(\sqrt{5} + 5 \right) I_1 R + \left(3\sqrt{5} + 5 \right) V \right] . \qquad (2.19)$$

Taking into account the equation on the last mesh, we also know $I_N = \frac{1}{3}I_{N-1}$, wherefrom I_N can be determined by substitution.

Example 2.53. Let us consider the electrical circuit in Fig. 2.13 where there are $2N$ resistors of resistance R and N DC sources of voltage V_k, $1 \leq k \leq N$. We want to determine the current I_k through the k-th horizontal resistor.

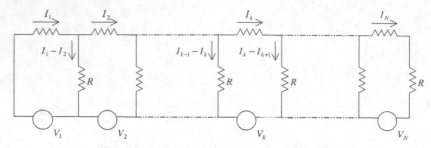

Fig. 2.13 Electrical circuit of Example 2.53

Proceeding as in the previous example, we obtain a two-step non-homogeneous linear equation:

$$I_k R + (I_k - I_{k+1}) R - (I_{k-1} - I_k) R = V_k$$

with "initial" and "final" conditions I_1 (to be determined subsequently) through

$$I_2 = I_1 - V_1/R \quad \text{and} \quad I_{N-1} = V_N/(2R) \, . \tag{2.20}$$

As the sorce term is generic, it is appropriate to use the $\mathcal{Z}$-transform method by setting $i(z) = \mathcal{Z}\{I_k\}$ and $v(z) = \mathcal{Z}\{V_k\}$:

$$\left(z^2 - 3z + 1\right) i(z) = I_1 z^2 + (I_2 - 3I_1) z + \frac{1}{R} v(z) \, .$$

By denoting J_k the right-hand side of (2.18) with the choices of A and B obtained by substituting (2.18) in system (2.20), and setting, for $k \geq 2$,

$$T_k = \mathcal{Z}^{-1}\left\{\frac{v(z)}{R\left(z^2 - 3z + 1\right)}\right\} = \frac{1}{\sqrt{5}R}\left(V * \left(\left(\frac{3+\sqrt{5}}{2}\right)^{k-1} - \left(\frac{3-\sqrt{5}}{2}\right)^{k-1}\right)\right)_k$$

we obtain

$$I_k = J_k + T_k \, . \qquad \square$$

Remark 2.54. Another method to solve difference equations, substantially identical to the one of the $\mathcal{Z}$-transform, is that of associating to each sequence X its **generating function**:

$$\sum_{k=0}^{+\infty} X_k \, z^k = x\left(\frac{1}{z}\right) \, .$$

The two methods are obviously equivalent. We have chosen to present the $\mathcal{Z}$-transform because it is most commonly used in signal processing and other engineering applications.

Exercise 2.19. Solve the following problem

$$\begin{cases} X_{k+2} - X_{k+1} - 2X_k = k & k \in \mathbb{N} \\ X_0 = 0 \qquad X_1 = 1 \end{cases}$$

Exercise 2.20. Solve the following problem (with a pulse signal on the right-hand side):

$$\begin{cases} X_{k+2} - 5X_{k+1} + 6X_k = K_k^0 & k \in \mathbb{N} \\ X_0 = X_1 = 0 \end{cases}$$

Exercise 2.21. Solve the following problem (with a Kronecker signal on the right-hand side):

$$\begin{cases} X_{k+1} - 2X_k = 3K_k^4 & k \in \mathbb{N} \\ X_0 = 1 \end{cases}$$

Exercise 2.22. Solve the equation $2X_k = 2 + 32 \sum_{s=0}^{k-1} (k-s-1) X_s, \ k \in \mathbb{N}$.

Exercise 2.23. Compute the $\mathcal{Z}$-transform of $X_k = k^3$, $k \in \mathbb{N}$, without using tables, by deducing it from the one of $Y_k = k(k+1)(k+2)$.

2.6 Linear equations with non-constant coefficients

In this section we examine some simple examples of linear difference equations with nonconstant coefficients.

Definition 2.55. *Given a sequence $A = \{A_k\}$, we call **first order linear homogeneous difference equation, with non-constant coefficients**, the set of equations*

$$X_{k+1} = A_k X_k \qquad \forall k \in \mathbb{N}. \tag{2.21}$$

Theorem 2.56. *For each initial datum X_0 the unique solution of (2.21) is:*

$$X_k = \left(\prod_{s=0}^{k-1} A_s \right) X_0. \tag{2.22}$$

Here and henceforth the symbol $\prod_{s=0}^{h} A_s$ will denote the product of the numbers $A_0, A_1, \ldots, A_h$.
We omit the proof of the theorem since it is completely analogous to the one of Theorem 2.2.

Definition 2.57. *Given two sequences A and B, we call **first order linear non-homogeneous difference equation, with non-constant coefficients**, the set of equations*

$$\boxed{X_{k+1} = A_k X_k + B_k \qquad \forall k \in \mathbb{N}} \tag{2.23}$$

Theorem 2.58. *For each initial value X_0 equation (2.23) has the unique solution:*

$$
X_k = X_0 \prod_{s=0}^{k-1} A_s + B_{k-1} + \sum_{r=0}^{k-2} B_r \left(\prod_{s=r+1}^{k-1} A_s \right) \qquad k \geq 1 \qquad (2.24)
$$

Proof. Set $G_0 = 1$ and $G_k = \prod_{s=0}^{k-1} A_s$ when $k > 0$, and divide both sides of (2.23) by G_{k+1}:

$$
\frac{X_{k+1}}{G_{k+1}} = \frac{X_k}{G_k} + \frac{B_k}{G_{k+1}}. \qquad (2.25)
$$

Moreover, if we set

$$
Z_k = \frac{X_k}{G_k} \qquad \beta_k = \frac{B_k}{G_{k+1}}
$$

equation (2.25) can be rewritten as $Z_{k+1} = Z_k + \beta_k$, whose solution is given by

$$
Z_k = Z_0 + \sum_{r=0}^{k-1} \beta_r = X_0 + \sum_{r=0}^{k-1} \frac{B_r}{G_{r+1}}.
$$

At this point, by substituting and using the distributive property, we obtain

$$
X_k = G_k Z_k = \left(\prod_{s=0}^{k-1} A_s \right) \left(X_0 + \sum_{r=0}^{k-1} \frac{B_r}{G_{r+1}} \right) = X_0 \prod_{s=0}^{k-1} A_s + \left(\prod_{s=0}^{k-1} A_s \right) \left(\sum_{r=0}^{k-1} \frac{B_r}{G_{r+1}} \right)
$$

from which (2.24) follows, thanks to the identity

$$
\left(\prod_{s=0}^{k-1} A_s \right) \left(\sum_{r=0}^{k-1} \frac{B_r}{G_{r+1}} \right) = B_{k-1} + \sum_{r=0}^{k-2} B_r \left(\prod_{s=r+1}^{k-1} A_s \right). \qquad \square
$$

Remark 2.59. The steps of the proof are correct only if $A_k \neq 0$, $\forall k$. However, claim (2.24) is valid even if some of the A_k vanishes, as can be verified by substitution. Moreover if some A_k vanishes then (2.24) simplifies as follows:

$$
\text{if } \exists \overline{k} \in \mathbb{N}: A_{\overline{k}} = 0 \quad \text{then} \quad X_k = B_{k-1} + \sum_{r=\overline{k}}^{k-2} B_r \left(\prod_{s=\overline{k}+1}^{k-1} A_s \right), \quad \forall k > \overline{k},
$$

e.g., for $k > \overline{k}$, the system exhibits loss of memory about the previous history: values X_k with $k > \overline{k}$ do not depend on values X_k, $k \leq \overline{k}$.

Remark 2.60. It should be noted that the number of operations necessary for calculating the term X_k of a first order linear homogeneous difference equation with non constants coefficients is k, with both formulas (2.21) and (2.22). Instead, if we consider a non-homogeneous equation, while the calculation of X_k by (2.23) needs $2k$ operations, the use of (2.24) requires an amount of operations of the order of k^2: therefore, the explicit formula for X_k (depending on k and the initial condition X_0) gives information on the solution but does not guarantee any computational advantage over the use of the recursive formula in calculating the values X_k.

Remark 2.61. In the particular case where $A_k = 1$ for each k, namely $X_{k+1} = X_k + B_k$, formula (2.24) is simplified as follows:

$$X_k = X_0 + \sum_{r=0}^{k-1} B_r .$$

We note in passing that a model with non-constant increments and $A_k = 1$ is typical of the sequence of partial sums of a series. In fact, given a sequence $X = \{X_k\}$, we define the sequence S of **partial sums** S_k associated to the series $\sum_{n=0}^{+\infty} X_n$ as the sequence whose general term is obtained by adding to the previous one the general term of X, that is

$$S_k = \sum_{n=0}^{k} X_n .$$

With these notations, the following recursive relationship holds:

$$S_{k+1} = S_k + X_{k+1} .$$

We observe that this problem can be solved in several cases with the techniques proposed in section 2.4 and related exercises.

Example 2.62. The number X_k of equilateral triangles with side $1/k$, obtained by splitting each side of an equilateral triangle of unit side in k equal parts and joining these points by segments parallel to the sides of the big triangle, satisfies the relation (see Exercise 1.12)

$$\begin{cases} X_{k+1} = X_k + 2k + 1 \\ X_1 = 1 . \end{cases}$$

By exploiting (2.24) and recalling the sum of an arithmetic progression, the computation gives:

$$X_k = 1 + \sum_{r=1}^{k-1} (2r + 1) = 1 + k - 1 + 2\frac{k(k-1)}{2} = k^2. \qquad \square$$

Let us see how one can extend the above considerations and results to multi-step equations.

Definition 2.63. *Given n sequences $A^0, A^1, \ldots, A^{n-1}$, s.t. $\exists \tilde{k}$ with $A^0_{\tilde{k}} \neq 0$ we call **linear homogeneous equation of order** n the equation*

$$X_{k+n} + A^{n-1}_k X_{k+n-1} + \cdots + A^1_k X_{k+1} + A^0_k X_k = 0. \tag{2.26}$$

The theorems that follow provide a description of the solutions set of (2.26). It is immediate to verify, by iteratively computing the values of X_k, that for each n-tuple of initial data $X_0, X_1, \ldots, X_{n-1}$, there exists a unique solution of (2.26).

With considerations similar to the case of constant coefficients, one proves the following theorem.

Theorem 2.64. *(**Solutions space**) The set of solutions of the homogeneous equation (2.26) is an n-dimensional vector space.*

Definition 2.65. *Given n sequences[7] $X^1, X^2, \ldots, X^n$, we call **Casorati matrix of index** k the $n \times n$ matrix*

$$\mathbf{W}_k = \begin{bmatrix} X^1_k & X^2_k & \cdots & X^n_k \\ X^1_{k+1} & X^2_{k+1} & \cdots & X^n_{k+1} \\ \vdots & \vdots & \ddots & \vdots \\ X^1_{k+n-1} & X^2_{k+n-1} & \cdots & X^n_{k+n-1} \end{bmatrix}.$$

*The determinant $\det \mathbf{W}_k$ is called **Casorati determinant**.*

Theorem 2.66. *Given n solutions $X^1, \ldots, X^n$ of the homogeneous equation (2.26) with $A^0_k \neq 0 \, \forall k \in \mathbb{N}$, the following statements are equivalent:*

i)   *the sequences $X^1, \ldots, X^n$ are linearly dependent;*
ii)   *there exists $k \in \mathbb{N}$ such that $\det \mathbf{W}_k = 0$;*
iii) *$\det \mathbf{W}_k = 0$ for each $k \in \mathbb{N}$.*

Proof. If i) holds then there exists a non-null vector $\mathbf{c}$ of constants such that $\mathbf{W}_k \mathbf{c} = \mathbf{0}$. So we have a linear homogeneous system of n algebraic equations in n unknowns that admits a non-trivial solution. By Cramer's Theorem, $\det \mathbf{W}_k = 0$.

If k_0 is such that $\det \mathbf{W}_{k_0} = 0$, then there exists a non-null vector $\mathbf{c}$ of constants such that $\mathbf{W}_{k_0} \mathbf{c} = \mathbf{0}$. Set $X_k = c_1 X^1_k + c_2 X^2_k + \cdots + c_n X^n_k$, $\forall k$, and conclude that X_k is a solution of the homogeneous equation (see Theorem 2.64) with initial condition $X_{k_0} = \cdots = X_{k_0+n-1} = 0$. From the existence and uniqueness theorem it follows that $X_k = 0$ for each k and then the sequences $X^1, \ldots, X^n$ are linearly dependent. $\square$

Theorem 2.67. *If $X^1, \ldots, X^n$ are linearly independent solutions of the homogeneous equation (2.26), then each solution of this equation is a linear combination of these solutions.*

---

[7] The reader has to pay attention to the position of the indices: the number X_1 is part of the initial datum, the sequence X^1 is a solution, X^1_1 is the term with index 1 of this sequence.

The nonhomogeneous case of (2.26) may be treated in the usual way by adding to a particular solution all the solutions of the homogeneous equation.

Exercise 2.24. Set

$$I_k = \int_0^{+\infty} e^{-x} x^k dx \qquad k \in \mathbb{N}$$

and determine a recursive relationship between I_k and I_{k-1}. Use Theorem 2.58 to derive an explicit expression of I_k.

Exercise 2.25. Solve the following equations (Hint: guess the expression of X_k by computing a few iterations, and verify it by induction):

(1) $X_{k+1} = \dfrac{k}{k+1} X_k$ (2) $X_{k+1} = \dfrac{3k+1}{3k+7} X_k$

(3) $X_{k+1} = e^{3k} X_k$ (4) $X_{k+1} = e^{\cos 2k} X_k$.

Exercise 2.26. Make explicit the general solution of the one-step equation

$$X_{k+1} = k X_k + 1.$$

Then discuss the coherence of the result with Theorems 2.64 e 2.66.

2.7 Examples of one-step nonlinear equations which can be turned into linear ones

In case of non-linear difference equations, general techniques for finding explicit solutions are not available. However, in some special cases it is possible to make a change of variables to transform the nonlinear equation into a linear one which is solvable by using the techniques of the previous sections. The inverse change of variable will then give the closed form solution of the original problem.

In this section we shall see that this method can always be applied to the case of difference equations of the first order in normal form, when the right-hand side is a linear rational function in the unknown.

In section 3.10 we will see other techniques for solving nonlinear equations.

Example 2.68. Let us consider the difference equation

$$X_{k+1} = -\frac{1}{X_k}.$$

We notice immediately that, unlike the linear case, it is not even evident that there is a solution: in fact, if $X_k = 0$ for some k, then it is impossible to proceed with the iterations.

However, if there is a solution, then an explicit formula for X_k is immediately found by noticing that, by setting $f(x) = -1/x$, we have $f(f(x)) = x$, hence

$$X_{k+2} = X_k \qquad \forall k \in \mathbb{N}.$$

Then the sequence $X_0, -X_0^{-1}, X_0, -X_0^{-1}, X_0, \ldots$ is the solution.   □

Example 2.69. Let us consider the difference equation

$$X_{k+1} = \frac{X_k}{1 + X_k}. \tag{2.27}$$

Even in this case, the solution might not exist.
Starting from a value X_0 for which the solutions exists, we proceed formally and, by setting $Y_k = 1/X_k$ for every k and substituting, we obtain

$$\frac{1}{Y_{k+1}} = \frac{1/Y_k}{1 + 1/Y_k} \qquad \text{that is} \qquad Y_{k+1} = Y_k + 1.$$

This problem is affine and can be solved by the method shown in Sect. 2.1: $Y_k = Y_0 + k$ from which

$$X_k = \frac{1}{Y_k} = \frac{1}{Y_0 + k} = \frac{1}{X_0^{-1} + k} = \frac{X_0}{1 + kX_0}. \tag{2.28}$$

We observe that the substitution carried out makes sense only if $X_k \neq 0$ for each k and in particular $X_0 \neq 0$. However, the formula obtained gives the solution X_k (check it by substitution) also if $X_0 = 0$! The formula applies to the calculation of the meaningful iterations also when there is no solution. For instance, if $X_0 = -1$, then the formula stops before the calculation of $X_1 = -1/0$; if instead $X_0 = -1/2$, then $X_1 = -1$ but X_2 can not be calculated since $X_2 = -1/0$, and so on.
Anyway, if $X_0 > 0$, then the solution X_k of (2.27) is well defined and positive for each k, therefore

$$X_k = \frac{X_0}{1 + kX_0} \qquad \forall k \in \mathbb{N}.$$

This conclusion remains true for all $X_0 < 0$ that do not lead to -1 in a finite number of steps.
In order to determine the forbidden initial values we can reverse the iteration of the function $f(x) = x/(1+x)$, i.e. iterate $f^{-1}(z) = z/(1-z)$ and then compute the values Z_k such that

$$\begin{cases} Z_{k+1} = \dfrac{Z_k}{1 - Z_k} \\ Z_0 = -1 \end{cases} \tag{2.29}$$

Notice that this is a nonlinear equation, but, as we will see shortly, its solution can be made explicit. Furthermore, it is clear that the forbidden values for X_0 are the values $-1/k$ with $k = 1, 2, \ldots$ as can be verified by induction. Observe that both (2.27) and (2.29) have rational right-hand sides.   □

Definition 2.70. *A **linear rational function** is a function defined as*

$$f : \mathbb{R} \setminus \{-d/c\} \to \mathbb{R} \qquad f(x) = \frac{ax + b}{cx + d} \qquad \text{where} \quad (c, d) \neq (0, 0).$$

Definition 2.71. *A **Möbius transformation** is a function defined as:*

$$f : \mathbb{C} \setminus \{-d/c\} \to \mathbb{C} \qquad f(z) = \frac{az + b}{cz + d} \qquad \text{where} \quad ad - bc \neq 0.$$

Remark 2.72. All Möbius transformations are compositions of translations, homotheties with center the origin and inversions $z \mapsto 1/z$. Furthermore, these transformations map every circle or straight line of the complex plane to a circle or a straight line in the complex plane.

Example 2.73. The function $f(z) = \dfrac{z - i}{z + i}$ defined on $\mathbb{C} \setminus \{-i\}$ transforms the real axis $\{x + iy : y = 0\}$ into the unit circle $\{x + iy : x^2 + y^2 = 1\}$.   □

Let us consider now the general recursive equation

$$X_{k+1} = \frac{aX_k + b}{cX_k + d} \qquad a, b, c, d \in \mathbb{R} \tag{2.30}$$

(I) If $c = 0$ or $ad - bc = 0$, then (2.30) is an affine equation and the explicit expression of X_k is given by Theorem 2.5.

(II) If $c \neq 0$ and $ad - bc \neq 0$, equation (2.30) is nonlinear. However, it is of the type $X_{k+1} = f(X_k)$ where f is a Möbius transformation. So, by means of a suitable change of the variable (of which the one of Example 2.69 was a special case), it is possible to transform (2.30) into a linear equation, and then make the values of X_k explicit.
We impose

$$\alpha = \frac{a\alpha + b}{c\alpha + d}, \tag{2.31}$$

that is we look for the α that, chosen as initial datum X_0, generates a constant solution $X_k = \alpha$, $\forall k$. Explicitly (remember that $c \neq 0$)

$$c\alpha^2 + (d - a)\alpha - b = 0 \quad \Rightarrow \quad \alpha_{1,2} = \frac{a - d \pm \sqrt{(a - d)^2 + 4bc}}{2c}.$$

We observe that $ad - bc \neq 0$ ensures $\alpha_{1,2} \neq -d/c$ and $\alpha_{1,2} \neq a/c$ (in fact $x = -d/c$ and $y = a/c$ are the equations of the asymptotes of the hyperbola with equation $y = (ax + b)/(cx + d)$).

If $X_0 = \alpha_1$, then $X_k = \alpha_1$ for each k; if $X_0 = \alpha_2$, then $X_k = \alpha_2$ for each k. If $X_0 \neq \alpha_1$ and $X_0 \neq \alpha_2$, then choosing one of the two roots, we set

$$Y_k = \frac{1}{X_k - \alpha} \qquad (2.32)$$

and correspondingly

$$X_k = \alpha + \frac{1}{Y_k} \qquad (2.33)$$

Substituting (2.33) in (2.30)

$$\alpha + \frac{1}{Y_{k+1}} = \frac{a\alpha + \dfrac{a}{Y_k} + b}{c\alpha + \dfrac{c}{Y_k} + d}$$

$$\frac{1}{Y_{k+1}} = -\alpha + \frac{(a\alpha + b)\,Y_k + a}{(c\alpha + d)\,Y_k + c} = \boxed{\text{by (2.31)}}$$

$$= \frac{a - \alpha c}{(c\alpha + d)\,Y_k + c}$$

$$Y_{k+1} = \frac{c\alpha + d}{a - \alpha c}Y_k + \frac{c}{a - \alpha c}. \qquad (2.34)$$

To summarize, if $c \neq 0$ and $ad - bc \neq 0$, then:

(i)   if $\dfrac{c\alpha + d}{a - \alpha c} = 1$, then $\alpha_1 = \alpha_2 = \dfrac{a - d}{2c}$ and, by Theorem 2.5,

$$Y_k = \frac{1}{X_0 - \alpha} + k\frac{c}{a - \alpha c}$$

and, for all k for which X_k is defined, we obtain

$$X_k = \alpha + \frac{1}{Y_k} = \alpha + \left(\frac{1}{X_0 - \alpha} + k\frac{c}{a - \alpha c}\right)^{-1}$$

In particular, if X_k is defined for each k, then $c \neq 0$ forces the existence of the finite limit

$$\lim_k X_k = \alpha.$$

(ii)   If $\dfrac{c\alpha + d}{a - \alpha c} \neq 1$ (remember that under our assumptions it is nonzero, too), then $\alpha_1 \neq \alpha_2$, the equilibrium of (2.34) is, for $\alpha = \alpha_1$,

$$\gamma = -\frac{c}{\sqrt{(a-d)^2 + 4bc}} \neq 0 \qquad (2.35)$$

and, by Theorem 2.5 and (2.33), we obtain:

$$Y_k = (Y_0 - \gamma)\left(\frac{c\alpha + d}{a - \alpha c}\right)^k + \gamma \qquad (2.36)$$

$$X_k = \alpha + \left(\left(\frac{1}{X_0 - \alpha} - \gamma\right)\left(\frac{c\alpha + d}{a - \alpha c}\right)^k + \gamma\right)^{-1} \qquad (2.37)$$

We observe that

$$\text{if } \left|\frac{c\alpha + d}{a - \alpha c}\right| > 1 \quad \text{then} \quad \lim_k X_k = \alpha_1 \quad \text{exists}$$

$$\text{if } \left|\frac{c\alpha + d}{a - \alpha c}\right| < 1 \quad \text{then} \quad \lim_k X_k = \alpha_1 + \frac{1}{\gamma} = \alpha_2 \quad \text{exists}$$

If $\left|\dfrac{c\alpha + d}{a - \alpha c}\right| = 1$ then one can have periodic or very complicated behaviors, according to whether the phase divided by 2π is, respectively, rational or not.

Notice that when $a, b, c, d \in \mathbb{R}$ but $\alpha \notin \mathbb{R}$ (i.e. $\Delta = (a-d)^2 + 4bc < 0$) then necessarily

$$\left|\frac{c\alpha + d}{a - \alpha c}\right| = 1;$$

in fact

$$\left|\frac{c\alpha + d}{a - \alpha c}\right| = \left|\frac{a + d \pm i\sqrt{-\Delta}}{a + d \mp i\sqrt{-\Delta}}\right| = 1.$$

This agrees with the fact that if $a, b, c, d \in \mathbb{R}$ but $\alpha \notin \mathbb{R}$ then X_k has no limit, neither finite, nor infinite (in fact, X_k is always real, so it cannot converge neither to α_1 nor to α_2).

Remark 2.74. Notice that in (2.33) and (2.35) we chose α_1, so $\alpha_1 + 1/\gamma = \alpha_2$.
By choosing α_2 one would obtain $\gamma = c/\sqrt{(a-d)^2 + 4bc}$ and for $\left| \dfrac{c\alpha + d}{a - \alpha c} \right| < 1$
one would have $\lim_{k} X_k = \alpha_2 + 1/\gamma = \alpha_1$.

Remark 2.75. Another method to make X_k explicit is described in Exercise 2.28. One can obtain qualitative information on the asymptotic behavior of X_k in a more simple way with the general techniques of Chap. 3 (see Exercise 3.40).

Remark 2.76. To find all initial data X_0 that generate the entire sequence by iterating (2.30) it is necessary to determine the values of X_0 that do not lead to $-d/c$ in a finite number of steps, i.e. it is necessary to exclude all values Z_k such that

$$\begin{cases} Z_{k+1} = f^{-1}(Z_k) \\ Z_0 = -\dfrac{d}{c} \end{cases}$$

where $f(x) = (ax + b)/(cx + d)$. By setting $z = (ax + b)/(cx + d)$, one obtains

$$x = f^{-1}(z) = (b - dz)/(cz - a).$$

Notice that when $a + d = 0$ (and then $cZ_0 = a$), the unique forbidden value for X_0 is $-d/c$. In fact, in this case, $X_0 \neq -d/c$ implies

$$X_{k+1} = \frac{aX_k + b}{cX_k - a} \neq \frac{a}{c} = -\frac{d}{c} \qquad \forall k.$$

Exercise 2.27. We define *golden section of a segment*, the longest of the two parts obtained by dividing the given segment in two, in such a way that the square of the length of the longest part is equal to the product of the lengths of the given segment times the length of the remaining part.
Compute the length of golden section in terms of the entire segment.

Exercise 2.28. The golden section of a segment of unit length has length x equal to the positive solution of

$$x = \frac{1}{1 + x}$$

and "by substituting the value of x in the right-hand side"

$$x = \frac{1}{1 + \dfrac{1}{1 + x}}.$$

By iterating the procedure (deduce x by the first equation and substitute it in the last found equation):

$$x = \cfrac{1}{1 + \cfrac{1}{1 + \cfrac{1}{1 + \cdots}}}$$

This last right-hand side is also called *continued fraction*.
This complicated (and vague ...) expression can be made precise by constructing a sequence X_k such that

$$X_{k+1} = \frac{1}{1 + X_k}$$

and checking if, for appropriate initial data X_0, X_k approximates x.
(1)   Calculate the explicit form of X_k.
(2)   Show that X_k tends to the golden section x for every initial datum $X_0 \neq -1$.

Exercise 2.29. Solve the equation $X_{k+1}X_k + 2X_{k+1} + 4X_k + 9 = 0$.

2.8 The Discrete Fourier Transform

The discrete Fourier transform (DFT) is a very useful tool to treat periodic sequences.

Definition 2.77. *A sequence X is said to be **periodic** with period N, where $N \in \mathbb{N}$, if $X_k = X_{k+N}$ holds for every k in $\mathbb{N}$.*

We observe that a periodic sequence X with period N is completely described by a finite number of consecutive terms, that is by the N-dimensional vector $\mathbf{X} = \begin{bmatrix} X_0 & X_1 & \cdots & X_{N-1} \end{bmatrix}^T$.

Example 2.78. The N-th complex roots of unity $\{z \in \mathbb{C} : z^N = 1\}$ are exactly N in number and are expressed as $e^{2\pi ki/N}$ with $k = 0, 1, \ldots, N-1$. Therefore, by setting $\omega = e^{2\pi i/N}$, the sequence X defined by $X_k = \omega^k$ is N-periodic and covers a trajectory that has exactly N points on the unit circle: the vertices of an N-sided regular polygon centered at the origin.   □

Remark 2.79. If w is an N-th complex root of unity (i.e. $w \in \mathbb{C}$ and $w^N = 1$), then

$$\sum_{k=0}^{N-1} w^k = \begin{cases} N & \text{if } w = 1 \\ 0 & \text{if } w \neq 1. \end{cases}$$

In fact, the case $w = 1$ is trivial and the other ones follow by

$$0 = w^N - 1 = (w - 1)\left(w^{N-1} + w^{N-2} + \cdots + w + 1\right).$$ □

If f is a real function in one real variable which is continuous, differentiable with continuous derivative[8] and periodic with period 2π, then, for suitable numbers c_n,

$$f(x) = \sum_{n \in \mathbb{Z}} c_n e^{inx}.\tag{2.38}$$

---

[8] Actually it is sufficient that f is 2π periodic, continuous and that it is possible to divide $[0, 2\pi]$ in a finite number of intervals $I_k = [a_k, b_k]$ such that f has continuous derivative in each I_k and $f'_+(a_k)$ and $f'_-(b_k)$ exist (finite) for each k.

The right-hand side of the previous equality is called **Fourier series** expansion of the function f. The **Fourier coefficients** c_n are computed by

$$c_n = \frac{1}{2\pi} \int_0^{2\pi} f(t) e^{-int} dt \,. \tag{2.39}$$

Assume now that N is an even number (actually, the best situation from the viewpoint of the numerical calculations is when N is a power of 2: $N = 2^p$, $p \in \mathbb{N} \setminus \{0\}$). In this case we carry out a numerical approximation of the integral in (2.39) that requires the knowledge of the values of f only at N equally spaced points: we use the **trapezoidal rule** [9], taking into account that $f(0) = f(2\pi)$,

$$C_n = \frac{1}{2\pi} \frac{2\pi}{N} \sum_{k=0}^{N-1} f(2k\pi/N) e^{-in2k\pi/N} = \frac{1}{N} \sum_{k=0}^{N-1} f(2k\pi/N) \left(e^{2\pi i/N} \right)^{-kn}$$

$$\tag{2.40}$$

To improve the notation we set

$$y_k = f(2k\pi/N) \qquad \omega = e^{2\pi i/N} = \cos \frac{2\pi}{N} + i \sin \frac{2\pi}{N} \,.$$

Since ω is an N-th root of 1 and its powers ω^k, $k = 0, 1, \ldots N - 1$, are all N-th complex roots of unity (see Example 2.77), we can write

$$C_n = \frac{1}{N} \sum_{k=0}^{N-1} y_k \, \omega^{-kn} \,. \tag{2.41}$$

We now give a (very useful) interpretation of the approximate Fourier coefficients C_n.

We look for a **trigonometric polynomial** p (a finite sum instead of a series of oscillating terms as in (2.38)) which interpolates f in the following sense:

$$p(x) = \sum_{h=-N/2}^{N/2-1} \gamma_h \, e^{ihx}, \quad p(2k\pi/N) = f(2k\pi/N) = y_k \,, \quad k = 0, 1, \ldots, N-1 \,.$$

$$\tag{2.42}$$

The above conditions are equivalent to the following linear algebraic system of N equations and just as many unknowns γ_h

$$\sum_{h=-N/2}^{N/2-1} \gamma_h \, e^{ih(2k\pi/N)} = y_k \qquad k = 0, 1, \ldots, N-1$$

---

[9] The error made by calculating the integral of a function f on a bounded and closed interval with the trapezoidal rule is small for large N, and can be estimated: if f' is continuous, then $|c_n - C_n| \leq \dfrac{2\pi^2}{N} \max |f'|$; moreover if f'' is continuous, then $|c_n - C_n| \leq \dfrac{\pi^3}{3N^2} \max |f''|$.

which can be rewritten in the form:

$$\sum_{h=-N/2}^{N/2-1} \gamma_h \, \omega^{hk} = y_k \qquad k = 0, 1, \ldots, N-1 \,.$$ (2.43)

We change the unknowns so that the index of summations takes the values $0, 1, \ldots, N-1$, instead of $-N/2, -N/2+1, \ldots, N/2-1$, by setting

$$Y_h = \gamma_h \quad \text{if } 0 \le h < N/2 \,, \qquad Y_h = \gamma_{h-N} \quad \text{if } N/2 \le h < N \,.$$

We observe that the vector $\mathbf{Y} = \begin{bmatrix} Y_0 \ Y_1 \ \cdots \ Y_{N-1} \end{bmatrix}^T$ implicitly defines a periodic sequence Y with period N.

Thanks to the N periodicity of the sequence $\{\omega^k\}$, we have $\omega^{hk} = \omega^{(h-N)k}$, therefore system (2.43) becomes

$$\sum_{h=0}^{N-1} Y_h \, \omega^{hk} = y_k \qquad k = 0, 1, \ldots N-1$$ (2.44)

The $N \times N$ matrix $\mathbb{F}$ with coefficients $\omega^{hk} = \mathbb{F}_{h\,k}$

$$\mathbb{F} = \begin{bmatrix} 1 & 1 & 1 & 1 & \cdots & 1 \\ 1 & \omega & \omega^2 & \omega^3 & \cdots & \omega^{N-1} \\ 1 & \omega^2 & \omega^4 & \omega^6 & \cdots & \omega^{2(N-1)} \\ \vdots & \vdots & \vdots & \vdots & \ddots & \vdots \\ 1 & \omega^{N-1} & \omega^{2(N-1)} & \omega^{3(N-1)} & \cdots & \omega^{(N-1)(N-1)} \end{bmatrix}$$ (2.45)

is a symmetric Vandermonde matrix (see Exercise 2.44): its rows (and columns) are the powers of the distinct numbers $1, \omega, \omega^2, \ldots, \omega^{N-1}$ with exponents $0, 1, \ldots, N-1$. Then $\mathbb{F}$ is invertible, as $\det \mathbb{F} \ne 0$. The inverse matrix can be easily expressed exploiting Remark 2.78: by choosing $w = \omega^k$, one obtains

$$\sum_{h=0}^{N-1} \omega^{hk} = \begin{cases} N & \text{if } k = 0 \text{ or } k \text{ is an integer times } N \\ 0 & \text{otherwise.} \end{cases}$$

As $\overline{(\omega^k)} = \omega^{-k}$, one deduces

$$\left(\mathbb{F} \overline{\mathbb{F}} \right)_{nm} = \sum_{h=0}^{N-1} \omega^{nh} \omega^{-hm} = \sum_{h=0}^{N-1} \omega^{h(n-m)} = \begin{cases} N & \text{if } n = m \\ 0 & \text{otherwise,} \end{cases}$$

e.g. $\mathbb{F} \overline{\mathbb{F}} = N \, \mathbb{I}_N$. Therefore $\mathbb{F}^{-1} = \dfrac{1}{N} \overline{\mathbb{F}}$ and we can explicitly solve system (2.44):

$$Y_n = \frac{1}{N} \sum_{k=0}^{N-1} y_k \, \omega^{-kn} \,.$$ (2.46)

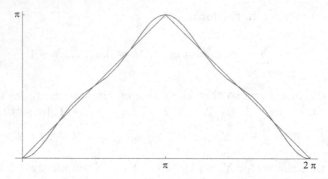

Fig. 2.14 Trigonometric polynomial p that interpolates f (Example 2.79)

By matching (2.40) with (2.44) we find that the Fourier coefficients approximated by the trapezoidal rule coincide with the coefficients of the trigonometric polynomial (2.42) interpolating at the points $X_k = 2k\pi/N$. If we consider the vectors $\mathbf{y} = \begin{bmatrix} y_0\, y_1\, \cdots\, y_{N-1} \end{bmatrix}^T$ and $\mathbf{Y} = \begin{bmatrix} Y_0\, Y_1\, \cdots\, Y_{N-1} \end{bmatrix}^T$, then systems (2.44) and (2.46), with reference to (2.45), are written in the form

$$\mathbf{y} = \mathbb{F}\mathbf{Y} \qquad \mathbf{Y} = \mathbb{F}^{-1}\mathbf{y} = \frac{1}{N}\overline{\mathbb{F}}\mathbf{y}\,.$$

The vector $\mathbf{Y}$ is called **Discrete Fourier Transform** of $\mathbf{y}$, while $\mathbf{y}$ is called inverse discrete Fourier transform of $\mathbf{Y}$:

$$\mathbf{Y} = \mathrm{DFT}\,\mathbf{y}\,, \qquad \mathbf{y} = (\mathrm{DFT})^{-1}\,\mathbf{Y}\,.$$

Therefore, the transformation amounts to left-multiplication by the matrix $\frac{1}{N}\overline{\mathbb{F}}$, while the inverse transformation consists in the left-multiplication[10] by the matrix $\mathbb{F}$ (where $\mathbb{F}$ is defined in (2.45)).

In analogy with what was said for $\mathbf{Y}$, the vector $\mathbf{y}$ implicitly defines a periodic sequence y with period N:

$$y = \{y_h\}_{h\in\mathbb{N}} \quad \text{such that} \quad y_{h+N} = y_h \quad \forall h \in \mathbb{N}\,.$$

Example 2.80. Let us consider the 2π periodic function defined by $f(x) = |x|$ if $|x| < \pi$, and extended periodically to the whole $\mathbb{R}$. Then, in $[0, 2\pi]$ we have $f(x) = x$ if $0 \leq x \leq \pi$ and $f(x) = 2\pi - x$ if $\pi \leq x \leq 2\pi$. We choose

---

[10] In the applications the following variant of the definition is often encountered, which is substantially equivalent: the DFT transformation amounts to left-multiplying by $\frac{1}{\sqrt{N}}\mathbb{F}$, while its inverse transformation amounts to left-multiplying by $\frac{1}{\sqrt{N}}\overline{\mathbb{F}}$.

$N = 8$, $X_k = k\pi/4$, $y_k = f(X_k)$ with $k = 0, \ldots, 7$. We obtain $\omega = e^{i\pi/4}$,

$$\mathbf{y} = \left[\, 0 \quad \frac{\pi}{4} \quad \frac{\pi}{2} \quad \frac{3}{4}\pi \quad \pi \quad \frac{3}{4}\pi \quad \frac{\pi}{2} \quad \frac{\pi}{4} \,\right]^T$$

and the discrete Fourier transform $\mathbf{Y} = \mathrm{DFT}\,\mathbf{y}$ is obtained from (2.46):

$$\mathbf{Y} = \left[\, 0 \quad \frac{-2+\sqrt{2}}{16}\pi \quad 0 \quad \frac{-2-\sqrt{2}}{16}\pi \quad \frac{\pi}{2} \quad \frac{-2-\sqrt{2}}{16}\pi \quad 0 \quad \frac{-2+\sqrt{2}}{16} \,\right]^T .$$

Furthermore, thanks to the identity $\left[Y_0\ Y_1\ \cdots\ Y_7\right]^T = \left[\gamma_{-4}\ \gamma_{-3}\ \cdots\ \gamma_3\right]^T$ we obtain the trigonometric polynomial that interpolates f over nine equally spaced points in $[0, 2\pi]$:

$$p(x) = \frac{\pi}{2} - \frac{\pi}{8}\left(2+\sqrt{2}\right)\cos x + \frac{\pi}{8}\left(\sqrt{2}-2\right)\cos 3x .\qquad \square$$

Let us consider, more in general, the polynomial

$$q(z) = \frac{1}{8}\sum_{k=0}^{7} y_k\, z^k = \frac{1}{8}\left(y_0 + y_1 z + y_2 z^2 + \cdots + y_7 z^7\right) \qquad z \in \mathbb{C}.$$

The actual calculation of each element Y_n of $\mathbf{Y} = \mathrm{DFT}\,\mathbf{y}$ has the **computational complexity**[11] of the evaluation of $q(\omega)$ with $\omega = e^{2\pi i/N}$ and q polynomial of degree $N - 1$.
We evaluate Y_n (of Example 2.79) with the **Ruffini-Horner method**:

i) $q = y_{N-1}$;
ii) for $k = 6, 5, 4, \ldots, 1, 0$;
iii) $q = qz + y_k$.

In the three previous steps for the computation algorithm the notation $a = b$ means the value b is assigned to a (according to the usual convention of computer programs). Therefore, for polynomials of degree 8 the algorithm needs 7 additions and 7 multiplications. In general, it requires as many additions and multiplications as the degree of the polynomial (i.e. $N - 1$). Moreover, $\mathbf{Y}$ has N components, so the total number of multiplications required to calculate the DFT is of the order of N^2. This number can be prohibitive when N is large.

In 1965 J.W.Cooley and J.Tukey devised a faster algorithm, which reduces the number of required multiplications: if $N = 2^p$, then $N\log_2 N = Np$ multiplications are needed, a considerable reduction compared to N^2 (for instance, if one chooses $N = 2^{12} = 4096$, then $p = 12$, $N^2 = 16.777.216$ while $Np = 49.152$).

---

[11] Number of elementary operations to be performed.

The Cooley-Tukey algorithm is called **Fast Fourier Transform** (FFT). Before going into the details we will explain the underlying idea.

Referring to the problem of calculating the DFT, suppose one needs to calculate N values of a polynomial of degree $N - 1$:

$$q(z) = \frac{1}{N} \sum_{k=0}^{N} y_k z^k = \frac{1}{N} \left(y_0 + y_1 z + y_2 z^2 + \cdots + y_{N-1} z^{N-1} \right) \qquad z \in \mathbb{C}.$$

One can exploit the fact that $z = \omega^{-n} = e^{-2\pi i n / N}$ and N is an even number of the kind $N = 2^p$. In fact, under such assumptions, the sought complex values (where q has to be evaluated) come in pairs z_n, $-z_n$ with the same square:

$$z_n = e^{-2\pi i n / N}$$

$$z_n = -z_{n+N/2} \qquad z_n^2 = \left(z_{n+N/2} \right)^2 .$$

In the case of Example 2.79 ($p = 3$, $N = 8$) q can be described by splitting the terms of even and odd degree:

$$q(z) = q_{even}\left(z^2\right) + q_{odd}\left(z^2\right) =$$

$$= \left(y_0 + y_2 z^2 + y_4 z^4 + y_6 z^6 \right) + z \left(y_1 + y_3 z^2 + y_5 z^4 + y_7 z^6 \right)$$

In order to calculate $q(z_n)$ and $q(z_{n+4})$, $n = 1, 2, 3, 4$, having set $u_n = z_n^2 = z_{n+4}^2$ we compute $q_{even}(u_n)$ and $q_{odd}(u_n)$ with $3 + 4 = 7$ multiplications. So

$$q(z_n) = q_{even}(u_n) + q_{odd}(u_n)$$

$$q(z_{n+4}) = q_{even}(u_n) - q_{odd}(u_n),$$

are evaluated by 7 multiplications. Since we have 4 pairs of points, we have a total of $4 \cdot 7 = 28$ multiplications instead of the $8 \cdot 7 = 56$ ones required by the Ruffini-Horner method. Moreover, since $N = 2^p$ the squares also come in opposite pairs with the same square. Then the polynomials q_{even} and q_{odd} can be computed in z_k with the same technique, and so on. This explains the numerical effectiveness of the DFT algorithm.

2.9 Fast Fourier Transform (FFT) Algorithm

Let us consider two vectors $\mathbf{y}$ and $\mathbf{Y}$ with N components, such that $\mathbf{Y} = DFT\mathbf{y}$:

$$y_k = \sum_{n=0}^{N-1} Y_n \omega^{kn} \qquad k = 0, 1, \ldots, N - 1$$

$$\tag{2.47}$$

$$Y_n = \frac{1}{N} \sum_{k=0}^{N-1} y_k \omega^{-kn} \qquad n = 0, 1, \ldots, N - 1$$

Assume N is decomposable into a product of integers: $N = p_1 p_2$ with $p_1 \neq 1 \neq p_2$. Hence the indices n and k can be written in the form:

$$\begin{cases} n = n_1 p_2 + n_0 & n_0 = 0, 1, \ldots, p_2 - 1 & n_1 = 0, 1, \ldots, p_1 - 1 \\ k = k_1 p_1 + k_0 & k_0 = 0, 1, \ldots, p_1 - 1 & k_1 = 0, 1, \ldots, p_2 - 1 \end{cases}$$

Taking into account the identity $\omega^{-N} = 1$ we deduce

$$\omega^{-n(k_1 p_1 + k_0)} = \omega^{-n k_0} \omega^{-(n_1 p_2 + n_0) k_1 p_1} = \omega^{-n k_0} \omega^{-n_0 k_1 p_1}$$

and substituting in (2.47)

$$\begin{aligned} Y_n &= \frac{1}{N} \sum_{k_1=0}^{p_2-1} \sum_{k_0=0}^{p_1-1} y_{k_1 p_1 + k_0} \, \omega^{-n(k_1 p_1 + k_0)} = \\ &= \frac{1}{p_1} \sum_{k_0=0}^{p_2-1} \omega^{-n k_0} \left(\frac{1}{p_2} \sum_{k_1=0}^{p_1-1} y_{k_1 p_1 + k_0} \, \omega^{-n_0 k_1 p_1} \right) = \qquad (2.48) \\ &= \frac{1}{p_1} \sum_{k_0=1}^{p_2-1} \omega^{-n k_0} \, \widetilde{y}_{k_0, n_0} \end{aligned}$$

where $\widetilde{y}_{k_0, n_0} = \dfrac{1}{p_2} \sum_{k_1=0}^{p_1-1} y_{k_1 p_1 + k_0} \, \omega^{-n_0 k_1 p_1}$.

In the terminology of computer science a multiplication followed by an addition corresponds to a **flops**[12]. The direct calculation of **Y** starting from **y** by using the second of (2.47) would require N^2 flops for each component. We observe that (2.48) has reduced the number of flops under the assumption $N = p_1 p_2$: p_2 operations are needed to compute $\widetilde{y}$, and with the substitution in (2.48) we have p_1 more. In conclusion, to compute each Y_n, $n = 0, \ldots, N-1$, we perform $N(p_1 + p_2)$ flops.

If N is further factorized $N = p_1 p_2 \cdots p_s$, then the number of flops reduces to $N(p_1 + p_2 + \cdots + p_s)$. If $N = p^s$ we obtain $Nps = pN \log_p N$ flops. In particular:

$$\boxed{\text{if } N = 2^s \quad \text{then FFT corresponds to} \quad 2N \log_2 N \text{ flops}}$$

Example 2.81. With reference to the terminology of Signal Theory, we consider a **discrete signal**, namely a sequence of numbers having integer [13] indices $X = \{X_k\}_{k \in \mathbb{Z}}$.

---

[12] More precisely, flops (**fl**oating **op**erations **p**er **s**econd) is the number of operations corresponding to the FORTRAN instruction $A(I, J) = A(I, J) + T * A(I, K)$: it includes an addition and a multiplication in floating-point arithmetic, some calculations of indexes and some references to memory. It also meets the modern definition of flops as a floating-point operation; in this way a flops in the first definition corresponds to two flops in the second.

[13] Throughout the book, except that in Example 2.81, the index of a sequence is always natural ($n \in \mathbb{N}$). However, to describe the action of a causal filter of a periodic signal the choice of indices $k \in \mathbb{Z}$ is the natural one.

Let us consider a **system** or **causal filter**, that is a transformation that associates to the discrete input signal X the output signal $Y = \{Y_k\}_{k \in \mathbb{Z}}$, according to the law:

$$Y_n = \sum_{m=-\infty}^{+\infty} K_{n-m} X_m \qquad n \in \mathbb{Z} \qquad\qquad (2.49)$$

$$X \quad \rightarrow \quad \boxed{\text{filter}} \quad \rightarrow \quad Y$$

where $K = \{K_m\}_{m \in \mathbb{Z}}$ is a given sequence that verifies $\sum_m |K_m| < +\infty$ and $K_m = 0$ if m is less than zero[14]. We observe first that, if the input X is N periodic, then also the output Y is N periodic. In fact, $\forall n \in \mathbb{Z}$:

$$Y_{n+N} = \sum_{m=-\infty}^{+\infty} K_{n+N-m} X_m = \sum_{m=-\infty}^{+\infty} K_{n-m} X_{m-N} = \sum_{m=-\infty}^{+\infty} K_{n-m} X_m = Y_n.$$

Actually the N-periodic signals X and Y are described by vectors with N components which we denote by $\mathbf{X} = [X_0\ X_1\ \cdots\ X_{N-1}]^T$ and $\mathbf{Y} = [Y_0\ Y_1\ \cdots\ Y_{N-1}]^T$. An important problem is that of solving the filter (2.49), which mean determining $\mathbb{D}$ such that $\mathbf{X} = \mathbb{D}\mathbf{Y}$.

System (2.49) can be represented by a linear map of $\mathbb{R}^n$ to itself (see Appendix D): instead of calculating the series in (2.49) one computes finite sums that correspond to the multiplication of a matrix by a vector. Therefore, we try to determine whether there exists a matrix $\mathbb{K}$ of order $N \times N$ such that $\mathbf{Y} = \mathbb{K}\mathbf{X}$ and, if so, compute it: if $\mathbb{D}$ exists then it must equal to $\mathbb{K}^{-1}$. For this purpose we define[15] the vector $\mathbf{K}$ of components κ_n, with

$$\kappa_n = \sum_{j=-\infty}^{+\infty} K_{n+jN} \qquad n = 0, \ldots, N-1 .$$

Then

$$Y_n = \sum_{m=-\infty}^{+\infty} K_{n-m} X_m = \sum_{p=-\infty}^{+\infty} \sum_{l=0}^{N-1} K_{n-(l-pN)} X_{l-pN} =$$

$$= \sum_{p=-\infty}^{+\infty} \sum_{l=0}^{N-1} K_{n+pN-l} X_l = \sum_{l=0}^{N-1} X_l \sum_{p=-\infty}^{+\infty} K_{n+pN-l} = \sum_{l=0}^{N-1} \kappa_{n-l} X_l.$$

Therefore, (2.49) has an N dimensional formulation when X is N periodic. Let $\mathbb{K}$ be an $N \times N$ matrix such that $\mathbb{K}_{nm} = \kappa_{n-m}$. Then:

$$\mathbf{Y} = \mathbb{K}\mathbf{X} \qquad\qquad (2.50)$$

---

[14] This last property characterizes the casuality of the filter: to a (non-periodic) input X such that $X_k = 0$ if $k < 0$, there corresponds an output Y such that $Y_k = 0$ if $k < 0$.

[15] We assume $\sum_{m=-\infty}^{+\infty} |K_m| < +\infty$, a necessary and sufficient condition for bounded input signals to give bounded output signals (a property called stability of the filter).

which is a system of N equations and can be further simplified by transforming it into N decoupled scalar equations.

We observe that $\mathbb{K}$ is a **circulant matrix**, for each row is a cyclic permutation of the previous row.

We take the inverse Discrete Fourier Transform (DFT^{-1}) of both sides of (2.50) and, by setting $\mathbf{x} = (\text{DFT})^{-1}\,\mathbf{X}$, $\mathbf{y} = (\text{DFT})^{-1}\,\mathbf{Y}$, $\mathbf{k} = (\text{DFT})^{-1}\,\mathbf{K}$, we obtain

$$
\begin{aligned}
y_n &= \sum_{k=0}^{N-1}\left(\sum_{m=0}^{N-1}\kappa_{k-m}\,X_m\right)\omega^{kn} = \\
&= \sum_{k=0}^{N-1}\sum_{m=0}^{N-1}\kappa_{k-m}\,\omega^{(k-m)n}\,X_m\,\omega^{mn} = \\
&= \sum_{m=0}^{N-1}X_m\,\omega^{mn}\sum_{k=0}^{N-1}\kappa_{k-m}\,\omega^{(k-m)n} = \\
&= x_n\,\mathsf{K}_n \qquad\qquad n = 0,\dots,N-1
\end{aligned}
$$

where $\mathbf{x} = [\,x_0\ x_1\ \cdots\ x_{N-1}\,]^T$, $\mathbf{y} = [\,y_0\ y_1\ \cdots\ y_{N-1}\,]^T$, $\mathbf{k} = [\,\mathsf{K}_0\ \mathsf{K}_1\ \cdots\ \mathsf{K}_{N-1}\,]^T$. The N scalar equations $y_n = x_n\,\mathsf{K}_n$ are decoupled: if $\mathsf{K}_n \neq 0$, $\forall n$, one obtains

$$
x_n = \frac{y_n}{\mathsf{K}_n},
$$

and, by computing the transform $(X = \text{DFT}\,x)$ we get:

$$
X_n = \frac{1}{N}\sum_{h=0}^{N-1}\frac{y_h}{\mathsf{K}_h}\,\omega^{-hn}\ .
$$

Since $\mathbf{y} = (\text{DFT})^{-1}\mathbf{Y}$, X_n can be expressed in terms of Y_n as:

$$
X_n = \frac{1}{N}\sum_{h=0}^{N-1}\frac{\displaystyle\sum_{j=0}^{N-1}Y_j\omega^{jh}}{\mathsf{K}_h}\,\omega^{-hn}
$$

i.e.,

$$
\begin{aligned}
\mathbf{X} &= \frac{1}{N}\overline{\mathbb{F}}\left(\text{diag}\left(\mathsf{K}_0^{-1},\mathsf{K}_1^{-1},\dots,\mathsf{K}_{N-1}^{-1}\right)\right)\mathbb{F}\,\mathbf{Y} = \\
&= \text{DFT}\left(\text{diag}\left(\mathsf{K}_0^{-1},\mathsf{K}_1^{-1},\dots,\mathsf{K}_{N-1}^{-1}\right)(\text{DFT})^{-1}\mathbf{Y}\right)
\end{aligned}
$$

where

$$
\text{diag}\left(\mathsf{K}_0^{-1},\mathsf{K}_1^{-1},\dots,\mathsf{K}_{N-1}^{-1}\right) =
\begin{bmatrix}
\mathsf{K}_0^{-1} & 0 & \cdots & 0 \\
0 & \mathsf{K}_1^{-1} & \cdots & 0 \\
\vdots & \vdots & \ddots & \vdots \\
0 & 0 & \cdots & \mathsf{K}_{N-1}^{-1}
\end{bmatrix}.
$$

In conclusion:

$$
\mathbb{D} = \frac{1}{N}\overline{\mathbb{F}}\left(\text{diag}\left(\mathsf{K}_0^{-1},\mathsf{K}_1^{-1},\dots,\mathsf{K}_{N-1}^{-1}\right)\right)\mathbb{F}\ .
$$

Eventually, we observe that when N is large, instead of calculating the matrix product, it is more convenient to compute transforms and inverse transforms by FFT. □

Definition 2.82. *We call **circular convolution of two** N**-dimensional** vectors* $\mathbf{x} = \begin{bmatrix} x_0 \; x_1 \; \cdots \; x_{N-1} \end{bmatrix}^T$ *and* $\mathbf{y} = \begin{bmatrix} y_0 \; y_1 \; \cdots \; y_{N-1} \end{bmatrix}^T$ *the sum:*

$$(\mathbf{x} * \mathbf{y})_k = \sum_{j=0}^{N-1} x_j y_{k-j} \qquad k = 0, 1, \ldots, N-1$$

where y_{k-j} *conventionally denotes the value* y_{k-j+N} *if* $k - j < 0$.

We emphasize the analogy between the circular convolution of vectors and the discrete convolution of sequences (see Definition 2.45 and Exercise 2.47). Although these operations are different, the choice of denoting both by the same symbol does not create ambiguity because these operations are performed on objects of different nature.

We illustrate some useful relations between circular convolution and DFT.

Theorem 2.83. *Given two* N*-dimensional vectors* $\mathbf{x}$ *and* $\mathbf{y}$ *with discrete Fourier transforms* $\widehat{\mathbf{x}} = DFT\mathbf{x}$, $\widehat{\mathbf{y}} = DFT\mathbf{y}$, *we have:*

$$
\begin{array}{l}
(I) \quad \mathbf{x} = N \overline{\widehat{\overline{\widehat{\mathbf{x}}}}} \\[2mm]
(II) \quad \widehat{\mathbf{x} * \mathbf{y}} = N \widehat{\mathbf{x}} \widehat{\mathbf{y}} \\[2mm]
(III) \quad \widehat{\mathbf{x}\mathbf{y}} = \widehat{\mathbf{x}} * \widehat{\mathbf{y}}
\end{array}
$$

Proof. **(I)** $\mathbf{x} = (DFT)^{-1} \widehat{\mathbf{x}} = \mathbb{F}\widehat{\mathbf{x}} = \overline{\overline{\mathbb{F}\widehat{\mathbf{x}}}} = N \overline{\widehat{\overline{\widehat{\mathbf{x}}}}}$.

(II) In all sums the index runs from 0 to $N - 1$:

$$\left(\widehat{\mathbf{x} * \mathbf{y}} \right)_k = \frac{1}{N} \sum_h \omega^{-kh} \sum_l x_l y_{h-l} = \frac{1}{N} \sum_{h,l} \omega^{-k(h-l)} y_{h-l} \omega^{-kl} x_l =$$

$$= \frac{1}{N} \sum_l \omega^{-kl} x_l \sum_h \omega^{-k(h-l)} y_{h-l} =$$

$$= N \left(\frac{1}{N} \sum_l \omega^{-kl} x_l \right) \left(\frac{1}{N} \sum_h \omega^{-k(h-l)} y_{h-l} \right) =$$

$$= N \left(\widehat{\mathbf{x}} \right)_k \left(\widehat{\mathbf{y}} \right)_k .$$

(III) By choosing $\mathbf{x} = \overline{\overline{\mathbf{u}}}$, $\mathbf{y} = \overline{\overline{\mathbf{v}}}$ in (II), combining and taking (I) into account, we obtain:

$$\overline{\overline{\widehat{\mathbf{u}}}} * \overline{\overline{\widehat{\mathbf{v}}}} = N \overline{\overline{\widehat{\mathbf{u}}}} \, \overline{\overline{\widehat{\mathbf{v}}}} = \frac{1}{N} \mathbf{u}\mathbf{v} .$$

By transforming $\mathbf{u}\mathbf{v} = N\overline{\widehat{\mathbf{u}}} * \overline{\widehat{\mathbf{v}}}$, again taking (I) into account, we obtain:

$$\widehat{\mathbf{u}\mathbf{v}} = \overline{\overline{\widehat{\mathbf{u}}}} * \overline{\overline{\widehat{\mathbf{v}}}} = \widehat{\mathbf{u}} * \widehat{\mathbf{v}} . \qquad \square$$

2.10 Summary exercises

Exercise 2.30. Compute the minimum number of moves X_k described recursively in the solution of Exercises 1.9 and 1.10.

Exercise 2.31. Given $\lambda \in \mathbb{R}$, find the solution of the problem

$$\begin{cases} X_{k+1} = \dfrac{1}{3}X_k - 2 \\ X_0 = \lambda \end{cases}$$

Study the asymptotic behavior of X_k for $\lambda \in \mathbb{R}$, and plot the sequence for $\lambda = 0$.

Exercise 2.32. Given $\lambda \in \mathbb{R}$, determine the solution of the equation

$$\begin{cases} X_{k+1} = -X_k + 3 \\ X_0 = \lambda \end{cases}$$

Study the asymptotic behavior for $\lambda \in \mathbb{R}$. Plot the sequence for $\lambda = 0, 5$.

Exercise 2.33. Determine the general solution of the following homogeneous difference equations:

1) $X_{k+3} + X_k = 0 \quad \forall k \in \mathbb{N}$ 2) $X_{k+4} - X_k = 0 \quad \forall k \in \mathbb{N}$

3) $X_{k+4} + X_k = 0 \quad \forall k \in \mathbb{N}$ 4) $X_{k+4} + 8X_{k+2} + 16X_k = 0 \quad \forall k \in \mathbb{N}$.

Exercise 2.34. Solve the problem

$$\begin{cases} X_{k+2} - X_{k+1} - 2X_k = k \quad k \in \mathbb{N} \\ X_0 = 0, \quad X_1 = 1 \end{cases}$$

Exercise 2.35. Solve the problem:

$$\begin{cases} X_{k+2} - 5X_{k+1} + 6X_k = K_k^0 \\ X_0 = X_1 = 0 \end{cases} \quad \forall k \in \mathbb{N}$$

where K_k^0 denotes the *Kronecker impulse* at 0: $K_0^0 = 1$, $K_k^0 = 0$, $\forall k > 0$.

Exercise 2.36 (Gambler's ruin). A player makes a sequence of bets, each one with fixed value (for instance, one Euro) and winning probability $p \in (0, 1)$ (the probability p is known and independent of the bet sequence).

The player begins with k Euros and the game ends when the player loses everything or if he gather s Euros (where s is known and $0 \le k \le s$). Compute the probability R_k that starting with k Euros the player loses all them, having set s Euros as target.

Exercise 2.37 (Critical analysis of the cobweb model). In the model of Example 2.8 assume more realistically that the supply function is

$$Q_{k+1}^o = -c + dP_{k+1}^e$$

where P_{k+1}^e is the *expected price* at time $k+1$ based on *expectations* that are formed in the following way:

1. (*normal price*) the expected price depends on from what the current price deviates from the normal price P_N which, for simplicity, we assume is equal to the

equilibrium price $P_E = (a + c) / (b + d)$, that the producer thinks that sooner or later will be the price at which it will sell his product:

$$P_{k+1}^e = P_k + \delta (P_N - P_k) \qquad \delta \in (0, 1) \, ;$$

2. (*adaptive expectations*) expectations are adjusted each period according to the discrepancy between what is expected and what is observed:

$$P_{k+1}^e - P_k^e = \beta (P_k - P_k^e) \qquad \beta \in (0, 1) \, .$$

In both cases, determine the closed-form expression of the dynamic equation of the price, comparing the results obtained with those of Example 2.8.

Exercise 2.38. Express the coefficients of the power series of the particular solutions identified in Exercise 1.14 for the equations of Bessel, Hermite, Laguerre and Legendre. Express the related solutions.

Exercise 2.39. Consider a financial transaction involving k payments of a constant (unit) amount to k annual maturities (starting from the end of the first year). Compute the *present value* A_k of the financial transaction, that is the equivalent of an immediate payment, in case of constant annual compound rate of interest r. Write a recursive relationship for A_{k+1} and A_k, determining the closed-form solution.
Answer the questions above in the case the rate r_k depends on k.

Exercise 2.40. A mortgage of amount M is borrowed and will be repaid in k constant payments. Assuming the annual compound interest rate r is constant and the payments are made at date 1, 2, ..., k, compute the amount Q of each payment.

Exercise 2.41. A **reverse-floater** bond gives coupons that vary over time according to a formula, which is established at issue time but depends on parameters related to the performance of economic indicators. Consider for example a bond over a period of 10 years which provides an annual coupon of $6,5\%$ for the first two years and thereafter an interest rate equal to

$$r_k = \text{maximum between } \{1, 5 \text{ and } 15 - 2\tau_k\}$$

where τ_k is the Euribor 12 months.

a) Calculate the maximum and the minimum annual coupon interest rate achievable in each year.

b) The bond considered has a high component of risk/opportunity to gain, not less than that of an equity. In particular, the price of the bond on the secondary market[16] will be subject to significant fluctuations. To illustrate this situation, evaluate the price P_k to which the bond can be sold in year k to a buyer who expects a coupon payment of not less than τ_k, and assume that there are no changes in the Euribor until maturity.

b) Based on the assumption of the previous point, rate the balance of the transaction supposing that the bond is sold in year k (without having in the meantime reinvested the coupons received).

---

[16] In the secondary market are dealt bonds before their maturity.

b)   If the Euribor index takes values 5, 5.1, 5.2, 4.9, 4, 3, 2.5, 2.6, 4, 7.5 in the years of life of the bond, what is the balance of the transaction if you sell the bond at maturity?

Exercise 2.42. Prove Theorem 2.46 about the $\mathcal{Z}$-transform of the discrete convolution.

Exercise 2.43. Show that the following relation holds

$$S_{k+1} = \frac{1}{k+2}\left[(n+1)^{k+2} - (n+1) - \binom{k}{2}S_k - \cdots - \binom{k+2}{k+1}S_1\right],$$

having set $S_k = 1^k + 2^k + 3^k + \cdots + n^k$.

Exercise 2.44. The proof of Theorem 2.15 exploits the knowledge of the Vandermonde determinant. Check the correctness of the expression used.

Exercise 2.45. Evaluate the $\mathcal{Z}$-transform of the sequence $X_k = \dfrac{1}{k!}$.

Exercise 2.46 (Elastic bar supported by pillars). Consider a metal bar made of homogeneous material and uniform cross-section, resting on $N-1$ equally spaced pillars (Fig. 2.15) and subject to loads W at both endpoints.
In the absence of other loads, it can be shown that the bending moments at the supports verify the two-steps equation, called *equation of three moments* (see Example 1.22),

$$M_{k-1} + 4M_k + M_{k+1} = 0 \qquad 1 \le k \le N-1$$

and that the bending moments at the ends are given by

$$M_0 = M_N = -Wd.$$

Determine the bending moments at each contact point with a pillar (note that instead of two initial conditions, we have an initial condition and a final one, also called *boundary conditions*).

Exercise 2.47. Consider the following two vectors $\mathbf{X} = \begin{bmatrix} X_0 \ X_1 \ \cdots \ X_{N-1} \end{bmatrix}^T$ and $\mathbf{Y} = \begin{bmatrix} Y_0 \ Y_1 \ \cdots \ Y_{N-1} \end{bmatrix}^T$ as well as two sequences X and Y whose first N terms coincide respectively with the components of $\mathbf{X}$ and $\mathbf{Y}$, while $X_k = Y_k = 0$ if $k \ge N$. Prove that $(X * Y)_k = 0$ if $k > 2N-2$ and

$$(\mathbf{X} * \mathbf{Y})_k = (X * Y)_k + (X * Y)_{k+N} \qquad k = 0, 1, \ldots, N-1.$$

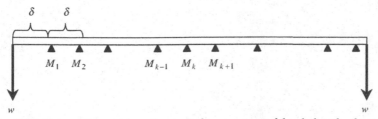

Fig. 2.15 Bar resting on equally spaced supports and loaded at both ends

3

Discrete dynamical systems: one-step scalar equations

In this chapter we tackle the study of nonlinear problems. We deal with models describing sequences of measurements, which are spaced by discrete steps of prescribed but not necessarily constant length; often, but not always, the steps stand for time-steps and the sequences are the related time series. For this reason these models are called discrete dynamical systems.

It is useful to study such models without prescribing the initial condition, since in many cases we have only an approximate knowledge of it. Moreover it is interesting to understand the qualitative behavior of all orbits, and to find whether they share common structural properties or whether they can be partitioned into subsets characterized by different structural properties.

A relevant difference between the linear and the nonlinear context is the fact that it is not possible to give an explicit formula of the general solution in the nonlinear case, except in few particular cases. Anyway it is convenient to perform a local comparison with "close" linear discrete dynamical systems (linearization method) with the aim of acquiring information about the qualitative behavior of the orbits. Due to the remarkable underlying complexity of nonlinear models, here we deal only with discrete dynamical systems which are one-step, scalar-valued, and autonomous.

3.1 Preliminary definitions

A **discrete dynamical system**, denoted by DDS for short, is the formal description of an evolutive phenomenon in terms of a map whose image is contained in its domain: starting from any admissible initial value, a sequence of values is generated by the iterated computation of the given map.

We introduce a formal definition of discrete dynamical system, which includes all the scalar models we plan to study in this chapter.

E. Salinelli, F. Tomarelli: *Discrete Dynamical Models.*
UNITEXT – La Matematica per il 3+2 76
DOI 10.1007/978-3-319-02291-8_3, © Springer International Publishing Switzerland 2014

Definition 3.1. *Assume $I \subset \mathbb{R}$ is an interval containing at least two distinct points and $f : I \to I$ is a continuous function. Then the pair $\{I, f\}$ is called* **discrete dynamical system in I, of the first order, autonomous, in normal form.**

Remark 3.2. Notice that, given a DDS $\{I, f\}$ and an initial value $X_0 \in I$, a sequence $\{X_k\}$ is uniquely defined by the iterative evaluation of f:

$$X_{k+1} = f(X_k) \qquad\qquad k \in \mathbb{N},$$

and the sequence fulfils

$$
\begin{aligned}
X_1 &= f(X_0) \\
X_2 &= f(X_1) = f(f(X_0)) \\
X_3 &= f(X_2) = f(f(f(X_0))) \\
\cdots &= \cdots \\
X_k &= f(X_{k-1}) = f(f(\ldots(f(X_0)))) = \underbrace{f \circ f \circ \cdots \circ f}_{k \text{ times}}(X_0) .
\end{aligned}
$$

All the above computations are well defined since $f(I) \subseteq I$.
On the other hand, any event described by a recursive law can be understood as a sequence generated by a DDS starting from a particular initial value.

Example 3.3. We list some examples of DDS $\{I, f\}$ together with the related recursive law:

$$f(x) = \frac{1}{2}x \qquad I = [0, 1] \qquad\qquad X_{k+1} = \frac{1}{2}X_k$$

$$f(x) = 5x \qquad I = \mathbb{R} \qquad\qquad X_{k+1} = 5X_k$$

$$f(x) = 3x + 2 \qquad I = \mathbb{R} \qquad\qquad X_{k+1} = 3X_k + 2$$

$$f(x) = \frac{x}{1+x} \qquad I = (0, +\infty) \qquad\qquad X_{k+1} = \frac{X_k}{1+X_k}. \qquad\qquad \square$$

To denote the **k-th iterate** of f we adopt the following notation

$$f^0 = \text{ identity map}$$
$$f^k = \underbrace{f \circ f \circ \cdots \circ f}_{k \text{ times}} .$$

Thus, referring to the DDS $\{I, f\}$, we have $X_k = f^k(X_0)$, $\forall k \in \mathbb{N}$.

Remark 3.4. Maybe the reader is used to employ the (useful, but formally incorrect) abuse of notation which omits the parentheses to denote the product of a computed value times itself: for instance, $\sin^2 x$ instead of $(\sin x)^2$.

Nonetheless such habit should be limited to the cases where it does not create ambiguity with the composition, but it must be avoided in cases, like the present context, where $\sin^2 x$ more aptly denotes $\sin(\sin x)$.

Definition 3.5. *A DDS $\{I, f\}$ is called **linear** if f is linear; **affine** if f is affine; **non linear** if f is neither linear nor affine.*

Example 3.6. Referring to Example 3.3, the first two DDS are linear, the third one is affine, the fourth is nonlinear. □

We emphasize that (except in very peculiar cases, of which we will discuss just few examples) closed formulae for the solution of a nonlinear DDS are not available. Precisely one does not know a function dependent only on k and X_0 which provides the value of X_k. This fact explains the paramount importance of the qualitative analysis of the sequences fulfilling the recursive relationship, of their asymptotic behavior and of the connections between such behavior and the initial conditions.

Definition 3.7. *Given a DDS $\{I, f\}$ and $X_0 \in I$, the sequence defined by*

$$\{X_0, X_1, X_2, \ldots, X_k, \ldots\} = \{X_0, f(X_0), f^2(X_0), \ldots, f^k(X_0), \ldots\}$$

*is called **orbit** (or **trajectory**) of the DDS $\{I, f\}$ associated to the initial value X_0, and is denoted by $\gamma(I, f, X_0)$.*

Definition 3.8. *The whole set of orbits of a DDS $\{I, f\}$ associated to all $X_0 \in I$ is called **phase portrait**.*

In order to understand a DDS, it is very helpful to plot its phase portrait, where the dynamics is described by arrows associated to subsequent moves under the iteration of the function driving the DDS.

Example 3.9. Consider the DDS $\{\mathbb{R}, x^4\}$ and represent $\mathbb{R}$ with an (oriented) vertical line and draw the phase portrait. We notice that $f(0) = 0$, $f(1) = 1$ and $f(x) > x$ if $x > 1$, $0 < f(x) < x$ if $0 < x < 1$ and $f(x) > 0$ if $x < 0$. The graphical representation in Fig. 3.1 highlights that all orbits fall in the nonnegative region from the first iteration on; moreover, if $0 \le |x| < 1$ the orbit $f^k(x)$ tends to zero, while if $|x| > 1$ the orbit $f^k(x)$ converges to $+\infty$ (the values 0 and 1 match two constant orbits). □

Example 3.10. Figure 3.2 shows the phase portrait of the discrete dynamical system $\{[-\pi, \pi], \cos x\}$. □

Example 3.11. The set of infinitely many coupled equations of the kind

$$X_{k+1} - X_k = g(X_k) \qquad\qquad k \in \mathbb{N} \qquad\qquad (3.1)$$

is called *first order difference equation*. We notice that (3.1) is equivalent to the DDS $\{I, f\}$ whenever, after setting $f(x) = g(x) + x$, one can find an

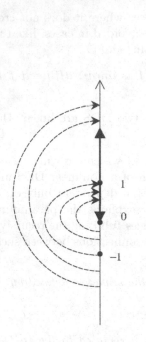

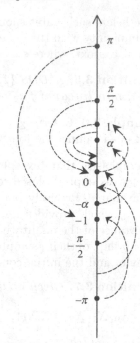

Fig. 3.1. DDS $\{\mathbb{R}, x^4\}$ **Fig. 3.2.** DDS $\{[-\pi, \pi], \cos x\}$

interval $I \subseteq \mathbb{R}$ such that $f(I) \subseteq I$: this means that under these assumptions the phase portrait of the DDS $\{I, f\}$ coincides with the set of all solutions of the difference equation (3.1) with initial value X_0 in I.   □

Remark 3.12. Unfortunately some discrete mathematical models are described by a DDS which is led by a function f whose image is not a subset of the domain of f (for instance $f(x) = \log x$), or by a function whose domain is not even an interval (for instance $f(x) = 1/x$). However, also in such cases, it is important to study the difference equation[1] $X_{k+1} = f(X_k)$, which is well-posed only for the choices of the initial value X_0 such that $f(X_k)$ belongs to the domain of f for all k.

Definition 3.13. *A number* $\alpha \in \mathbb{R}$ *is called* **equilibrium** *(or* **fixed point** *or* **stationary point***) of the DDS* $\{I, f\}$ *if*

$$\boxed{\quad \alpha = f(\alpha) \qquad \alpha \in I \quad}$$

If we start from an initial datum $X_0 = \alpha$ *which is an equilibrium of the DDS, then the resulting orbit is constant and is called* **stationary orbit***:*

$$\gamma(I, f, \alpha) = \{\alpha, \alpha, \alpha, \dots\}.$$

---

[1] Obviously equivalent to $X_{k+1} - X_k = g(X_k)$ where $g(x) = f(x) - x$.

*If, given $X_0 \in I$ there exists k such that $f^k(X_0) = \alpha = f(\alpha)$, correspondingly the trajectory is called **eventually stationary**:*

$$\gamma(I, f, X_0) = \left\{ X_0, f(X_0), f^2(X_0), \ldots, f^{k-1}(X_0), \alpha, \alpha, \alpha, \ldots \right\}.$$

Remark 3.14. The search of equilibria of a DDS $\{I, f\}$ is equivalent to looking for intersections of the line $y = x$ and the graph of f, contained in the subset $I \times I$ of the $x\,y$ plane.

Remark 3.15. If α is an equilibrium of the DDS $\{I, f\}$, then α is also an equilibrium of the DDS $\{I, f^k\}$ for all $k \in \mathbb{N}$.

Definition 3.16. *A **cycle of order** s (or **periodic orbit** of (least) **period** s, or s **cycle**) of the DDS $\{I, f\}$ is a set of s distinct points in the interval I $\{\alpha_0, \alpha_1, \ldots, \alpha_{s-1}\}$ that fulfill*

$$\alpha_1 = f(\alpha_0), \quad \alpha_2 = f(\alpha_1), \quad \ldots, \quad \alpha_{s-1} = f(\alpha_{s-2}), \quad \alpha_0 = f(\alpha_{s-1}).$$

*In this case s is called **period of the orbit** (or **order of the cycle**).*
If we choose as initial datum $X_0 = \alpha_0$, then the relative orbit $\gamma(I, f, \alpha_0)$ has an s periodic behavior:

$$\gamma(I, f, \alpha_0) - \{\alpha_0, \alpha_1, \alpha_2, \ldots, \alpha_{s-1}, \alpha_0, \alpha_1, \alpha_2, \ldots, \alpha_{s-1}, \alpha_0, \ldots\}.$$

*More broadly, an orbit X of $\{I, f\}$ is called **eventually periodic** if there exist $\alpha_0, \alpha_1, \ldots, \alpha_{s-1}$, different from each other, and $h \in \mathbb{N}$ fulfilling*

$$X_h = \alpha_0, \quad X_{h+1} = \alpha_1, \quad \ldots, \quad X_{h+s} = \alpha_0.$$

Periodic orbits of least period s (if they exist) are made of values which solve the equation $\{x \in I : f^s(x) = x\}$ but are not solutions of the $s - 1$ equations $\{x \in I : f^h(x) = x\}$, $h = 1, 2, \ldots, s - 1$ (see Remark 3.15); actually it is enough to exclude the solutions of equations with h that divides s.

Example 3.17. The linear DDS $\{\mathbb{R}, 2x\}$ has exactly one equilibrium ($\alpha = 0$) and no periodic orbits. □

Example 3.18. The linear DDS $\{\mathbb{R}, -x\}$ has one equilibrium ($\alpha = 0$) and infinitely many periodic orbits of period 2: $\{X_0, -X_0\}$ (see Fig. 2.2). □

Example 3.19. The nonlinear DDS $\{(0, +\infty), 1/x\}$ has only one equilibrium ($\alpha = 1$) and infinitely many periodic orbits of period 2: $\{X_0, X_0^{-1}\}$ (see Fig. 3.3). □

Example 3.20. The nonlinear DDS $\{[0, 1], a(x - x^2)\}$ where $a \in \mathbb{R}$ shows an eventually stationary orbit if $X_0 = 1$: $\{1, , 0, 0, \ldots\}$. □

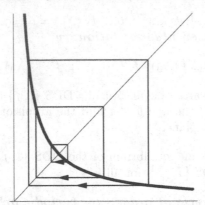

Fig. 3.3 Equilibrium and periodic orbits of $\{(0, +\infty), 1/x\}$

Example 3.21. The DDS $\{[0, 1], 4(x - x^2)\}$ has two equilibria, 0 and 3/4, that are the solutions of $\{x \in [0, 1] : 4(x - x^2) = x\}$, and only one 2 periodic orbit $\{(5 - \sqrt{5})/8, \ (5 + \sqrt{5})/8\}$, whose elements are obtained by solving

$$\left\{x \in [0, 1] : \ 4\left(4(x - x^2) - 16(x - x^2)^2\right) = x\right\} \tag{3.2}$$

and removing the values 0 and 3/4. Notice that solving the degree-four algebraic equation (3.2) is not an instant computation; nevertheless this difficulty is easily circumvented by gathering the factors x and $(x - 3/4)$ and obtaining an equation of degree-two. $\square$

Example 3.22. The continuous function

$$f(x) = \begin{cases} x + 2 & x < -1 \\ -x & -1 \le x \le 1 \\ x - 2 & x > 1 \end{cases}$$

produces the eventually 2 periodic trajectory $X : \ X_0 = 5, \ X_1 = 3, \ X_2 = 1, \ X_3 = -1, \ X_4 = 1 \ \dots$. $\square$

The next theorem provides a sufficient condition for the existence of fixed points.

Theorem 3.23. *Let $I = [a, b]$ be a bounded closed interval in $\mathbb{R}$ and $f : I \to I$ a continuous function. Then there exists $\alpha \in I$ such that $\alpha = f(\alpha)$.*

Proof. The function $g(x) = x - f(x)$ is continuous in I, as it is the difference of continuous functions, and

$$g(a) = a - f(a) \le 0 \le b - f(b) = g(b).$$

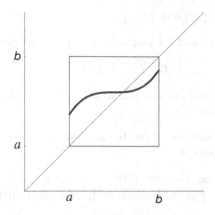

Fig. 3.4 Graph of a function fulfilling the conditions in Theorem 3.23

Therefore, either g vanishes at a or g vanishes at b or, if it does not vanish at the boundary then, by the Intermediate Value Theorem, g must vanish in the interior of the interval $[a, b]$. □

Corollary 3.24. *If $\{I, f\}$ is a DDS and I is a bounded closed interval, then there exists an equilibrium.*

If I is unbounded or open, then both Theorem 3.23 and Corollary 3.24 fail, as shown by the following examples:

$$I = \mathbb{R} \qquad f(x) = x + e^x$$

$$I = (0, 1) \qquad f(x) = x^2.$$

However the next result holds.

Theorem 3.25. *Let A be any nonempty subset of $\mathbb{R}$, $f \colon A \to A$ a continuous function, $X_0 \in A$ and $X_{k+1} = f(X_k)$, $k \in \mathbb{N}$. If $\lim_k X_k = L \in A$ exists, then L is a fixed point of f, that is $L = f(L)$.*

Proof. If $X_k \to L \in A$, we obtain $X_{k+1} \to L$. Then:

$$L = \lim_k X_k = \lim_k X_{k+1} = \boxed{\text{since } X_{k+1} = f(X_k)}$$

$$= \lim_k f(X_k) = \boxed{\text{since } f \text{ is continuous}}$$

$$= f\left(\lim_k X_k\right) =$$

$$= f(L). \qquad \square$$

Theorem 3.25 makes effective the subsequent strategy in the study of the asymptotic behavior of a DDS $\{I, f\}$:

if f is continuous, we look for real solutions of the equation

$$\{L \in I : \quad f(L) = L\},$$

and among these solutions we can find all finite limits (if they exist) of $X_{k+1} = f(X_k)$ as X_0 varies in I.

The monotonicity properties allow to ascertain whether divergent solutions exist when I is unbounded.

Example 3.26. We consider the DDS $\{I, f\}$ where $I = [-2, +\infty)$ and $f(x) = \sqrt{x+2}$. The function f is continuous and, recalling that the arithmetic root is nonnegative:

$$\sqrt{L+2} = L \quad \Rightarrow \quad \begin{cases} L \geq 0 \\ L+2 = L^2 \end{cases} \quad \Rightarrow \quad L = 2.$$

Due to Theorem 3.25, the possible finite limit is the equilibrium $\alpha = 2$. Moreover, $\lim_k X_k = +\infty$ is ruled out since $X_k \leq 2$ or $X_k > 2$ imply respectively $X_{k+1} \leq 2$ or $X_{k+1} < X_k$. $\qquad\square$

The knowledge of any admissible limits of a DDS is not conclusive for the asymptotic analysis: we must clarify whether the solutions $\{X_k\}$ do converge to a limit. To this aim the **theorem on the existence of limits of monotone sequences**[2] proves valuable:

any monotone increasing (or decreasing) sequence of real numbers tends to a limit; this limit is finite if and only if the sequence is bounded, is $+\infty$ ($-\infty$) if and only if the sequence is unbounded.

Often the induction principle is effective for proving monotonicity of a particular solution X_k of a given DDS.

Example 3.27. We consider the sequence $X_{k+1} = X_k^2$ with $X_0 = \lambda \in (0, 1)$ and show, by induction, that X_k is strictly decreasing.
By $\lambda \in (0, 1)$ we deduce $X_1 = \lambda^2 < \lambda = X_0 < 1$. Moreover, if $0 < X_k < X_{k-1}$ then $0 < X_k^2 < X_{k-1}^2$ that is $X_{k+1} < X_k$. Then X_k is strictly decreasing.
Notice that, since the sequence is bounded from below by 0, by the previous analysis and Theorem 3.25 we get $\lim_k X_k = 0$. $\qquad\square$

Exercise 3.1. Draw the phase diagrams of $\{I, f\}$ where:

$$1) \quad I = \mathbb{R}, \, f(x) = \arctan x \qquad\qquad 2) \quad I = (0, +\infty), \, f(x) = \frac{1}{x}.$$

Exercise 3.2. Exploit the graphical method to find all fixed points of $\{\mathbb{R}, \, e^x - 1\}$.

---

[2] We recall that a sequence X is called monotone increasing (respectively decreasing) if $X_{k+1} \geq X_k$ (respectively $X_{k+1} \leq X_k$) for any k.
Monotonicity is strict if the inequality is strict for any k.

Exercise 3.3. Find the 3 periodic orbits of $\{[0,1]\,,\ 1-2\,|x-1/2|\}$.

Exercise 3.4. Find equilibria and cycles (if any) of $\{\mathbb{R},\ x^2\}$.

Exercise 3.5. Find equilibria and cycles (if any) of $\{[-3,3]\,,\ \sqrt{9-x^2}\}$.

Exercise 3.6. Find equilibria and cycles (if any) of $\{\mathbb{R},\ (|x|-x)/2\}$.

Exercise 3.7. Example 3.17 shows that a DDS may have some equilibria and no 2 periodic orbit. Prove that if a DDS $\{I,f\}$ has a 2 periodic orbit then it must also have an equilibrium.

Exercise 3.8. Study the monotonicity of this sequence, as λ varies in $[-2,+\infty)$,

$$\begin{cases} X_{k+1} = \sqrt{X_k + 2} \\ X_0 = \lambda \end{cases} \tag{3.3}$$

Exercise 3.9. Prove that if the DDS $\{I,f\}$ (where $I \subset \mathbb{R}$ is an interval and f is continuous) has a 2 periodic orbit, then it has an equilibrium in the interval whose endpoints belong to this orbit.

3.2 Back to graphical analysis

The graphical method, that was introduced in Sec. 1.3 for studying the particular case of a DDS ruled by linear affine function f, actually is extremely useful for the analysis of general DDSs: the technique extends in a straightforward manner to any nonlinear function f.

The strength of this method relies not only on the help it supplies to the visual perception of trajectories, but also on the fact that it provides a simple algorithm whose implementation and iteration can be easily delegated to computer routines and hence strongly enforced. For instance, if $\{I,f\}$ is the DDS and X_0 the initial datum, then a computer program to evalute the first 100 iterations executes the subsequent steps

---

Algorithm for the graphical analysis

(1) $k = 0$
(2) $x = X_0$
(3) $y = f(x)$
(4) $x = y$
(5) $k = k + 1$
(6) if $k = 100$ stop, if $k < 100$ go back to (3)

Command (6) stops the iterations at a given value of k, otherwise the cycle would not end.

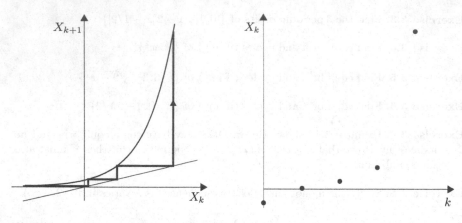

Fig. 3.5. Cobweb of $\{\mathbb{R}, e^x\}$, $X_0 = -2$ **Fig. 3.6.** Trajectory

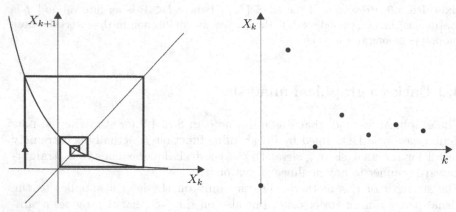

Fig. 3.7. Cobweb of $\{\mathbb{R}, e^{-x}\}$, $X_0 = -1$ **Fig. 3.8.** Trajectory

If we are interested in the values X_k, we can simply insert the command (3′) between (3) and (4):

$$(3')\quad \text{print } k \text{ and } x$$

in this way we get a chart of the trajectory:

$$
\begin{array}{cc}
\text{index} & 0 \ \ 1 \ \ 2 \ \ \dots \ \ k \ \dots \\
\text{related values} & X_0 \ X_1 \ X_2 \ \dots \ X_k \ \dots
\end{array}
$$

In order to add further steps and achieve additional clarity in the graphical representation, it is convenient to drop the vertical segments with endpoints $(X_k, 0)$ and (X_k, X_k) and the orizonthal segments with endpoints $(0, X_{k+1})$ and (X_k, X_{k+1}): in this way we obtain a polygonal chain that is usually called **cobweb**.

Exercise 3.10. Analyze the DDS in Example 3.26 with the graphical method: notice how conjecturing and proving the right monotonicity properties turns out easier this way. Moreover the graphs suggest a straightforward proof strategy: if $-2 \leq x \leq 2$, then $x < f(x) < 2$, so X_k is monotone and bounded. . . .

3.3 Asymptotic analysis under monotonicity assumptions

Given a DDS $\{I, f\}$, checking the monotonicity of the associated trajectories may be a difficult task. Much easier is to analyze the monotonicity of the function f first, then by this analysis, together with the results in the previous section, we can deduce hints about the asymptotic behavior of the trajectories. We collect all these facts in two flow-charts (listed below and labeled by Algorithm I and Algorithm II) that describe a practical approach to the qualitative study of all trajectories of $\{\mathbb{R}, f\}$ when f is monotone, without any additional condition or knowledge about its differentiability. The proof is left as an exercise to the reader.

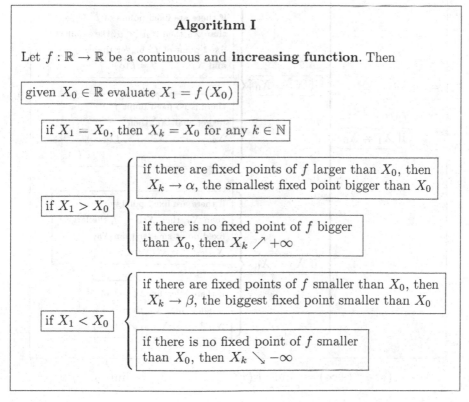

Algorithm I

Let $f : \mathbb{R} \to \mathbb{R}$ be a continuous and **increasing function**. Then

given $X_0 \in \mathbb{R}$ evaluate $X_1 = f(X_0)$

if $X_1 = X_0$, then $X_k = X_0$ for any $k \in \mathbb{N}$

if $X_1 > X_0$ $\begin{cases} \text{if there are fixed points of } f \text{ larger than } X_0, \text{ then} \\ X_k \to \alpha, \text{ the smallest fixed point bigger than } X_0 \\ \\ \text{if there is no fixed point of } f \text{ bigger} \\ \text{than } X_0, \text{ then } X_k \nearrow +\infty \end{cases}$

if $X_1 < X_0$ $\begin{cases} \text{if there are fixed points of } f \text{ smaller than } X_0, \text{ then} \\ X_k \to \beta, \text{ the biggest fixed point smaller than } X_0 \\ \\ \text{if there is no fixed point of } f \text{ smaller} \\ \text{than } X_0, \text{ then } X_k \searrow -\infty \end{cases}$

The symbols $\nearrow$ and $\searrow$ denote increasing and decreasing monotonicity respectively.

Notice that if f is decreasing, then f^2 is increasing, therefore in order to prove Algorithm II it is enough to apply Algorithm I separately to the subsequence of terms with even indexes and to the one with odd indexes.

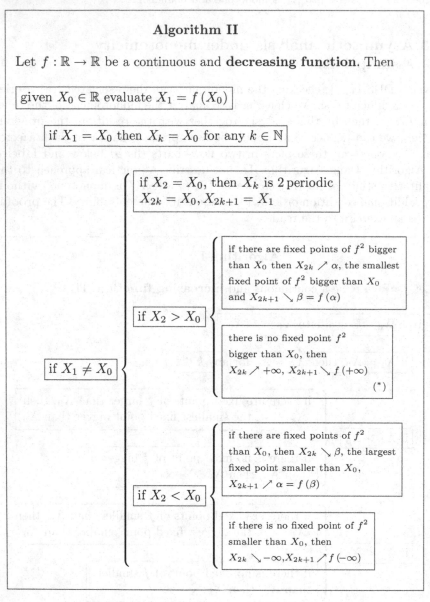

Algorithm II

Let $f : \mathbb{R} \to \mathbb{R}$ be a continuous and **decreasing function**. Then

given $X_0 \in \mathbb{R}$ evaluate $X_1 = f(X_0)$

if $X_1 = X_0$ then $X_k = X_0$ for any $k \in \mathbb{N}$

if $X_1 \neq X_0$

    if $X_2 = X_0$, then X_k is 2 periodic $X_{2k} = X_0, X_{2k+1} = X_1$

    if $X_2 > X_0$

        if there are fixed points of f^2 bigger than X_0 then $X_{2k} \nearrow \alpha$, the smallest fixed point of f^2 bigger than X_0 and $X_{2k+1} \searrow \beta = f(\alpha)$

        there is no fixed point f^2 bigger than X_0, then $X_{2k} \nearrow +\infty, X_{2k+1} \searrow f(+\infty)$ (*)

    if $X_2 < X_0$

        if there are fixed points of f^2 than X_0, then $X_{2k} \searrow \beta$, the largest fixed point smaller than X_0, $X_{2k+1} \nearrow \alpha = f(\beta)$

        if there is no fixed point of f^2 smaller than X_0, then $X_{2k} \searrow -\infty, X_{2k+1} \nearrow f(-\infty)$

$(*) \quad f(+\infty) = \lim_{x \to +\infty} f(x), \qquad f(-\infty) = \lim_{x \to -\infty} f(x)$

Remark 3.28. The monotonicity of f entails strong qualitative restrictions to the dynamics of the DDS $\{I, f\}$ (it is a straightforward consequence of the previous Algorithms):

- if f is increasing, then the DDS cannot have periodic orbits, though it can have equilibria;
- if f is decreasing, then the DDS can have only equilibria and periodic orbits of period 2; orbits with larger periods are not allowed in the dynamics.

Example 3.29. With the aim of analyzing the qualitative properties of orbits of $\{\mathbb{R}, x + \sin x\}$, we observe that $f(x) = x + \sin x$ is an increasing function (indeed $f'(x) = 1 + \cos x \geq 0$, $\forall x \in \mathbb{R}$) and we can apply Algorithm I. Thus:

- if $X_0 = k\pi$, then $X_k = X_0$ for any k;
- if there exists an integer number k such that $2k\pi < X_0 < (2k+1)\pi$, then $X_k \nearrow (2k+1)\pi$;
- if there exists an integer number k such that $(2k+1)\pi < X_0 < (2k+2)\pi$, then $X_k \searrow (2k+1)\pi$. $\square$

Example 3.30. We study the discrete dynamical system $\{\mathbb{R}, -x^3\}$. The function f is monotone strictly decreasing: we apply Algorithm II. For $X_0 \in \mathbb{R}$ we obtain

$$X_1 = f(X_0) = -X_0^3$$

and

$$X_1 = X_0 \qquad \Leftrightarrow \qquad -X_0^3 = X_0 \qquad \Leftrightarrow \qquad X_0 = 0.$$

Then, if $X_0 = 0$ we get $X_k = 0$ for any $k \in \mathbb{N}$.

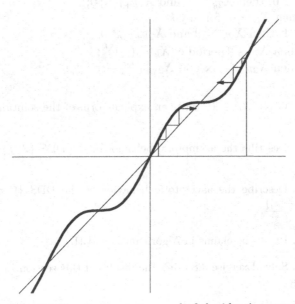

Fig. 3.9 Graph of $x + \sin x$ and of the identity map

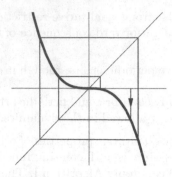
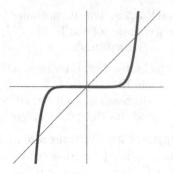

Fig. 3.10 $f(x) = -x^3$ **Fig. 3.11** $f^2(x) = x^9$

If $X_0 \neq 0$ we obtain $X_1 \neq X_0$ and now we have to evaluate X_2 :

$$X_2 = f(X_1) = f^2(X_0) = -\left(-X_0^3\right)^3 = X_0^9.$$

Since

$$X_2 = X_0 \qquad \Leftrightarrow \qquad X_0^9 = X_0 \qquad \Leftrightarrow \qquad X_0 = 0 \text{ or } X_0 = \pm 1,$$

we have the following conclusions:

- if $X_0 < -1$, then $X_{2k} \searrow -\infty$ and $X_{2k+1} \nearrow +\infty$;
- if $X_0 = -1$, then X_k is 2 periodic: $X_k = (-1)^{k+1}$;
- if $-1 < X_0 < 0$, then $X_{2k} \nearrow 0$ and $X_{2k+1} \searrow 0$;
- if $X_0 = 0$, then $X_k = 0$ for any k;
- if $0 < X_0 < 1$, then $X_{2k} \searrow 0$ and $X_{2k+1} \nearrow 0$;
- if $X_0 = 1$, then X_k is 2 periodic: $X_k = (-1)^k$;
- if $X_0 > 1$, then $X_{2k} \nearrow +\infty$ and $X_{2k+1} \searrow -\infty$.   □

Exercise 3.11. Given $X_0 \in \mathbb{R}$, write an explicit form of the solution of the DDS in Example 3.30.

Exercise 3.12. Describe the asymptotic behavior of the DDS $\{\mathbb{R}, f\}$ with $f(x) = (x + |x|)/4$.

Exercise 3.13. Describe the asymptotic behavior of the DDS $\{[-\pi, \pi], f\}$ with $f(x) = x + \dfrac{(\pi - |x|)}{2}$.

Exercise 3.14. Prove the claims in Algorithms I and II.

Exercise 3.15. Solve Exercise 3.8 using the results of this section.

3.4 Contraction mapping Theorem

This section, that can be omitted at first reading, presents a result of great theoretical and numerical relevance: the Contraction mapping Theorem. The theorem is stated here only in the particular case connected to the object of this text: discrete dynamical systems. This result allows to ascertain the existence and uniqueness of a fixed point for a DDS $\{I, f\}$ with f non-necessarily differentiable, and to perform a numerical approximation of such fixed point. This approximation is extremely useful when the equation $\{\alpha \in I : \ \alpha = f(\alpha)\}$ cannot be solved in closed form.

Definition 3.31. *Assume that $I \subseteq \mathbb{R}$ is an interval. A map $f : I \to I$ is a **contraction** if there exists a constant $\tau < 1$ such that*

$$\boxed{|f(x) - f(y)| \leq \tau |x - y| \qquad \forall x, y \in I}$$

Notice that a contraction is always continuous, precisely it is uniformly continuous.

Theorem 3.32 (Contraction mapping Theorem). *Assume I is a closed interval in $\mathbb{R}$ and $f : I \to I$ is a contraction mapping. Then the DDS $\{I, f\}$ has a unique equilibrium α.*
Moreover, for any initial datum $X_0 \in I$, after setting $X_{k+1} = f(X_k)$, we obtain:

$$\lim_k X_k = \alpha .$$

The convergence is very fast, as clarified by this estimate:

$$|X_k - \alpha| < \frac{|f(X_0) - X_0|}{1 - \tau} \tau^k .$$

Proof. We choose $X_0 \in I$ and we obtain, for $k = 1, 2, \dots$

$$|X_{k+1} - X_k| = |f(X_k) - f(X_{k-1})| \leq \tau |X_k - X_{k-1}|$$

hence, by iterating

$$|X_{k+1} - X_k| \leq \tau^k |X_1 - X_0| .$$

Exploiting the triangle inequality and the above inequality we obtain, for any $h \geq 1$,

$$|X_{k+h} - X_k| \leq \sum_{n=0}^{h-1} |X_{k+n+1} - X_{k+n}| \leq |X_1 - X_0| \tau^k \sum_{n=0}^{h-1} \tau^n =$$

$$= |X_1 - X_0| \tau^k \frac{1 - \tau^h}{1 - \tau} \leq \frac{|X_1 - X_0|}{1 - \tau} \tau^k .$$

Since $\tau < 1$, $\{X_k\}$ is a Cauchy sequence and, since I is closed, there exists $\alpha \in I$ such that $\lim_k X_k = \alpha$. Since f is continuous, we find

$$f(\alpha) = f(\lim_k X_k) = \lim_k f(X_k) = \lim_k X_{k+1} = \alpha\,.$$

If α_1, α_2 are both fixed points, we obtain $|\alpha_1 - \alpha_2| = |f(\alpha_1) - f(\alpha_2)| \leq \tau|\alpha_1 - \alpha_2|$, that is $\alpha_1 = \alpha_2$.

If we fix k and take the limit of $|X_{k+h} - X_k| \leq \dfrac{|X_1 - X_0|}{1 - \tau}\tau^k$ as $h \to +\infty$, we deduce

$$|\alpha - X_k| < \frac{|X_1 - X_0|}{1 - \tau}\tau^k\,. \qquad\qquad \square$$

Remark 3.33. The claim in Theorem 3.32 still holds if $I \subset \mathbb{R}$ is any nonempty closed set, as $\mathbb{R}$ is complete.

Remark 3.34. A contraction mapping could be non-differentiable (for instance: $f(x) = |x|/2$); however, thanks to Lagrange's Theorem, any $C^1(I)$ function f such that there exists τ fulfilling

$$|f'(x)| \leq \tau < 1 \qquad\qquad \forall x \in I$$

is a contraction.

Example 3.35. A remarkable application of the Contraction mapping Theorem is given by the search for zeroes of a function $g : I \to I$

$$\{x \in I : \quad g(x) = 0\}$$

under the conditions $g \in C^1(I)$ and $|1 + g'(x)| \leq \tau < 1$ or $|1 - g'(x)| \leq \tau < 1$. This problem is equivalent to searching fixed points of $f : I \to I$, where, respectively, $f(x) = x + g(x)$ or $f(x) = x - g(x)$. Indeed:

$$f(x) = x \qquad \Leftrightarrow \qquad g(x) = 0.$$

The conditions imply that such f is a contraction map, therefore g has a unique zero which can be approximated by the sequence $X_{k+1} = f(X_k)$ starting from whatever $X_0 \in I$ (or $X_{k+1} = -f(X_k)$). $\qquad \square$

Exercise 3.16. Study the dynamics of the DDS $\{\mathbb{R}, |x - 1|/2\}$.

3.5 The concept of stability

In this section we introduce several definitions with the aim of specifying formally the notion of stability.

Definition 3.36. *An equilibrium α of the DDS $\{I, f\}$ is called a **stable equilibrium** if, for any $\varepsilon > 0$, there exists a constant $\delta > 0$ such that $|X_0 - \alpha| < \delta$*

$and \, X_k = f^k(X_0)$, $k \in \mathbb{N}$, *together entail*

$$|X_k - \alpha| < \varepsilon \qquad \forall k \in \mathbb{N}.$$

Conversely, α *is called* **unstable equilibrium** *if it is not stable, that is if there exists* $\varepsilon_0 > 0$ *such that, for any* $\delta > 0$, *we can find* $X_0 \in I$ *and* $\widetilde{k} > 0$ *such that*

$$|X_0 - \alpha| < \delta \qquad\qquad |X_{\widetilde{k}} - \alpha| > \varepsilon_0.$$

The definition above is worth considering carefully. The stability of an equilibrium α means that: first, the initial datum $X_0 = \alpha$ produces a constant trajectory which coincides with the equilibrium; second, an initial datum X_0 "slightly different" from α produces a trajectory that "remains close" to the equilibrium forever.

The study of stability has extreme importance in the applications: in fact the initial datum is hardly ever known precisely, nevertheless often it is possible to bound the error of its measurement.

The reader is invited to verify that all equilibria in Examples 3.18 and 3.19 are stable (a suitable choice is $\delta = \varepsilon$ in the first case, $\delta = \varepsilon/(1-\varepsilon)$ in the second, obviously for $\varepsilon \in (0,1)$), while 0 is unstable for $\{\mathbb{R}, 2x\}$.

Definition 3.37. *An equilibrium* α *of the DDS* $\{I, f\}$ *is called a* **globally attractive equilibrium** *if, for any* $X_0 \in I$, *after setting* $X_k = f^k(X_0)$, *we have*

$$\lim_k X_k = \alpha.$$

Definition 3.38. *An equilibrium* α *of the DDS* $\{I, f\}$ *is called a* **locally attractive equilibrium** *if there exists* $\eta > 0$ *such that, for any* $X_0 \in I \cap (\alpha - \eta, \alpha + \eta)$, *after setting* $X_k = f^k(X_0)$, *we have*

$$\lim_k X_k = \alpha.$$

Notice that the equilibria in Examples 3.18 and 3.19 are not attractive (not even locally), though both are stable.

Definition 3.39. *An equilibrium* α *of the DDS* $\{I, f\}$ *is called a* **globally asymptotically stable equilibrium** *if these two conditions hold together:*

1) α *is stable;*
2) α *is globally attractive.*

α *is called a* **locally asymptotically stable equilibrium** *if it is stable and locally attractive.*

Remark 3.40. If f is a contraction in I, then Theorem 3.32 implies that $\{I, f\}$ is a DDS with a unique equilibrium and this equilibrium is globally asymptotically stable.

Example 3.41. In the linear case, $\{\mathbb{R}, ax + b\}$, if there is an attractive equilibrium (that is $|a| < 1$) then it is (unique and) globally asymptotically stable too. If f is nonlinear the issue is more subtle. $\square$

Exercise 3.17. To gain acquaintance with the various definitions, the reader is invited to prove that for the DDS $\{\mathbb{R}, x^3 + x/2\}$, the point 0 is a stable and locally attractive equilibrium, but it is not globally attractive. Then 0 is locally asymptotically stable.

To understand the next definition, recalling the notion of **distance of a point** x **from a closed set** C is well-timed:

$$\operatorname{dist}(x, C) \;=\; \min\{|x - c| \,:\, c \in C\}.$$

Definition 3.42. *A set $A \subset I$ is called **attractor** (or **well**, or **locally attractive set**) for a DDS $\{I, f\}$ if all these conditions hold:*

1) *A is closed, that is all the limit points of A belong to A;*
2) *A is **invariant**, that is $f(A) = A$;*
3) *there exists $\eta > 0$ such that, for any $x \in I$ fulfilling $\operatorname{dist}(x, A) < \eta$, we have*

$$\lim_k \operatorname{dist}\left(f^k(x), A\right) = 0;$$

4) *A is **minimal**, that is there are no proper subsets of A fulfilling 1), 2) and 3).*

Definition 3.43. *If A is an attractor of $\{I, f\}$, then the set*

$$\left\{ x \in I : \lim_k f^k(x) \in A \right\}$$

*is called **attraction basin** of A for the DDS $\{I, f\}$.*

Example 3.44. A locally attractive equilibrium is an attractor.   □

Example 3.45. The DDS $\{\mathbb{R}, x\,e^{1-x}\}$ has two equilibria: 0 and 1. The equilibrium 0 is unstable and non-attractive, 1 is locally asymptotically stable, the interval $[0, 1]$ is an invariant set but it is not an attractor, since it does not verify condition 4).   □

Remark 3.46. Given a DDS $\{I, f\}$, the set

$$T = \bigcap_{k \geq 1} f^k(I)$$

is always invariant, that is $f(T) = T$. In fact, if we take into account the inclusions $f^{k+1}(I) \subseteq f^k(I)$, we obtain:

$$T = \bigcap_{k \geq 1} f^k(I) = \bigcap_{k \geq 2} f\left(f^k(I)\right) = f\left(\bigcap_{k \geq 2} f^k(I)\right) = f\left(\bigcap_{k \geq 1} f^k(I)\right) = f(T).$$

Often, but not always, the set $T = \bigcap_{k \geq 1} f^k(I)$ is an attractor or, at least, the union of several attractors of $\{I, f\}$. This is a consequence of the following facts: for any $X_0 \in I$, the iterations $f^k(X_0)$ approximate T, that is $f^k (X_0) \in \bigcap_{j=1}^{k} f^j (I)$, and the inclusions $f^{k+1} (I) \subseteq f^k (I)$ entail that $\lim\limits_{k} f^k (X_0)$ belongs to the closure of T.

Definition 3.47. *A periodic orbit $\{\alpha_0, \alpha_1, \ldots, \alpha_{s-1}\}$ of least period s is called a **stable orbit** for the DDS $\{I, f\}$ if the points $\alpha_0, \alpha_1, \ldots, \alpha_{s-1}$ are stable equilibria of f^s.*

Definition 3.48. *(equivalent to the previous one)*
*A periodic orbit $\{\alpha_0, \alpha_1, \ldots, \alpha_{s-1}\}$, of least period s, is a **stable orbit** for the DDS $\{I, f\}$ if for any $\varepsilon > 0$ there exists $\delta > 0$ such that*

$$given \quad j \in \{0, \ldots, s-1\} \quad and \quad X_j \in I : \qquad |X_j - \alpha_j| < \delta$$

then

$$\left| f^{k-j} (X_j) - \alpha_{k(mod\, s)} \right| < \varepsilon \qquad \forall k > j$$

where $k\,(mod\,s)$ denotes the only integer in $\{0, 1, \ldots, s-1\}$ congruent to k modulo s (that is, the smallest integer that gives the same remainder as k when divided by s).

Definition 3.49. *A periodic **orbit** is called **locally asymptotically stable** if it is both a stable orbit and a locally attractive set.*

Example 3.50. Graphical analysis of $\{(0, 1), f\}$, $f(x) = 3.1(x - x^2)$.
Applying the Remark 3.14 to f^2 we can show that there is a unique 2 periodic orbit $\{\gamma, \beta\}$ that is locally asymptotically stable, but it is not globally asymptotically stable due to the existence of the two equilibria 0 and 21/31. These equilibria are neither stable, nor locally attractive.
Notice that the values γ or β (belonging to the periodic orbit) are not attractors if they are considered individually (see Figs. 3.12, 3.13). $\qquad\qquad \square$

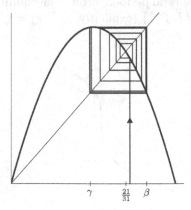

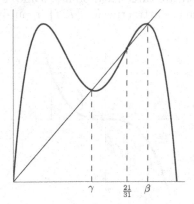

Fig. 3.12 $f(x) = 3.1(x - x^2)$ **Fig. 3.13** f^2 with $f(x) = 3.1(x - x^2)$

Definition 3.51. *A set $R \subset I$ is called* **repulsive set** *(or* **repeller***) for the DDS $\{I, f\}$ if these conditions hold:*

1) R *is closed;*
2) R *is invariant, that is $f(R) = R$;*
3) *there exists an open neighborhood U of R (that is $R \subset U$, $\mathbb{R} \backslash U$ is closed) such that, for any neighborhood V of R there exists $X_0 \in V \backslash R$ such that $f^k(X_0) \notin U$ for infinitely many values of k;*
4) R *is minimal, namely there is no proper subset of R fulfilling 1), 2), 3).*

The definition of repulsive set formalizes the case of a set R mapped into itself by f, with the additional property that in any neighborhood of R there are points that are repelled from R for infinitely many values of k.

Definition 3.52. *A* **repulsive equilibrium** *is a repeller made by only one point.*

Remark 3.53. An equilibrium is repulsive if and only if it is unstable.

A warning: being a repeller is not the opposite of being attractive. For instance, every periodic orbit of $\{\mathbb{R}, -x\}$ oscillates around 0, which is a stable equilibrium but it is neither attractive, nor repulsive.

Remark 3.54. If $f : I \to I$ is a **homeomorphism** (continuous, invertible with continuous inverse) and A is attractive for the DDS $\{I, f\}$, then A is a repeller for the DDS $\{I, f^{-1}\}$.
If $f : I \to I$ is a homeomorphism and A is a repeller for $\{I, f\}$ then A is attractive for $\{I, f^{-1}\}$.

We notice that if the trajectories $\{I, f\}$ describe the evolution in time of a physical quantity, then by reverting the time arrow we find the trajectories of $\{I, f^{-1}\}$.

Example 3.55. We consider the DDS $\{I, f\}$ with $I = [-1, 1]$ and $f(x) = x^3$. There are three fixed points: 0 and ± 1; there is no periodic orbit. The equilibrium 0 is attractive for $\{I, f\}$, repulsive for $\{I, f^{-1}\}$ (explicitly, $f^{-1}(x) = \sqrt[3]{x}$

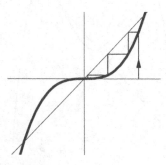

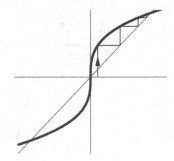

Fig. 3.14 $I = [-1, 1]$, $f(x) = x^3$ **Fig. 3.15** $I = [-1, 1]$, $f(x) = \sqrt[3]{x}\,\text{sign}(x)$

if $x \geq 0$, $f^{-1}(x) = -\sqrt[3]{-x}$ if $x < 0$). The equilibria -1 and $+1$ are repulsive for $\{I, f\}$, attractive for $\{I, f^{-1}\}$.

Precisely, 0 is locally asymptotically stable for $\{I, f\}$, whereas ± 1 are locally asymptotically stable for $\{I, f^{-1}\}$. $\square$

Remark 3.56. About the stability analysis of DDSs, we observe that the techniques based on monotonicity properties and the Contraction Mapping Theorem provide information of global nature and give an exhaustive description of the dynamics. Nevertheless in many circumstances these powerful tools do not apply. If this is the case a local analysis is performed: whenever the map associated to the DDS is differentiable this provides a lot of information, as shown in detail in the next section.

3.6 Stability conditions based on derivatives

When a function $f : I \to I$ has one or more continuous derivatives, it is possible to ascertain the stability and/or the asymptotic stability of equilibria for the DDS $\{I, f\}$ by performing simple tests.

Theorem 3.57. (*First order stability condition*)
If α is an equilibrium for the DDS $\{I, f\}$ and f belongs to C^1, then:

$$
\begin{array}{lll}
|f'(\alpha)| < 1 & \Rightarrow & \alpha \text{ locally asymptotically stable} \\
|f'(\alpha)| > 1 & \Rightarrow & \alpha \text{ unstable}
\end{array}
$$

Proof. Assume $|f'(\alpha)| < 1$. Due to the continuity of f' there are $d > 0$ and $r < 1$ such that $|f'(x)| \leq r < 1$ if $x \in (\alpha - d, \alpha + d)$. By the Mean Value Theorem, if $x, t \in (\alpha - d, \alpha + d)$, then there is x^* such that

$$
|f(x) - f(t)| = |f'(x^*)| \, |x - t| \leq r \, |x - t|.
$$

Therefore Definition 3.36 of stability is fulfilled with the choice $\delta = \min\{\varepsilon, d\}$. Moreover, if $x \in (\alpha - d, \alpha + d)$, then:

$$
\left| f^k(x) - \alpha \right| = \left| f^k(x) - f^k(\alpha) \right| \leq r \left| f^{k-1}(x) - f^{k-1}(\alpha) \right| \leq \cdots \leq r^k \, |x - \alpha|.
$$

Therefore, $\lim_k r^k = 0$ implies that α is locally attractive, hence it is locally asymptotically stable too.

Now assume $|f'(\alpha)| > 1$. Due to the continuity of f' there are $m > 1$ and $d > 0$ such that $|f'(x)| \geq m > 1$ for $x \in (\alpha - d, \alpha + d)$. Due to Mean Value Theorem, if $x \neq \alpha$ and $x \in (\alpha - d, \alpha + d)$, then $f^k(x)$ cannot belong to $(\alpha - d, \alpha + d)$ for every k, otherwise we would obtain

$$
\left| f^k(x) - \alpha \right| = \left| f^k(x) - f^k(\alpha) \right| \geq m \left| f^{k-1}(x) - f^{k-1}(\alpha) \right| \geq \cdots \geq m^k \, |x - \alpha|.
$$

This is a contradiction as $\lim_k m^k = +\infty$ and $|x - \alpha| \neq 0$. $\square$

Example 3.58 (Root-finding algorithm by successive approxima-tions). If $g : [a, b] \to [a, b]$ is a continuous function, the problem of finding its zeroes is equivalent to the problem of finding fixed points in $[a, b]$ of the function f defined by: $f(x) = x + g(x)$. In fact, $g(\alpha) = 0$ if and only if $f(\alpha) = \alpha$. If α is an attractive equilibrium then, no matter how X_0 is chosen in the basin of α, the sequence defined by $X_{k+1} = X_k + g(X_k)$ converges to α.

For instance, if $g \in C^1$ and $|1 + g'(\alpha)| < 1$, Theorem 3.57 ensures the convergence of the method. If $f(x) = x + g(x)$ is a contraction (see Section 3.4) the algorithm works efficiently for any choice of X_0 in $[a, b]$.

Sometimes, even if the condition $|1 + g'(x)| < 1$ is not fulfilled, we can adapt the method by substituting f with

$$\varphi(x) = x + \psi(x) g(x) ,$$

where $\psi \in C^1$ never vanishes in $[a, b]$ and ψ is suitably chosen in order to have $|\varphi'(\alpha)| < 1$; in such case the approximations are generated by the iterations $Y_{k+1} = Y_k + \psi(Y_k) g(Y_k)$.   □

Definition 3.59. *Given $\{I, f\}$ with f in $C^1(I)$, an **equilibrium** α for this DDS is called **hyperbolic** if $|f'(\alpha)| \neq 1$, **superattractive** if $f'(\alpha) = 0$ and **neutral** if $|f'(\alpha)| = 1$.*

Going back to the stability analysis based on the derivatives, we emphasize that the case $|f'(\alpha)| = 1$ is left open by Theorem 3.57 and actually it may be connected to dynamics extremely different from each other.

If the test of the first derivative of f at the equilibrium is inconclusive, then it is necessary to deepen the analysis, as we will show in the sequel. To this aim we state beforehand the next definition.

Definition 3.60. *An **equilibrium** α for the DDS $\{I, f\}$ is called*

- **semistable from above** *if for any $\varepsilon > 0$ there exists $\delta > 0$ such that $\alpha \leq x < \alpha + \delta$ implies $|f^k(x) - \alpha| < \varepsilon$ for any $k \in \mathbb{N}$;*
- **attractive from above** *if there exists $\eta > 0$ such that $\alpha \leq x < \alpha + \eta$ implies $\lim_k f^k(x) = \alpha$;*
- **asymptotically stable from above** *if it is semistable from above and attractive from above.*
- **unstable from above** *(or **repulsive from above**) if it is not semistable from above.*

The previous definitions corresponds to a restriction of the properties formalized in Definitions 3.36-3.39 to initial values bigger than α. The reader can easily adapt these definitions to a left neighborhood of α, and formulate the definitions of semistable, attractive, asymptotically stable and repulsive from below.

Theorem 3.61. *Assume that α is an equilibrium of the DDS $\{I, f\}$, with $f \in C^1$ and $f'(\alpha) = 1$. Then*

---

f convex in a neighborhood of α	$\Rightarrow$	*α semistable from below*
f concave in a neighborhood of α	$\Rightarrow$	*α semistable from above*

If the convexity (respectively, the concavity) of f is strict, then α is asymptotically stable from below and repulsive from above (respectively, asymptotically stable from above and repulsive from below).

Proof. Confining ourselves to the case of convex functions (in such case we know that $f(x) \geq f(\alpha) + f'(\alpha)(x - \alpha) \quad \forall x$), it is sufficient noticing that there exists $\delta > 0$ such that
$$x - \delta \leq x \leq \alpha \quad \Rightarrow \quad x \leq f(x) \leq \alpha.$$
This chain of inequalities improves as $x < f(x) < \alpha$ if the convexity is strict. $\square$

In Fig. 3.16 and 3.17 we show with graphical examples the meaning of the previous theorem.

Theorem 3.61 holds even if f is neither C^1 nor convex (or concave), provided the inequalities highlighted in the proof hold true: the argument works whenever $\alpha = f(\alpha)$ and $x \leq f(x)$ (respectively $x \geq f(x)$) in a left (respectively right) neighborhood of α.

Theorem 3.62 (Stability condition of the second order). *Assume that α is an equilibrium for the DDS $\{I, f\}$, with $f \in C^2$ and $f'(\alpha) = 1$. Then:*

$f''(\alpha) > 0 \quad \Rightarrow$	α	$\begin{cases} \textit{asymptotically stable from below} \\ \textit{repulsive from above} \end{cases}$
$f''(\alpha) < 0 \quad \Rightarrow$	α	$\begin{cases} \textit{asymptotically stable from above} \\ \textit{repulsive from below} \end{cases}$

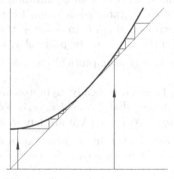

Fig. 3.16 f convex

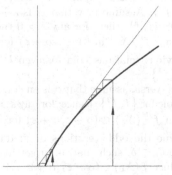

Fig. 3.17 f concave

If $f''(\alpha)$ vanishes, we can go on with the test based on convexity or concavity of f separately in a right or left neighborhood of α, and achieve the next result.

Theorem 3.63 (Stability condition of the third order). *Assume that α is an equilibrium for the DDS $\{I, f\}$ with $f \in C^3$, $f'(\alpha) = 1$ and $f''(\alpha) = 0$. Then:*

$$
\begin{array}{lll}
f'''(\alpha) > 0 & \Rightarrow & \alpha \ unstable \\[2mm]
f'''(\alpha) < 0 & \Rightarrow & \alpha \ locally \ asymptotically \ stable
\end{array}
$$

The reader is invited to formulate and prove a statement for a stability condition when $f \in C^4$ and

$$
f'(\alpha) = 1 \qquad f''(\alpha) = 0 \qquad f'''(\alpha) = 0.
$$

We are left to consider the case

$$
f(\alpha) = \alpha \qquad f'(\alpha) = -1. \tag{3.4}
$$

Before tackling this problem, we recommend to study with the graphical method some examples where conditions (3.4) are fulfilled (like $f(x) = 3x^2 - x$) and take time to think about the previously studied affine linear case (for instance, $f(x) = 2 - x$).

In the affine linear case we obtain stability but not asymptotic stability: precisely, infinitely many 2 periodic orbits oscillate around the equilibrium. This property suggests to study the oscillating behavior of a general DDS, by considering separately the two sequences defined by terms labeled with even and odd indices: $\{X_{2k}\}_{k \in \mathbb{N}}$, $\{X_{2k+1}\}_{k \in \mathbb{N}}$.

Lemma 3.64. *Assume that α is an equilibrium for the DDS $\{I, f\}$. Then α is a locally asymptotically stable equilibrium for $\{I, f\}$ if and only if α is a locally asymptotically stable equilibrium for $\{I, f^2\}$.*

Proof. If α is an equilibrium for $\{I, f\}$ then (Remark 3.15) α is an equilibrium for $\{I, f^2\}$. Assume now that α is a locally asymptotically stable equilibrium for the DDS $\{I, f\}$; then, for any $\varepsilon > 0$ there is $\delta > 0$ such that if $X_0 \in (\alpha - \delta, \alpha + \delta)$ we obtain $f^k(X_0)$ in $(\alpha - \varepsilon, \alpha + \varepsilon)$ for any k and $\lim_k f^k(X_0) = \alpha$. In particular, the previous relations hold for even k; therefore α is locally asymptotically stable for $\{I, f^2\}$.

Vice-versa, assume that α is an equilibrium for $\{I, f\}$ and α is locally asymptotically stable for $\{I, f^2\}$. Hence: for any $\varepsilon > 0$ there is $\delta > 0$ such that if $X_0 \in (\alpha - \delta, \alpha + \delta)$; then $f^{2k}(X_0) \in (\alpha - \varepsilon, \alpha + \varepsilon)$ for any k and $\lim_k f^{2k}(X_0) = \alpha$. We are left to examine the odd iterations of f: due to the continuity of f in α there exists δ_1, $0 < \delta_1 \leq \delta$, such that, after setting $\varepsilon_1 = \min(\varepsilon, \delta)$, from $X_0 \in (\alpha - \delta_1, \alpha + \delta_1)$ we deduce $f(X_0) \in (\alpha - \varepsilon_1, \alpha + \varepsilon_1)$; then

$$f^{2k+1}(X_0) = f^{2k}(f(X_0)) \in (\alpha - \varepsilon, \alpha + \varepsilon); \quad \lim_k f^{2k+1}(X_0) = \lim_k f^{2k}(f(X_0)) = \alpha. \qquad \square$$

Theorem 3.65. *If $f : I \to I$ with $f \in C^3$, $f(\alpha) = \alpha$ and $f'(\alpha) = -1$, then*

$$
\begin{aligned}
2f'''(\alpha) + 3\left(f''(\alpha)\right)^2 > 0 \quad &\Rightarrow \quad \alpha \text{ locally asymptotically stable} \\
2f'''(\alpha) + 3\left(f''(\alpha)\right)^2 < 0 \quad &\Rightarrow \quad \alpha \text{ unstable}
\end{aligned}
$$

Proof. After setting $g(x) = f^2(x)$, $x \in I$, we consider the two DDSs $\{I, f\}$ and $\{I, g\}$. Due to Lemma 3.64 and Theorem 3.63, it is sufficient to verify that:

(i) $g'(\alpha) = 1$;   (ii) $g''(\alpha) = 0$;   (iii) $g'''(\alpha) = -2f'''(\alpha) - 3\left(f''(\alpha)\right)^2$.

Checking the three conditions is a delicate and instructive exercise on the differentiation of composite functions: the reader is invited to check them by himself before comparing the results with the computations below.

$$g'(x) = (f(f(x)))' = f'(f(x)) f'(x)$$

$$g'(\alpha) = f'(f(\alpha)) f'(\alpha) = f'(\alpha) f'(\alpha) = (-1)(-1) = 1$$

$$g''(x) = (f'(f(x)) f'(x))' = f''(f(x))(f'(x))^2 + f'(f(x)) f''(x)$$

$$g''(\alpha) = f''(\alpha)(f'(\alpha))^2 + f'(\alpha) f''(\alpha) = 0$$

$$g'''(x) = \left(f''(f(x))(f'(x))^2 + f'(f(x)) f''(x)\right)' =$$
$$= f'''(f(x))(f'(x))^3 + 3f''(f(x)) f''(x) f'(x) + f'(f(x)) f'''(x)$$

$$g'''(\alpha) = -2f'''(\alpha) - 3\left(f''(\alpha)\right)^2. \qquad \square$$

Remark 3.66. If the function f is a polynomial of second degree and α is a fixed point of f such that $f'(\alpha) = -1$, then α is locally asymptotically stable for $\{\mathbb{R}, f\}$.
In fact, $f''' \equiv 0$ and the claim follows from Theorem 3.65.
Prove that the attractivity is global.

Exercise 3.18. Find the equilibria, if any, of these discrete dynamical systems, and study their stability with tests based on derivatives:

1) $\{\mathbb{R}, (x^3 + x)/2\}$,   2) $\{\mathbb{R}, x^3 + x^2\}$,   3) $\{\mathbb{R}, xe^{1-x}\}$,

4) $\{\mathbb{R}, x^3 + x^2 + x\}$,   5) $\{\mathbb{R}, e^x - 1\}$,   6) $\{\mathbb{R}, x - x^3\}$,

7) $\{\mathbb{R}, x^3 + x\}$,   8) $\{\mathbb{R}, x^2 - x\}$,   9) $\{\mathbb{R}, x^3 - x\}$.

Table 3.1 Outline for the analysis of the stability of an equilibrium α when f has some derivatives

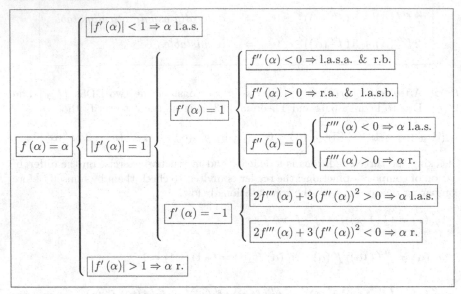

Abbreviations: l.a.s. = locally asymptotically stable,
l.a.s.a. = locally asymptotically stable from above,
l.a.s.b. = locally asymptotically stable from below,
r. = repulsive, r.a. = repulsive from above,
r.b. = repulsive from below.

3.7 Fishing strategies

In this section we examine a relevant example: we apply the previous discrete dynamical systems techniques to optimize activities like fishing or hunting.

Let us consider a fish species whose population dynamics, when left undisturbed by external agents and up to suitable normalization, is described by the iterative law

$$X_{k+1} = 1.5X_k - 0.5X_k^2 \,.$$

The steps of this DDS encapsulate the breeding seasons that succeed each other: we are assuming that the logistic dynamics (see Example 1.13) with the choice of parameters 1.5 and 0.5 gives a satisfactory description of reproduction and intra-specific competition.

We study the effects on this species of two different fishing strategies.

First strategy – Fixed amount of fishing harvest (at every step):
after every breeding season, a *fixed amount* b of fish is caught. The related population dynamics is modified as follows

$$X_{k+1} = 1.5X_k - 0.5X_k^2 - b \,.$$

If $0 < b < 1/8$, then it is easy to verify (by using the first-order stability condition of Theorem 3.57) that there are two positive equilibria, $\alpha_1 \geq \alpha_2 > 0$,

$$\alpha_{1,2} = \frac{1}{2} \pm \frac{1}{2}\sqrt{1 - 8b}$$

with α_1 stable and attractive, α_2 repulsive. To avoid extinction, b has to be chosen in such a way that $\alpha_2 = \alpha_2(b) < X_0$, that is (since X_0 is a datum whereas b is the value that we can choose) b must fulfill:

$$\begin{cases} 0 < b < \dfrac{1}{8} & \text{if } X_0 \geq \dfrac{1}{2}, \\ b < \dfrac{1}{2}\left(X_0 - (X_0)^2\right) & \text{if } 0 < X_0 < \dfrac{1}{2}. \end{cases} \tag{3.5}$$

If $b = 1/8$, there is a unique equilibrium $\alpha_1 = \alpha_2 = 1/2$. Moreover $f'(\alpha_1) = 1$ and f is concave. Therefore the equilibrium is semistable from above and unstable from below. If $X_0 \geq 1/2$ then there is no extinction. We emphasize that we fish in a dangerous manner: if $X_0 < 1/2$ extinction takes place. Moreover, if $X_0 \geq 1/2$, first the dynamics rapidly evolves toward the equilibrium, thereafter we run the risk of plummeting in the basin of $-\infty$ as a consequence of any small perturbation.

If $b > 1/8$ there are no equilibria and all trajectories lead the population to extinction: $f(x) < x$ for any x entails $X_{k+1} < X_k$ for any k, then the limit exists and it cannot be finite since there are no equilibria (see Theorem 3.25). Then, any $b \leq 1/8$ is a **sustainable amount of fishing** for the strategy planning a fixed amount of fishing in every season (i.e. time-step), provided b is not too big with respect to the initial population.

Nevertheless the fixed-amount strategy may jeopardize the fishery in the future, because the more we come closer to the maximum sustainable amount (that is b approaches $1/8$ and consequently α_1 approaches both α_2 and $1/2$) the higher is the risk that *a slight fluctuation depending on factors beyond our control leads the population toward extinction*: if $b < (X_0 - X_0^2)/2$, then the whole dynamics takes place in (α_2, α_1), yet a small perturbation may lead to the region $\{x < \alpha_2\}$ that is contained in the basin of $-\infty$.

Second strategy – Proportional amount of fishing harvest (at every step):
we plan to catch a *fixed fraction* r of the fish population, with the parameter r suitably chosen in the interval $(0, 1)$.
Consequently the DDS that describes the population dynamics becomes $X_{k+1} = 1.5X_x - 0.5X_k^2 - rX_k$, that is:

$$X_{k+1} = (1.5 - r)X_k - 0.5X_k^2.$$

The equilibria of this DDS are the solutions of $0.5\alpha^2 = (0.5 - r)\alpha$, so

$$\alpha_1 = 1 - 2r \qquad\qquad \alpha_2 = 0.$$

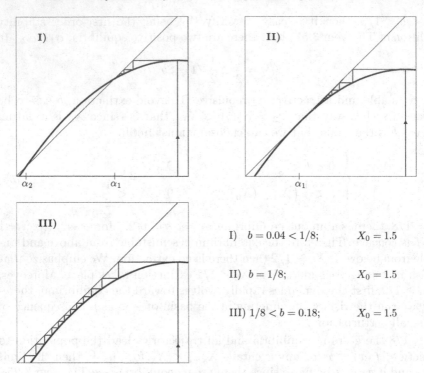

Fig. 3.18 First fishing strategy: fixed amount

I) $b = 0.04 < 1/8$; $X_0 = 1.5$

II) $b = 1/8$; $X_0 = 1.5$

III) $1/8 < b = 0.18$; $X_0 = 1.5$

After setting $f(x) = (1.5 - r)x - 0.5x^2$, we find $f'(x) = (1.5 - r) - x$ and $f'(0) = 1.5 - r$, that is 0 is stable and attractive if and only if $0.5 < r < 2.5$. Recalling that we have a restriction on the parameter r, that is $r \in (0, 1)$, in this range we find that 0 is unstable (this is a nice property for the strategy under examination!) if $0 < r \le 0.5$; notice that the instability if $r = 0.5$ follows from Theorem 3.62.

We study the other equilibrium $\alpha_1 = 1 - 2r$. By $f'(1 - 2r) = 0.5 + r$ we get

- if $0 < r < 0.5$ then α_1 is stable;
- if $r > 0.5$ then α_1 is repulsive;
- if $r = 0.5$ then $\alpha_1 = \alpha_2 = 0$ is a stable equilibrium from above and repulsive from below.

We look for positive stable equilibria, so for $0 < r < 0.5$, $\alpha_1 = 1 - 2r$ is the only positive equilibrium and is stable. Since α_1 is attractive, the X_k converge to $\alpha_1 > 0$ and the fishing harvest P (equal to rX_k) tends to stabilize in the long range at the value

$$P(r) = r(1 - 2r) = r - 2r^2.$$

The choice of r that maximizes the harvest P in the long range is $r_{\max} = 1/4$ with $P_{\max} = 1/8$.

Then $1/8$ is the maximum sustainable harvest when applying the strategy with fixed fraction and this maximum guarantees also the stability.

Comparison of the two fishing strategies.

The maximum sustainable harvest is the same in both cases. Nevertheless there is a huge difference:

harvesting this maximum leads to an unstable situation in the first case, whereas it leads to a stable one in the second case.

Applying the second strategy, if the harvest exceeds the value $1/8$ or if the population decreases for whichever different cause, then the amount of harvest decreases too, allowing the population to recover.

Then the **strategy with fixed fraction is certainly better**. Nevertheless it is difficult to implement due to difficulties in estimating the population size X_k and, consequently, the optimal amount $X_k/4$ of harvest in any season k.

In practice, one tries to induce a constant "effectiveness" of fishing by introducing suitable regulations: for instance, allowing to fish only in certain days of the week (and only in the fishing season, well away from the reproduction season of the population) should give a harvest proportional to X_k, whose exact value is difficult to evaluate.

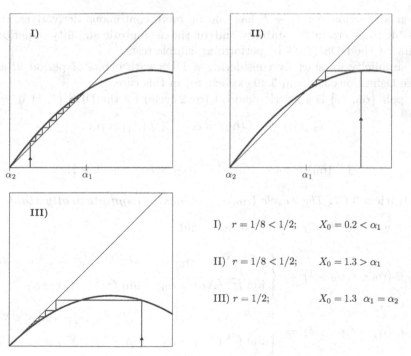

I) $r = 1/8 < 1/2$; $X_0 = 0.2 < \alpha_1$

II) $r = 1/8 < 1/2$; $X_0 = 1.3 > \alpha_1$

III) $r = 1/2$; $X_0 = 1.3$ $\alpha_1 = \alpha_2$

Fig. 3.19 Second fishing strategy: fixed fraction

Moreover we notice that, with the parameters we chose for the undisturbed population dynamics, the maximum sustainable harvest P_{max} equals 25% of the population. In general even with different choices, if one tries to achieve a harvest bigger than the maximum sustainable with the appropriate parameters, then the outcome is a dramatic drop in the population and consequently an equally dramatic drop in the fishing.

Exercise 3.19. Verify that with different parameters, for instance 1.7 and 0.7, or, more in general, with $1 + a$ and a, with $0 < a < 2$, we obtain the same qualitative results of the example previously examined.

Exercise 3.20. We notice that a more realistic fishing strategy model should take into account of the fact that reproduction and fishing activity take place at different times. For instance, fixed proportion before reproduction:
$$Y_{k+1} = 1.5 \left(1 - r\right) Y_k - 0.5 \left(1 - r\right)^2 \left(Y_k\right)^2 ;$$
or, fixed proportion after reproduction:
$$Z_{k+1} = 1.5 \left(1 - r\right) Z_k - 0.5 \left(1 - r\right) \left(Z_k\right)^2 .$$
Try a comparison of the related dynamics. Describe a model of the first strategy (fixed amount) with fishing activity before or after reproduction.

3.8 Qualitative analysis and stability of periodic orbits

When a function $f : I \to I$ has one or more continuous derivatives, it is possible to ascertain the stability and/or the asymptotic stability of periodic orbits for the DDS $\{I, f\}$ by performing simple tests.
For simplicity we start by considering a DDS with orbits of period 2, and make Definitions 3.48 and 3.49 explicit for in this case.
The pair $\{\alpha_0, \alpha_1\}$ is a 2 periodic orbit (or 2 cycle) for the DDS $\{I, f\}$ if

$$\alpha_0 \neq \alpha_1 \qquad f\left(\alpha_0\right) = \alpha_1 \qquad f\left(\alpha_1\right) = \alpha_0$$

hence

$$f^{2k}\left(\alpha_0\right) = \alpha_0 \qquad f^{2k+1}\left(\alpha_0\right) = \alpha_1 \qquad k \in \mathbb{N}.$$

Definition 3.67. *The 2 cycle* $\{\alpha_0, \alpha_1\}$ *is **locally asymptotically stable** if*

for any $\varepsilon > 0$ *there exists* $\delta > 0$ *such that*

$$X_0 \in (\alpha_0 - \delta, \alpha_0 + \delta) \Rightarrow \begin{cases} \left|f^{2k}\left(X_0\right) - \alpha_0\right| < \varepsilon, & \left|f^{2k+1}\left(X_0\right) - \alpha_1\right| < \varepsilon \; \forall k \\ \lim_k f^{2k}\left(X_0\right) = \alpha_0 & \lim_k f^{2k+1}\left(X_0\right) = \alpha_1 \end{cases}$$

$$X_1 \in (\alpha_1 - \delta, \alpha_1 + \delta) \Rightarrow \begin{cases} \left|f^{2k}\left(X_1\right) - \alpha_1\right| < \varepsilon, & \left|f^{2k+1}\left(X_1\right) - \alpha_0\right| < \varepsilon \; \forall k \\ \lim_k f^{2k}\left(X_1\right) = \alpha_1 & \lim_k f^{2k+1}\left(X_1\right) = \alpha_0 \end{cases}$$

that is, if $|X_0 - \alpha_0| < \delta$ and $|X_1 - \alpha_1| < \delta$, set $X_k = f^k(X_0)$, $k \in \mathbb{N}$, then:

- *even-index terms X_{2k} stay uniformly close to α_0 and converge to α_0;*
- *odd-index terms X_{2k+1} stay uniformly close to α_1 and converge to α_1.*

*A 2 cycle is called **repulsive** if it is a repulsive set.*

To ascertain the stability of a 2 cycle $\{\alpha_0, \alpha_1\}$ for the DDS $\{I, f\}$ we study the DDS $\{I, g\}$ where $g = f^2$, in analogy to the analysis performed for an equilibrium α with $f'(\alpha) = -1$ (Lemma 3.64 and Theorem 3.65).

Since any equilibrium of f^2 corresponds to an equilibrium or 2 cycle for f, with a proof similar to the one of Lemma 3.64, we can prove the next lemma.

Lemma 3.68. *Assume $f : I \to I$ is continuous.*
Then $\{\alpha_0, \alpha_1\}$ is a 2 cycle for the DDS $\{I, f\}$ if and only if

$$\alpha_0 \text{ and } \alpha_1 = f(\alpha_0) \text{ are equilibria of } \{I, f^2\} \text{ but not of } \{I, f\}.$$

A 2 cycle $\{\alpha_0, \alpha_1\}$ for $\{I, f\}$ is stable (respectively attractive, repulsive) if and only if both equilibria α_0 and α_1 are stable (respectively, attractive, repulsive) for $\{I, f^2\}$.

Hence, applying Theorem 3.57 to the equilibria of f^2, we obtain that

if $\{\alpha_0, \alpha_1\}$ is a 2 cycle of $\{I, f\}$ s.t.

$$\left|(f^2)'(\alpha_0)\right| < 1 \text{ and } \left|(f^2)'(\alpha_1)\right| < 1$$

then $\{\alpha_0, \alpha_1\}$ is a locally asymptotically stable 2 cycle;

$$(3.6)$$

if $\{\alpha_0, \alpha_1\}$ is a 2 cycle of $\{I, f\}$ s.t.

$$\left|(f^2)'(\alpha_0)\right| > 1 \text{ or } \left|(f^2)'(\alpha_1)\right| > 1$$

then $\{\alpha_0, \alpha_1\}$ is a repulsive 2 cycle.

Notice that conditions (3.6) are hard to test in the examples since they deal with long and useless computations. Conversely the next statement, in whole equivalent to (3.6), is more explicit and can be useful in the applications.

Theorem 3.69. *Assume that $f \in C^1(I)$ and $\{\alpha_0, \alpha_1\}$ is a 2 cycle for the DDS $\{I, f\}$. Then*

$$\begin{array}{ll}
|f'(\alpha_0) f'(\alpha_1)| < 1 & \Rightarrow \quad \{\alpha_0, \alpha_1\} \text{ locally asymptotically stable} \\
|f'(\alpha_0) f'(\alpha_1)| > 1 & \Rightarrow \quad \{\alpha_0, \alpha_1\} \text{ repulsive}
\end{array}$$

Proof. We apply (3.6) together with the identity

$$(f^2)'(\alpha_0) = f'(f(\alpha_0)) f'(\alpha_0) = f'(\alpha_1) f'(\alpha_0) = (f^2)'(\alpha_1). \qquad \square$$

Example 3.70. We look for the 2 periodic orbits of the DDS, if any, $\{[0, 1], 13x(1 - x)/4\}$. To this aim, after noticing that

$$\frac{13}{4}x(1 - x) = x \qquad \Leftrightarrow \qquad x = 0 \text{ or } x = \frac{9}{13}$$

we write explicitly the double iteration of f

$$f^2(x) = \frac{13}{4}\left[\frac{13}{4}x(1-x)\right]\left[1 - \frac{13}{4}x(1-x)\right] = \frac{169}{16}x(1-x)\left(1 - \frac{13}{4}x + \frac{13}{4}x^2\right)$$

The fixed points of f^2, different from 0 and 9/13, are the solutions of $f^2(x) = x$, such that $f(x) \neq x$, that is

$$-\frac{2197}{64}x^2 + \frac{2873}{64}x - \frac{221}{16} = 0 \quad \Leftrightarrow \quad x_{1,2} = \frac{17 \pm \sqrt{17}}{26}.$$

By $f'(x) = 13(1-2x)/4$, we deduce:

$$\left| f'\left(\frac{17-\sqrt{17}}{26}\right) f'\left(\frac{17+\sqrt{17}}{26}\right) \right| = \frac{1}{16}\left| (-4+\sqrt{17})(-4-\sqrt{17}) \right| = \frac{1}{16} < 1$$

and, thanks to Theorem 3.69, we can conclude that the 2 periodic orbit $\left\{ \frac{17-\sqrt{17}}{26}, \frac{17+\sqrt{17}}{26} \right\}$ is locally asymptotically stable (see Fig. 3.20).   □

Example 3.71. Given the DDS $\{[0,1], f\}$, with $f(x) = (1+2\sqrt{2})(x-x^2)$, we look for 2 periodic orbits and study their the stability.
Following Lemma 3.68, we look for fixed points of

$$f^2(x) = f(f(x)) = (1+2\sqrt{2})\left[(1+2\sqrt{2})x(1-x)\right]\left[1 - (1+2\sqrt{2})x(1-x)\right] =$$
$$= (9+4\sqrt{2})x(1-x)\left(1 - (1+2\sqrt{2})x + (1+2\sqrt{2})x^2\right)$$

that are not fixed points of f: $(1+2\sqrt{2})x(1-x) \neq x$, that is to say different from

$$0 \quad \text{and} \quad \frac{8-2\sqrt{2}}{7}.$$

$f(x) = \frac{13}{4}(x - x^2)$ $\qquad\qquad\qquad\qquad$ $f(x) = (1+2\sqrt{2})(x - x^2)$

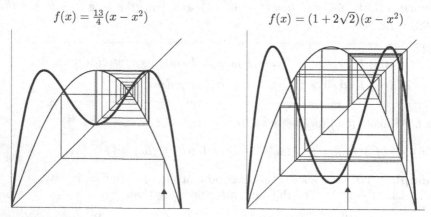

Fig. 3.20 Graphs of f and its iteration f^2 in the interval $[0,1]$ (two examples)

Under these conditions, with some computation, we obtain

$$f^2(x) = x \qquad \Leftrightarrow \qquad x_1 = \frac{4 - \sqrt{2}}{7} \quad \text{or} \quad x_2 = \frac{2 + 3\sqrt{2}}{7} .$$

To analyze the stability of the periodic orbit $\{(4 - \sqrt{2})/7, (2 + 3\sqrt{2})/7\}$ we observe that $f'(x) = (1 + 2\sqrt{2})(1 - 2x)$ entails

$$\left| f'\left(\tfrac{4 - \sqrt{2}}{7}\right) f'\left(\tfrac{2 + 3\sqrt{2}}{7}\right) \right| = |-3| > 1$$

and, due to Theorem 3.69, we conclude that the orbit is repulsive. $\qquad\square$

Remark 3.72. If $\{\alpha_0, \alpha_1, \ldots, \alpha_{s-1}\}$ is an s cycle for the DDS $\{I, f\}$ with $f \in C^1(I)$, then, for $j = 0, 1, \ldots, s - 1$, the subsequent identity holds

$$(f^s)'(\alpha_j) = f'\left(f^{s-1}(\alpha_j)\right) f'\left(f^{s-2}(\alpha_j)\right) \cdots f'(\alpha_j) =$$
$$= f'(\alpha_0) f'(\alpha_1) \cdots f'(\alpha_{s-1}) ,$$

that is the number $(f^s)'(\alpha_j)$ does not depend only on α_j but (as it is natural) it depends on the whole cycle, moreover it can be evaluated without any iteration but only computing the product of the derivatives at all points belonging to the orbit.

Hence, by the same argument used in proving Theorem 3.69, we obtain a characterization of the stability for s cycles.

Theorem 3.73. *Assume that* $f : I \to I$, $f \in C^1(I)$ *and* $\{\alpha_0, \alpha_1, \ldots, \alpha_{s-1}\}$ *is an s cycle for the DDS $\{I, f\}$. Then*

$$
\begin{array}{lll}
|f'(\alpha_0) f'(\alpha_1) \cdots f'(\alpha_{s-1})| < 1 & \Rightarrow & s \text{ loc. asymptot. stable cycle} \\
|f'(\alpha_0) f'(\alpha_1) \cdots f'(\alpha_{s-1})| > 1 & \Rightarrow & s \text{ repulsive cycle}
\end{array}
$$

Theorems 3.69 and 3.73 are very easy to implement if the cycle is explicitly known. But the main difficulty consists in establishing whether the s cycle exists and in computing it. To understand this it is enough to think that even when the map is a polynomial of degree 2, for instance if $I = [0, 1]$ and $f(x) = 4(x - x^2)$, in order to find a 4 cycle one has to compute the fixed points of f^4, that is solve $f^4(x) - x = 0$, a 16-degree equation!

3.9 Closed-form solutions of some nonlinear DDS

As already pointed out, there are no general techniques for solving in closed form a nonlinear recursive law. Nevertheless, whenever by a suitable (necessarily non linear) change of variable one achieves a linear problem referred to the transformed variable, it is possible to make the solution explicit using the theory of linear DDSs (a particular case has been studied in Section 2.7). We consider the sequence X which is given by a recursive law

$$X_{k+1} = f(X_k)$$

where $X_0 \in I$ is prescribed, I is an interval, and $f : I \to I$ is not an affine function.

Then X is a trajectory of the DDS $\{I, f\}$.

Theorem 3.74. *Assume $f : I \to I$ is a continuous function with continuous derivative, b is a real constant different from 0 and 1, $g : I \to \mathbb{R}$ is a continuous function fulfilling*

$$b\, g(f(x)) = g(x)\, f'(x) \tag{3.7}$$

and the condition [3]

$$g(t) \neq 0 \qquad \forall t \in I.$$

If the function $\psi : I \to \mathbb{R}$ is defined by

$$\psi(x) = \int_{X_0}^{x} \frac{dt}{g(t)} + c \tag{3.8}$$

where $\; c = \dfrac{1}{b-1} \displaystyle\int_{X_0}^{X_1} \dfrac{dt}{g(t)}, \;$ we write the explicit form for every trajectory of the DDS $\{I, f\}$:

$$\boxed{ X_k = \psi^{-1}\left(b^k \psi(X_0)\right) } \tag{3.9}$$

where ψ^{-1} denotes the inverse function of ψ.

Remark 3.75. Notice that the Fundamental Theorem of Calculus entails $\psi \in C^1(I)$ and the sign of $\psi'(x) = 1/g(x)$ never changes in I since g is continuous and different from 0 in the interval I. Then ψ is strictly monotone, hence invertible.

Hence the identity (3.7) can be fulfilled only in intervals where f is strictly monotone.

Remark 3.76. The main difficulty to face, when applying the previous theorem to the relevant examples, consists in solving the functional equation (3.7) in the unknown g.

---

[3] The claim of Theorem 3.74 holds also in a more general context, even if g vanishes at finitely many points in I, provided g does not change in I and the function $1/g$ is integrable over I in the generalized sense.

Proof of Theorem 3.74. Since $\psi, f \in C^1(I)$ (see Remark 3.75), applying the chain-rule when differentiating the composition, we obtain

$$\frac{d}{dx}\psi\left(f\left(x\right)\right) = \psi'\left(f\left(x\right)\right)f'\left(x\right) = \frac{f'\left(x\right)}{g\left(f\left(x\right)\right)} = \boxed{\text{due to (3.7)}}$$

$$= \frac{b}{g\left(x\right)} = b\,\psi'\left(x\right).$$

Integrating from X_0 to x and setting equal to zero the indeterminate constant, $\psi\left(f\left(X_0\right)\right) - b\psi\left(X_0\right) = 0$, we find

$$\psi\left(f\left(x\right)\right) = b\psi\left(x\right). \tag{3.10}$$

Hence, recalling $X_1 = f\left(X_0\right)$, we get $c = \psi\left(X_0\right) = \psi\left(X_1\right)/b$ and $c = \psi\left(X_1\right) - \int_{X_0}^{X_1}\frac{dt}{g\left(t\right)}$. Therefore

$$c = \frac{1}{b-1}\int_{X_0}^{X_1}\frac{dt}{g\left(t\right)}.$$

Substituting X_k to x in (3.10), taking into account that $X_{k+1} = f\left(X_k\right)$, we obtain

$$\psi\left(X_{k+1}\right) = b\psi\left(X_k\right) \qquad k \in \mathbb{N}$$

and with the substitution

$$Y_k = \psi\left(X_k\right) \qquad k \in \mathbb{N}$$

we write

$$\begin{cases} Y_{k+1} = bY_k & k \in \mathbb{N} \\ Y_0 = \psi\left(X_0\right) \end{cases}$$

whose explicit solution is

$$Y_k = b^k Y_0 = b^k \psi\left(X_0\right)$$

hence

$$X_k = \psi^{-1}\left(b^k \psi\left(X_0\right)\right). \qquad \square$$

Example 3.77. Analysis of logistic dynamics with parameter a: $h_a\left(x\right) = a\left(x - x^2\right)$. In this case the functional equation (3.7) corresponds to looking for b and g such that $g\left(x\right) \neq 0$, $b \neq 0$, $b \neq 1$ and

$$b\,g\left(ax - ax^2\right) = g(x)\,a\left(1 - 2x\right). \tag{3.11}$$

An attempt to solve (3.11) consists in looking for an affine linear function g, that is: $g\left(x\right) = \gamma x + \delta$. Substituting g in formula (3.11), we obtain

$$-b\gamma ax^2 + b\gamma ax + b\delta = -2a\gamma x^2 + \left(a\gamma - 2a\delta\right)x + a\delta.$$

By matching the coefficients of the same powers of x, we obtain

$$b = a = 2 \qquad \gamma = -2\delta.$$

Then the choice of a linear affine g allows to solve **only the case** $a = 2$.
We choose $\gamma = -1$, $\delta = 1/2$, and obtain $b = 2$ and $g(x) = -x + 1/2$. We notice
that g vanishes at $x = 1/2$. Nevertheless it is enough to restrict the analysis to the interval $(-\infty, 1/2)$ where g neither vanishes nor changes sign since
$\max_{\mathbb{R}} h_2(x) = 1/2 = h_2(1/2)$. In fact, the trajectories of the DDS that are
different from the constant trajectory $(X_k \equiv 1/2)$ achieve only values strictly
less than $1/2$.
Substituting in (3.8), with $X_0 \in (-\infty, 1/2)$, we find

$$\psi(x) = \int_{X_0}^{x} \frac{dt}{\frac{1}{2} - t} + \int_{X_0}^{X_1} \frac{dt}{\frac{1}{2} - t} = \ln \frac{(1 - 2X_0)^2}{(1 - 2x)(1 - 2X_1)} = -\ln(1 - 2x)$$

$$x = \frac{1}{2}\left(1 - e^{-\psi}\right)$$

that is $\psi^{-1}(s) = \frac{1}{2}(1 - e^{-s})$. Now formula (3.9) gives

$$X_k = \frac{1}{2} - \frac{1}{2}\exp\left(-2^k\left(-\ln(1 - 2X_0)\right)\right) = \frac{1}{2}\left(1 - (1 - 2X_0)^{2^k}\right).$$

Summarizing,

the orbits of the logistic map h_2 with initial datum $X_0 \in \mathbb{R}$ are

$$X_k = \frac{1}{2}\left(1 - (1 - 2X_0)^{2^k}\right) \qquad k \in \mathbb{N}$$

(3.12)

The reader is invited to verify that this explicit solution holds even if $X_0 \geq 1/2$
and fulfills all monotonicity and convergence properties than can be easily deduced by a qualitative analysis performed with the tools of Sects. 3.1, 3.2 and
3.3. $\square$

Theorem 3.74 works under rather restrictive conditions: basically it requires
monotonicity of f over I (see Remark 3.75). To deal with the case of non-monotone f (for instance, the logistic map) some adjustment is required whenever, contrarily to the previous example, we do not know "a priori" that the
trajectories of the DDS eventually are contained in an interval of monotonicity for f.
When $0 \in I$ and 0 is an equilibrium of the DDS (that is $f(0) = 0$) we can
repeat the argument in the proof of Theorem 3.74, in every monotonicity interval for f: we introduce a piece-wise constant function $b = b(x)$, precisely
$|b|$ is constant and $\mathrm{sign}(b(x)) = \mathrm{sign}(f'(x))$, say b is constant in the intervals
I_j where f is monotone.
In every I_j we can define a function ψ_j and its inverse ψ_j^{-1}. Set Ψ as the function defined by glueing the ψ_j's and set Φ as the function defined by gluing

the ψ_j^{-1}'s. Then (changing variables) the DDS $\{I, \Phi \circ f\}$, though nonlinear, still has a simple enough structure to obtain information:

$$Y_{k+1} = b\,(Y_k)\,Y_k$$

The modulus of the solution is easily found: $|Y_k| = |b|^k\,|Y_0|$.
If, in addition Φ is even, then

$$X_k = \Phi\,(Y_k) = \Phi\,(|Y_k|) = \Phi\,\left(\left|b^k\Psi\,(X_0)\right|\right)\,.$$

We clarify the previous considerations by studying a relevant example.

Example 3.78. More about the logistic map h_a.
We consider solutions of (3.11) of the kind

$$g\,(t) = \sqrt{t\,(1-t)}\,,$$

that is a positive function in $(0,1)$. We find

$$b\sqrt{a\,(x - x^2)\,(1 - a\,(x - x^2))} = \sqrt{x\,(1-x)}\,a\,(1 - 2x)$$

equivalent for $x \in (0,1)$ to

$$\frac{b}{\sqrt{a}}\sqrt{1 - ax + ax^2} - 1 - 2x$$

that is **an identity in x only for** (we refer to the statement of Theorem 3.74):

$$a = 4 \quad \text{and} \quad b = 2, \qquad \text{if} \quad x \in [0, 1/2)$$

$$a = 4 \quad \text{and} \quad b = -2, \quad \text{if} \quad x \in (1/2, 1]$$

$$\forall a, b \quad \text{with} \quad a > 0 \qquad \text{if} \quad x = 1/2, 0, 1.$$

Since $0 \in I$ and $f\,(0) = 0$, we obtain

$$\psi_1\,(x) = \int_0^x \frac{dt}{\sqrt{t\,(1-t)}} = \int_0^x \frac{dt}{\sqrt{t}\sqrt{1-t}} = 2\arcsin\sqrt{x} \qquad x \in [0, 1/2]$$

Hence

$$\psi_1^{-1}\,(y) = \left(\sin\left(\tfrac{1}{2}y\right)\right)^2 \qquad y \in [0, \pi/2]\,.$$

Moreover

$$\psi_2\,(x) = \frac{\pi}{2} + \int_{1/2}^x \frac{-1}{\sqrt{t\,(1-t)}}dt = \pi - 2\arcsin\sqrt{x} \qquad x \in [1/2, 1]$$

$$\psi_2^{-1}\,(y) = \left(\cos\left(\tfrac{1}{2}y\right)\right)^2$$

The conclusion is attained by substituting b, ψ and ψ^{-1} in formula (3.9).

$$
\boxed{\begin{array}{c}
\text{The orbits of the logistic map } h_4 \\
\text{with initial datum } X_0 \in [0,1] \text{ are} \\[2mm]
X_k = \left(\sin\left(2^k \arcsin\sqrt{X_0}\right)\right)^2 \qquad \forall k \in \mathbb{N}
\end{array}} \tag{3.13}
$$

We check the correctness of the formula we have found, by setting $A = \arcsin\sqrt{X_0}$ for better readability:

$$
f(X_k) = 4\left(\sin\left(2^k A\right)\right)^2\left(1 - \left(\sin\left(2^k A\right)\right)^2\right) = 4\left(\sin\left(2^k A\right)\right)^2\left(\cos\left(2^k A\right)\right)^2 =
$$
$$
= \sin\left(2^{k+1} A\right) = X_{k+1} \ .
$$

Summarizing, formula (3.13) describes explicitly the trajectories. Notwithstanding the deceptive simplicity of this formula the trajectories of the DDS $\{[0,1],\ h_4\}$ exhibit a surprising variety of complex structures, as it will be discussed in Chap. 4. In particular, for almost every initial value in $[0,1]$, the trajectory $\{X_k\}$ is erratic and not periodic. $\qquad\square$

3.10 Summary exercises

Exercise 3.21. Using the techniques of Section 3.6, ascertain the stability of equilibria, if any, of the dynamical system $\{\mathbb{R}, ax + b\}$.

Exercise 3.22. Consider the sequence

$$
\begin{cases}
Y_{k+1} = 1 - \dfrac{4}{\pi}\arctan Y_k & \forall k \in \mathbb{N}\setminus\{0\} \\
Y_0 = a
\end{cases}
$$

If $a = 4$ find the behavior of the sequence, prove that it has limit as $k \to +\infty$ and evaluate such limit.
Which are the other real values of a such that the sequence has a limit?

Exercise 3.23. Study the asymptotic behavior of the sequence, by varying the non-negative real parameter a:

$$
\begin{cases}
Y_{k+1} = \dfrac{1}{3}Y_k^2 + \dfrac{4}{3} \\
Y_0 = a
\end{cases}
$$

Exercise 3.24. Study the asymptotic behavior of the sequence, by varying the non-negative real parameter a:

$$
\begin{cases}
Y_{k+1} = \dfrac{1}{5}Y_k^2 + \dfrac{6}{5} \\
Y_0 = a
\end{cases}
$$

Exercise 3.25. Find the limit of the sequence, by varying the nonnegative real parameter $a > 0$:

$$\begin{cases} Y_{k+1} = 2Y_k \exp\left(-\frac{1}{2}Y_k\right) & \forall k \in \mathbb{N} \\ Y_0 = a \end{cases}$$

Exercise 3.26. Given two positive real numbers a and b such that $a > \sqrt{b} > 0$, examine the sequence Y defined by the iterative formula:

$$\begin{cases} Y_{k+1} = \frac{1}{2}\left(Y_k + \frac{b}{Y_k}\right) & \forall k \in \mathbb{N} \\ Y_1 = a \end{cases}$$

Show that Y_k is defined for any k, it is monotone decreasing, Y_k converges to some limit L as $k \to +\infty$ and compute L. What can be said in the case $Y_1 = \sqrt{b}$?

After setting $\epsilon_k = |Y_k - L|$, show that

$$\epsilon_{k+1} = \frac{\epsilon_k^2}{2\sqrt{b}}.$$

Evaluate $\sqrt{3}$ up to an error not exceeding 10^{-7}.

Exercise 3.27. Prove that $\sqrt{6 + \sqrt{6 + \sqrt{6 + \cdots}}} = 3$.

Exercise 3.28. Given a function $g : \mathbb{R} \to \mathbb{R}$ that has no fixed point, after setting $f(x) = g(x) - x$ and $Y_{k+1} = Y_k + f(Y_k)$, prove that if the absolute maximum of f is strictly negative, then $Y_k \to -\infty$, whereas if the absolute minimum of f is strictly positive then $Y_k \to +\infty$.

Exercise 3.29. A kind of bacterium either undergoes cell division by splitting in two new bacteria both with the same features of the first one or dies. Assume p is the probability of cell division for each one of such bacteria (and of the whole offspring). Which is the probability s that the progeny of this bacterium never becomes extinct?

Exercise 3.30. Perform a graphical analysis of the DDS $\{I, f\}$ with $I = [0, 1]$ and $f(x) = \frac{1}{2\pi} \sin(\pi x)$.

Exercise 3.31. Perform a graphical analysis of the DDS $\{I, f\}$ with $I = [0, 1]$ and $f(x) = \sin(\pi x)$.

Exercise 3.32. Prove that there exists $\alpha \in (0, 1)$ such that $\lim_{k \to +\infty} \cos^k x = \alpha$ for any $x \in \mathbb{R}$.
Perform a numerical computation of the value $\alpha = 0.73909\ldots$.
Hint: show that every orbit of the DDS $\{I, \cos x\}$ is attracted by the unique equilibrium α.
Compute such equilibrium and show that it is stable.

Exercise 3.33. Find the equilibria (if any) and study their stability for the DDS $\{\mathbb{R}, x + x^3\}$.

Exercise 3.34. Find the equilibria (if any) and study their stability for the DDS $\{\mathbb{R}_+, x^{-2}\}$.

Exercise 3.35. Study this sequence defined by an iterative formula, as the parameter β varies in $\mathbb{R}$,

$$X_{k+1} = f_\beta(X_k) \quad \text{with} \quad f_\beta(x) = \beta + \sqrt{x}.$$

Exercise 3.36. Given the DDS $\{[0,1], T\}$ where

$$T(x) = \begin{cases} 2x & 0 \le x < 1/2 \\ 2(1-x) & 1/2 \le x \le 1 \end{cases}$$

show that the initial datum $X_0 = 1/7$ induces an eventually 3 periodic trajectory and that the DDS exhibits two periodic orbits of period 3, both unstable (hint: graphical analysis of the equation $T^3(x) = x$).

Exercise 3.37. Given $f \in C^1(I)$, with $I \subset \mathbb{R}$ interval and $1 + f'(x) \ne 0$ for any $x \in I$, show that the DDS $\{I, f\}$ has no orbit of least period 2.

Exercise 3.38. (a)   Given the real number α, find a non-recursive formula for X_k

$$\begin{cases} X_{k+2} = X_{k+1} + X_k & k \in \mathbb{N} \\ X_1 = 1, \quad X_2 = \alpha \end{cases}$$

(b)   Study the behavior of the sequence Y_k defined by

$$\begin{cases} Y_{k+1} = \dfrac{1}{1+Y_k} & k \ge 2 \\ Y_2 = 1 \end{cases}$$

(c)   Exploiting item (a), write a non-recursive formula for Y_k.
(d)   Study the discrete dynamical system, on varying the parameter $\beta \in \mathbb{R}$:

$$\begin{cases} Z_{k+1} = \dfrac{1}{1+Z_k} & k \ge 2 \\ Z_k = \beta \end{cases}$$

Exercise 3.39. Study the DDS defined by $X_{k+2} = (X_{k+1})^5 / (X_k)^6$.

Exercise 3.40. a)   After choosing the initial datum $X_0 > 0$, study the DDSs

$$X_{k+1} = \sqrt{1+X_k}, \qquad X_{k+1} = 1 + \frac{1}{X_k}.$$

b)   Prove that all the trajectories of the two DDS converge to the same equilibrium (monotonically in the first case, oscillating in the second), that is the subsequent relation between an iterated square root and a continuous fraction holds:

$$\sqrt{1 + \sqrt{1 + \sqrt{1 + \sqrt{1 + \cdots}}}} = 1 + \cfrac{1}{1 + \cfrac{1}{1 + \frac{1}{1+\cdots}}}.$$

4

Complex behavior of nonlinear dynamical systems: bifurcations and chaos

This chapter is mainly devoted to the qualitative analysis of the discrete dynamical system associated to the logistic growth. This is a fundamental issue in biology and an archetypal mathematical model exhibiting a nontrivial behavior, notwithstanding its elementary formulation; this behavior is shared by many other unimodal nonlinear DDSs: in its study we will come across equilibria, periodic orbits, bifurcations and strongly erratic trajectories that qualify as chaotic dynamics.

Several basic notions of topology, which are useful tools for reading this chapter, can be found in Appendix E.

4.1 Logistic growth dynamics

We consider a relevant example of one-step nonlinear discrete mathematical model: the dynamics of a population with constant growth rate together with the simplest "social" correction term. Empirical observations suggest that there are no examples of unbounded growth for biological species. As we already noticed in the first chapter (Examples 1.12, 1.13), the Malthus model of growth population described by the linear dynamical system $X_{k+1} = aX_k$ leads to trajectories with unbounded exponential growth, hence it requires some adjustment since, even neglecting the competition with different species and assuming unbounded availability of food resources in the environment, the overcrowding would always entail a tendency toward intra-specific competition. To take into account the social relations (increased aggressiveness and reduced attitude toward breeding) the model requires correction terms preventing unlimited growth and keep the population within a sustainable size.

The simplest possible correction to the Malthus model is the one introduced by Verhulst, and it relies on the substitution of the linear law with a polynomial of degree two, leading to the **logistic growth** model with discrete

E. Salinelli, F. Tomarelli: *Discrete Dynamical Models.*
UNITEXT – La Matematica per il 3+2 76
DOI 10.1007/978-3-319-02291-8_4, © Springer International Publishing Switzerland 2014

time:

$$X_{k+1} = aX_k (1 - X_k) \qquad 0 \le a \le 4 \qquad (4.1)$$

The polynomial $ax (1 - x)$ is called **logistic map** or **logistic function** and its graph is called **logistic parabola**. In this model (4.1), if the population is small the linear term with positive coefficient prevails, whereas for big populations the quadratic term with negative coefficient prevails (obviously small and big have a relative meaning with respect to the parameters and can be quantitatively specified). The law (4.1) is equivalent to

$$X_{k+1} = h_a (X_k)$$

together with $h_a (x) = ax - ax^2$, $0 \le a \le 4$; moreover, if $x \in [0, 1]$ then $h_a (x) \in [0, 1]$.

We emphasize that a normalization of measure units is always understood, only for convenience in the computations: we consider only values of the population in the range $[0, 1]$. Hence the coefficient a (positive due to its biological meaning) cannot exceed the value 4 in order for the h_a-image of $[0, 1]$ to be contained in $[0, 1]$. We can thus evaluate all the iterations for any choice of the initial datum $X_0 \in [0, 1]$ and obtain that any term of the sequence stays in the prescribed range: $X_k \in [0, 1]$ for any $k \in \mathbb{N}$. Also in the logistic growth, as in Malthus model, as soon as the initial value of the population is chosen, then the whole evolution is univocally determined: we say that they are **deterministic models**. Nevertheless, the dynamics of the logistic growth can no longer match a simple qualitative description for the behavior of X_k as a function of k: least of all we can hope to find either unbounded growth, or extinction for any initial datum and for any choice of the positive parame-

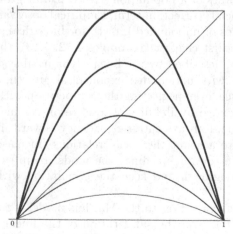

Fig. 4.1 Graphs of h_a, $a = 1/2, 1, 2, 3, 4$: the curve's thickness increases with a

ter $a \in [0,4]$; on the contrary monotonicity, oscillations and other complicate phenomena coexist by simple experiments with the graphical method, we can observe many different kinds of qualitative behavior. The reader is invited to sketch pictures or create computer plots, before proceeding (a Notebook for generating the cobwebs of a DDS is proposed in Appendix H).

Let us try to give an idea of the variety of possible dynamics, without claiming to be complete: notwithstanding the elementary formulation of the model, the whole phase portrait exhibits far from elementary objects whose complete description poses awkward problems.

We focus on the DDS $\{[0,1], h_a\}$ with $a \in [0,4]$.

One can see by the graphical method (and also prove analytically) that the qualitative behavior of trajectories changes dramatically as the parameter a grows: we look for an overall picture of all trajectories and for any value of a. The analysis is elementary for small values of a, as we will discuss in detail later, but the phase portrait is much more complex and requires the definition of new tools as a increases.

The graph of h_a is a concave parabola with vertex at $(1/2, a/4)$, and symmetric with respect to the vertical line of equation $x = 1/2$. Solving the equation $h_a(x) = x$ we deduce that the DDS $\{[0,1], h_a\}$ always has the equilibrium 0: this equilibrium is unique if $0 \le a \le 1$, while, if $1 < a \le 4$, h_a exhibits also the fixed point $\alpha_a = (a-1)/a$. Moreover, initial values 1 and (if $1 \le a \le 4$) $1/a$ lead to eventually constant trajectories, since $h_a(1) = 0$ and $h_a(1/a) = \alpha_a$. Starting from the previous considerations we sketch a preliminary qualitative analysis of the phase portrait of $\{[0,1], h_a\}$ as a varies.

(I) If $a \in [0,1]$, there is the unique equilibrium 0; it is globally asymptotically stable, that is every trajectory with initial datum $X_0 \in [0,1]$ is attracted

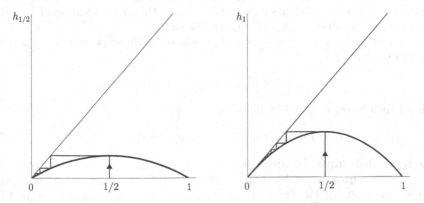

Fig. 4.2 Some cobwebs of h_a starting from $X_0 = 1/2$, for $a = 1/2$ and $a = 1$

by 0 which is a stable equilibrium. In fact $X_{k+1} < X_k$ and,

$$a \in [0, 1) \quad \Rightarrow \quad |h'_a(0)| = a < 1$$
$$a = 1 \quad \Rightarrow \quad h'_a(0) = 1, \ h''_a(0) = -2a < 0$$

and the conclusion follows by Theorems 3.25, 3.57 and 3.62.

(II)   If $1 < a \leq 4$ there are two equilibria: 0 and α_a. Precisely 0 is unstable $(h'_a(0) = a > 1)$ and:

- if $1 < a < 3$, then α_a is stable and attractive $(|h'_a(\alpha_a)| = |2 - a| < 1)$ and the attraction basin of α_a is $(0, 1)$;
- if $a = 3$, then $h'_3(\alpha_3) = -1$ so that α_3 is stable and attractive due to Example 3.66;
- if $3 < a \leq 4$ then α_a is unstable too $(|h'_a(\alpha_a)| = a - 2 > 1)$.

(III)   If $3 < a \leq 4$ then a 2 periodic orbit appears. In fact, $h_a^2(x) = x$ has four distinct solutions: two of them are 0 and α_a (the fixed point of h_a), the remaining two belong to the unique 2 periodic trajectory of the DDS.

(IV)   If $3 < a < 1 + \sqrt{6} = 3,44949...$, then the 2 periodic orbit is stable and attractive (the proof of its stability is postponed).

(V)   If $a \geq 1 + \sqrt{6}$, then the analysis is much more delicate and will be performed in the subsequent sections.

We state and prove a statement that clarifies how the nontrivial part of the dynamics associated to h_a takes place in $[0, 1]$.

Theorem 4.1. *If we consider the logistic map h_a in the whole $\mathbb{R}$ with $a > 1$, then every possible equilibrium and possible periodic orbit of period ≥ 2 of the DDS $\{\mathbb{R}, h_a\}$ is contained in $[0, 1]$.*

Proof. The statement about equilibria is a trivial consequence of the condition $a > 1$. If s is an integer ≥ 2, then, since 0 is a fixed point of h_a for any a and $h_a(1) = 0$, the claim about periodic orbits is proved if we show that every solution of $\{x \in \mathbb{R} : h_a^s(x) = x\}$ is contained in $[0, 1]$. We prove this fact.
If $x < 0$, then $h_a(x) < x < 0$, evaluating h_a again we obtain $h_a^2(x) < h_a(x) < x < 0$, and by iterating

$$h_a^s(x) < h_a^{s-1}(x) < \cdots < x \qquad \forall s \geq 1.$$

If $x > 1$ then $h_a(x) < 0$, and by iterating

$$h_a^s(x) < h_a^{s-1}(x) < \cdots < h_a(x) < 0 < x \qquad \forall s \geq 2. \qquad \square$$

The result just proved ensures that, if $a > 1$, then the study of the iterations of h_a can be performed in $[0, 1]$ only, without losing anything about the structure of the orbits of $\{\mathbb{R}, I\}$, since any solution that is not forever confined in I is eventually strictly decreasing.

4.2 Sharkovskii's Theorem

If we examine the DDS $\{\mathbb{R}, f\}$ with $f(x) = -3x^2 + 5x/2 + 1/2$, then it is easy to verify that $\{0, 1/2, 1\}$ is a periodic orbit with period 3: $f(0) = 1/2$, $f(1/2) = 1$, $f(1) = 0$.

The previous example is built by finding the unique function f of the kind $ax^2 + bx + x$ whose graph passes through $(0, 1/2)$, $(1/2, 1)$, $(1, 0)$. We could ask how many and which other periodic orbits there are for $\{\mathbb{R}, f\}$. A first answer to this question is provided by the next surprising result.

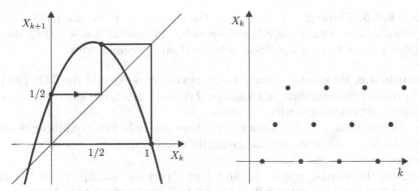

Fig. 4.3 Periodic orbit $\{0, 1/2, 1\}$ of the DDS $\{\mathbb{R}, -3x^2 + 5x/2 + 1/2\}$

Theorem 4.2. *If I is an interval contained in $\mathbb{R}$, $f : I \to I$ is continuous and $\{I, f\}$ has a 3 periodic orbit, then $\{I, f\}$ has an s periodic orbit too, for any positive integer s.*

Though Theorem 4.2 is already surprising by itself, actually it is only a small surprise among all the ones relative to periodic orbits: it is just a small piece of the information provided by the elegant result proved by O.M. Sharkovskii[1] in 1964, that is stated below. The most relevant issue about the structural complexity of the phase diagram of $\{I, f\}$ is that the claim follows by the continuity of f only.

Definition 4.3. *The subsequent ordering is called **Sharkovskii's ordering** of natural numbers:*

$$1 \prec 2 \prec 2^2 \prec 2^3 \prec 2^4 \prec \cdots \prec 9 \cdot 2^n \prec 7 \cdot 2^n \prec 5 \cdot 2^n \prec 3 \cdot 2^n \prec \cdots$$
$$\prec 9 \cdot 2^2 \prec 7 \cdot 2^2 \prec 5 \cdot 2^2 \prec 3 \cdot 2^2 \prec \cdots$$
$$\prec 9 \cdot 2 \prec 7 \cdot 2 \prec 5 \cdot 2 \prec 3 \cdot 2 \prec \cdots$$
$$\prec 9 \qquad \prec 7 \qquad \prec 5 \qquad \prec 3$$

---

[1] Oleksandr Mikolaiovich Sharkovskii, 1936 – .

The relation $a \prec b$ means that a precedes b in the ordering. The first dots stands for all powers of 2 in increasing order, the subsequent dots denote all powers of 2 in decreasing order multiplied by every odd number different from 1 in decreasing order (for instance, $7 \cdot 2^4 \prec 9 \cdot 2^3$).
Any natural number appears exactly once in the Sharkovskii ordering.

Theorem 4.4 (Sharkovskii). *If $f : I \to I$ is a continuous function on the interval $I \subseteq \mathbb{R}$ and there exists a periodic orbit of least period s for the DDS $\{I, f\}$, then the DDS $\{I, f\}$ has a periodic orbit of least period m for any $m \prec s$ according to Sharkovskii's ordering.*

Remark 4.5. Theorem 4.4 is sharp in the sense that there are DDS's with a 5 periodic orbit without any 3 periodic orbit; in general if $n \prec s$ then there are DDS's exhibiting an n periodic orbit and no s periodic orbit.

Example 4.6. By graphical analysis one can easily verify that the DDS $\{[0,1], h_{3,1}\}$ presents two equilibria and a unique 2 periodic orbit, but there is no other periodic orbit coherently with Theorem 4.4.
The first graph in Fig. 4.4, shows that there are only two equilibria: 0 and $21/31$ (there cannot be more than 2 equilibria since $h_{3,1}(x) = x$ is an equation of degree two).
Analyzing the second graph, we find four 2 periodic points: 2 are the already known equilibria (0 and $21/31$), the two others necessarily belong to a 2 periodic orbit.
Examining the third graph, we find only four intersections with the line bisecting the first quadrant, therefore there are no roots in addition to the four ones previously described. So there is no periodic orbit of least period four and no periodic orbit of any other period since 4 precedes every number different from 1 and 2 in the Sharkovskii ordering.   □

To make precise the analysis in the previous example we recall that, due to Theorem 4.1, if $a > 1$ then *every periodic point of h_a belong to the interval* $[0,1]$ so that we can omit the analysis of the graph in $\mathbb{R} \setminus [0,1]$.

For the complete proof of the Sharkovskii Theorem we refer to [6]; here we prove only a part of the general statement, that is Theorem 4.2. Beforehand we state two lemmas.

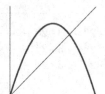 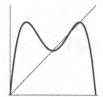 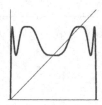

Fig. 4.4. Some iterations of the logistic map $h_{3.1}$: $h_{3.1}$, $(h_{3.1})^2$, $(h_{3.1})^4$

Lemma 4.7. *Given $I = [a, b]$ and $f : I \to \mathbb{R}$ continuous, if $f(I) \supset I$ then there exists a fixed point of f in I.*

Proof Since $f(I) \supset I$, there exist $c, d \in I$ such that $f(c) = a$ and $f(d) = b$. If $c = a$ or $d = b$ the lemma is proved.
Otherwise we have $c > a$ and $d < b$: in this case, after setting $g(x) = f(x) - x$ we obtain

$$g(c) = f(c) - c = a - c < 0 < b - d = f(d) - d = g(d) ;$$

due to the continuity of g and the Intermediate Value Theorem, there exists t in the interval (c, d) such that $g(t) = 0$ that is $f(t) = t$. $\square$

Lemma 4.8. *If U and V are bounded closed intervals and $f : U \to \mathbb{R}$ is continuous, then $V \subset f(U)$ implies the existence of an interval $U_0 \subset U$ such that $f(U_0) = V$.*

The proof is left to the reader.

Remark 4.9. Lemma 4.7 cannot be extended to functions of several variables: there are functions $f : B \to \mathbb{R}^n$, $B \subseteq \mathbb{R}^n$, with nonempty compact B such that $f(B) \supset B$ but f has no fixed point in B (see Example 4.10).
In particular Theorem 4.2 fails in the vector-valued case, where existence of a 3 periodic orbit does not necessarily implies the existence of every other integer period.

Example 4.10. Set $D = \{(x, y) \in \mathbb{R}^2 : x^2 + y^2 = 1, y \geq 0\}$ and define $f : D \to \mathbb{R}^2$ in polar coordinates as $f(\rho, \theta) = (\rho, 2\theta + \pi/2)$. Then $D \subset f(D)$ but f has no fixed point. $\square$

We give the details of the proof of Theorem 4.2, which can be omitted at first reading.

Proof of Theorem 4.2. Let $\{a, b, c\}$ be the 3 periodic orbit, with $a < b < c$. We assume $f(a) = b$, $f(b) = c$ and $f(c) = a$ (in the other case the proof goes in the same way).
Set $I_0 = [a, b]$ and $I_1 = [b, c]$. Then by the Intermediate Value Theorem

$$I_1 \subset f(I_0), \qquad I_0 \subset f(I_1), \qquad I_1 \subset f(I_1) .$$

Due to Lemma 4.7, the inclusion $I_1 \subset f(I_1)$ implies the existence of a fixed point in I_1, then there exists an equilibrium (that is a periodic orbit of period 1).
Assume that $n > 1$ and $n \neq 3$: we are going to prove that there exists an n periodic orbit. To this aim we define the closed intervals J_k, $k = 0, 1, \ldots, n$, such that

(1) $I_1 = J_0 \supset J_1 \supset J_2 \supset \cdots \supset J_n$
(2) $f(J_k) = J_{k-1}$ $k = 1, 2, \ldots, n - 2$

Fig. 4.5. Intervals I_0 and I_1

(3) $f^k (J_k) = I_1$ $k = 1, 2, \ldots, n - 2$
(4) $f^{n-1} (J_{n-1}) = I_0$
(5) $f^n (J_n) = I_1$.

First we show that the existence of intervals J_n with (1)-(5) implies the existence of a periodic orbit of least period n. Second, we will show the existence of the sets J_n. Property (5) and Lemma 4.7 entail the existence of a point $p \in J_n$ such that $f^n (p) = p$. We use (1)-(4) to show that p has least period n.

By (1) we get $p \in I_1 = [b, c]$; by (3) we obtain

$$f^k (p) \in I_1 \qquad k = 0, 1, \ldots, n - 2$$

and by (4) we deduce $f^{n-1} (p) \in I_0 = [a, b]$.

Necessarily $p \neq c$, otherwise $f (p) = f (c) = a \notin I_1$ and, since $f^{n-1} (p)$ is the unique element of the orbit not belonging to I_1, we would have $n = 2$, contradicting the fact that c has period 3.

Necessarily $p \neq b$, otherwise $p = b$ and $f^2 (p) = f^2 (b) = a \notin I_1$ would imply $n = 3$, a conclusion contradicting the condition $n \neq 3$.

Summarizing, $p \in (b, c)$, $f^{n-1} (p) \in I_0 = [a, b]$, that does not intersect (b, c). So $f^{n-1} (p) \neq p$, and p cannot have minimum period $n - 1$.

By contradiction, if the minimum period of p was strictly less than $n - 1$, then property (3) together with $p \neq b$ and $p \neq c$ would imply that the trajectory of p stays in (b, c) forever: a contradiction with (4). Hence the minimum period of p is n.

Second, we prove the existence of the intervals J_n.

Let $n > 1$ be fixed. We start by building the intervals for $k = 1, 2, \ldots, n - 2$: we set $J_0 = I_1$. Then by $f (I_1) \supset I_1$ we find $f (J_0) \supset J_0$ and by Lemma 4.8 there exists $J_1 \subset J_0$ such that $f (J_1) = J_0$. Therefore $J_1 \subset J_0$ implies $f (J_1) \supset J_1$ and we repeat the argument for J_k up to $k = n - 2$, thus obtaining (1) and (2) up to $n - 2$. To show (3) we notice that by (2) we get, for $k = 1, 2, \ldots, n - 2$,

$$f^2 (J_k) = f (f (J_k)) = f (J_{k-1}) = J_{k-2} \qquad (k \geq 2)$$
$$f^3 (J_k) = f (f^2 (J_k)) = f (J_{k-2}) = J_{k-3} \qquad (k \geq 3)$$
$$\cdots = \cdots$$
$$f^{k-1} (J_k) = f \left(f^{k-2} (J_k) \right) = f \left(J_{k-(k-2)} \right) = f (J_2) = J_1$$
$$f^k (J_k) = f \left(f^{k-1} (J_k) \right) = f (J_1) = J_0 = I_1.$$

When showing (4) we still have freedom in the choice of J_{n-1}. Now

$$f^{n-1} (J_{n-2}) = f \left(f^{n-2} (J_{n-2}) \right) = f (I_1)$$

and $f (I_1) \supset I_0$ entails $f^{n-1} (J_{n-2}) \supset I_0$; then by Lemma 4.8 there exists $J_{n-1} \subset J_{n-2}$ such that $f^{n-1} (J_{n-1}) = I_0$. Finally

$$f^n (J_{n-1}) = f \left(f^{n-1} (J_{n-1}) \right) = f (I_0)$$

and $f (I_0) \supset I_1$ implies $f^n (J_{n-1}) \supset I_1$, and again by Lemma 4.8 there exists $J_n \subset J_{n-1}$ such that $f^n (J_n) = I_1$, that is (5) is true. $\square$

Even omitting the whole proof of the Sharkovskii Theorem, we remark that, if the existence of a single periodic orbit for $\{I, f\}$ with least period $s > 1$ is known, then it is easy to prove the existence of a fixed point, with continuous f and I interval. In fact, if I is a closed bounded interval, the existence of the fixed point follows by Corollary 3.24, otherwise if we assume by contradiction $f(x) \neq x$ for any $x \in I$, then there are only two possibilities: either $f(x) > x$ for any x, or $f(x) < x$ for any x. If $\{\alpha, f(\alpha), f^2(\alpha), \dots, f^{s-1}(\alpha)\}$ is the s periodic orbit, with $s > 1$, then in the first case $\alpha < f(\alpha) < f^2(\alpha) < \cdots < f^s(\alpha) = \alpha$ whereas in the second one $\alpha > f(\alpha) > \cdots > f^s(\alpha) = \alpha$. Both the conclusions lead to a contradiction.

We close this section by stating a result whose meaning, roughly speaking, is as follows: if f "does not oscillate too much" then, for any fixed integer s, $\{I, f\}$ can have at most a finite number of s periodic orbits; nevertheless it may have infinitely many periodic orbits altogether, as it does happen in the case of the example at the beginning of this section.

Definition 4.11. *Let $f : I \to I$ be of class C^3.*
*We call **Schwarzian derivative of** f, and denote it by $\mathcal{D}f$, the subsequent function (the definition is understood at points $x \in I$ such that $f'(x) \neq 0$):*

$$(\mathcal{D}f)(x) = \frac{f'''(x)}{f'(x)} - \frac{3}{2}\left(\frac{f''(x)}{f'(x)}\right)^2.$$

Theorem 4.12 (D. Singer, 1978). *Let $\{I, f\}$ be a DDS with $f \in C^3(I)$ and f' vanishing at most at finitely many points $x_1, \dots, x_m$. Assume that $(\mathcal{D}f)(x) < 0 \; \forall x \in I \setminus \{x_1, \dots, x_m\}$. Then, for any $s \geq 1$, (I, f) can have only finitely many orbits with period s.*

Proof. Given $s \in \mathbb{N} \setminus \{0\}$, we study f^s. We find $f^s \in C^3$ and (see Exercise 4.2)

$$\mathcal{D}(f^s)(x) < 0 \qquad x \in I \setminus \{x_1, \dots, x_m\}.$$

It is sufficient to show that f^s has a finite number of fixed points, that is $f^s(x) = x$ has a finite number of solutions; by Rolle's Theorem, it suffices to verify that the equation $(f^s)'(x) = 1$ has finitely many solutions in I.

If this last claim is false, then again by Rolle's Theorem $(f^s)'$ must have infinitely many points with vanishing derivative and infinitely many local minima: at these points $\overline{x}$ we have

$$(f^s)''(\overline{x}) = 0 \qquad (f^s)'''(\overline{x}) \geq 0;$$

then $\mathcal{D}(f^s)(\overline{x}) < 0$ implies $(f^s)'(\overline{x}) < 0$ and $(f^s)'''(\overline{x}) > 0$. Thus $(f^s)'$ is negative at infinitely many points $\overline{x}$ alternating with infinitely many points $\overline{\overline{x}}$ where its value is 1.

We deduce the existence of infinitely many points in I where $(f^s)'$ vanishes, that is infinitely many points where f' vanishes: a contradiction with the previous assumptions. $\qquad \square$

We underline that the conditions and claim of Theorem 4.12 trivially hold for any polynomial f of degree two.

Remark 4.13. If $f : I \to I$, $f \in C^3$ with $f(\alpha) = \alpha$, $f'(\alpha) = -1$ and $(\mathcal{D}f)(\alpha) < 0$, then α is a locally asymptotically stable equilibrium. Indeed

$$2f'''(\alpha) + 3\left(f''(\alpha)\right)^2 = -2\left(\mathcal{D}f\right)(\alpha) > 0$$

and we can apply Theorem 3.65.

Exercise 4.1. Determine the parameters a, b and c in such a way that among the trajectories of $\{\mathbb{R}, f\}$ with $f(x) = ax^2 + bx + c$ there is the 3 cycle $\{1, 2, 3\}$.

Exercise 4.2. Verify that, if $f, g \in C^3$ then the Schwarzian derivative of the composition $f \circ g$ is

$$\mathcal{D}(f \circ g)(x) = (\mathcal{D}f)(g(x)) \cdot \left(g'(x)\right)^2 + (\mathcal{D}g)(x).$$

Exercise 4.3. Prove Lemma 4.8.

4.3 Bifurcations of a one-parameter family of DDS

By the expression **one-parameter family of functions** we denote a collection of functions f_a, each one of them labeled by a numerical value of the parameter a varying in a set of numbers denoted by A. Analogously, a **one-parameter family of discrete dynamical systems** is a collection $\{I, f_a\}$ of DDS's where $f_a : I \to I$ and $\{f_a\}_{a \in A}$ is a one-parameter family of functions with domain and image in the interval I.

For instance, the set of linear functions from $\mathbb{R}$ to $\mathbb{R}$ is a one-parameter family of functions $f_a(x) = ax$ where $a \in \mathbb{R}$ is the labeling parameter. The corresponding family of DDS's $\{\mathbb{R}, f_a\}$ has been studied in detail in Chap. 2.

The most important example is the already mentioned discrete dynamics of the logistic map h_a, with $a \in [0, 4]$: we plan to study it in further detail.

Often physical phenomena exhibit a continuous dependence on the parameters appearing in the laws ruling the phenomena. Nevertheless, in many situations relevant qualitative differences may develop as a consequence of slight variations of the parameters: this fact is very well described by the saying "the straw that broke the camel's back". Indeed the study of bifurcations deals with these special straws: that is, it classifies the parameters values that are associated to dramatic changes in the picture of the phase diagram.

Let us see through simple examples how the phase diagram of a one-parameter family of DDS's varies as the parameter changes. To this aim we try to summarize all the qualitative information in a single diagram, called **bifurcation diagram of** $\{I, f\}$:

> *in the set $A \times I$ we mark the fixed points of f_a and of its iterations as a function of a, we find the values of a such that this points are attractive and represent graphically this fact by vertical arrows pointing toward equilibria; the*

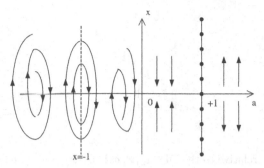

Fig. 4.6 Bifurcation diagram of $\{\mathbb{R}, ax\}$

*vertical arrows point away from the equilibria in the region where they are re-
pulsive; semistable equilibria are depicted with obvious changes; the diagram
is completed with arrows that are coherent with the dynamics in the regions
without fixed points.*

Example 4.14. We represent by a bifurcation diagram the dynamics associated to the family of linear functions $f_a(x) = ax$, $a \in \mathbb{R}$ (see Fig. 4.6). If
$|a| < 1$, the DDS has the unique equilibrium 0, that is stable and globally
attractive; if $|a| > 1$, then 0 is still the unique equilibrium point, but it is re-
pulsive; if $a = 1$, there are infinitely many stable equilibria, neither attractive,
nor repulsive; if $a = -1$, there are infinitely many periodic orbits $\{\alpha, -\alpha\}$.
As the value of a increases from $-\infty$ to -1 (except $a = -1$) the dynamics
stays qualitatively unchanged, the same happening if a varies in $(-1, 1)$ or if
a varies in $(1, +\infty)$; but whenever a crosses either the value -1 or the value
$+1$ there is a sudden change.   □

It is natural to look for a formal description of the existence and related consequences of a parameter value a_0 such that the dynamics for $a < a_0$ is different
from the dynamics for $a > a_0$.

Definition 4.15. *Let $f_a : I \to I$ be a one-parameter family of functions,
where $I \subset \mathbb{R}$ is an interval and $a \in A$. A parameter value a_0 in the interior
of A is called **bifurcation value** for $\{I, f_a\}$ if there exists $\varepsilon > 0$ such that
the number of equilibrium points plus the number of distinct points belonging
to periodic orbits is constant in $(a_0 - \varepsilon, a_0)$ and in $(a_0, a_0 + \varepsilon)$, but either the
constants are different or they are equal and either the stability or the kind of
these points are different on the two sides.*

Example 4.16. Assume $I = \mathbb{R}$, $f_a(x) = ae^x$, $a \in \mathbb{R}^+$. From the graphical
analysis we deduce:

- if $0 < a < e^{-1}$, then f_a has two fixed points x_0 and x_1, with
$$0 < f_a'(x_0) < 1 < f_a'(x_1),$$
  therefore x_0 is stable and attractive, x_1 is repulsive; the attraction basin of
  x_0 is $(-\infty, x_1)$;

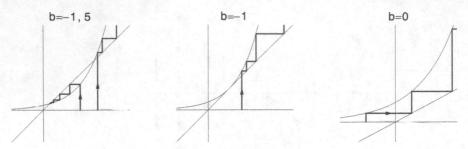

Fig. 4.7 Cobweb relative to the DDS's $\{\mathbb{R}, g_b\}$, $g_b(x) = e^{x+b}$, $b = -1.5;\ -1;\ 0$

- if $a = e^{-1}$, then f_a has a unique fixed point $\widetilde{x} = 1$, that turns out to be stable and attractive from below, unstable from above; the attraction basin of $\widetilde{x}$ is $(-\infty, 1)$;
- if $a > e^{-1}$, then there is no fixed point. Indeed $f(x) > x$ for any $x \in \mathbb{R}$ entails that for any real initial datum x we obtain $\lim_k f^k(x) = +\infty$, that is there are no fixed points and by Sharkovskii's theorem, if $a > e^{-1}$ there are not even periodic orbits.

Alternatively, the analysis can be done via the change of parameter $b = \ln a$ and the study of the family $g_b(x) = e^{x+b} = f_a(x)$ as b varies in $\mathbb{R}$. With this change $a = e^{-1}$ corresponds to $b = -1$. In this way, for any two values b_0 and b_1 of the parameter the graphs of g_{b_0} and g_{b_1} are swapped by a horizontal translation $b_1 - b_0$. Obviously this leads to the same results, up to the bifurcation value $b_0 = -1$ in place of the bifurcation value $a_0 = e^{-1}$. $\qquad\square$

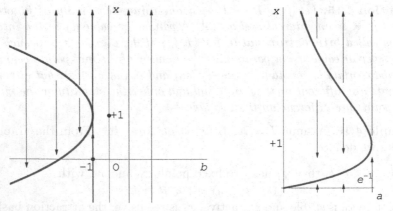

Fig. 4.8 Bifurcation diagrams: $\{\mathbb{R}, g_b\}$, $\{\mathbb{R}, f_a\}$, $g_b(x) = e^{x+b}$, $f_a(x) = ae^x$

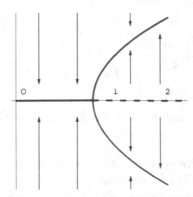

Fig. 4.9 Bifurcation diagram of $\{\mathbb{R}, a \arctan x\}$, $a \in [0, 2]$

Example 4.17. Set $f_a : \mathbb{R} \to \mathbb{R}$, $f_a(x) = a \arctan x$ with $a \in [0, 2]$.
If $0 \leq a \leq 1$, then $\alpha = 0$ is the unique equilibrium and it is globally asymptotically stable; if $1 < a \leq 2$ then 0 is still an equilibrium but loses its stability and becomes repulsive, moreover two additional equilibria appear, $\pm\alpha$, both locally asymptotically stable: the attraction basin of the positive one is $(0, +\infty)$, the attraction basin of the negative one is $(-\infty, 0)$. $\qquad\square$

To perform an elementary analysis of the bifurcation points of a DDS $\{I, f\}$, we consider the fixed points and the periodic orbits as a subset of $\Lambda \times I$:

$$\bigcup_{s=1}^{\infty} \{(a, \alpha) \in A \times I : f_a^s(\alpha) = \alpha\}.$$

Under reasonable assumptions on the dependence of f_a on a, this set is the union of smooth curves, whose intersections are isolated points and exhibit distinct tangents. In this context the bifurcation points can be subdivided in three classes, based on a local analysis of the intersection type:

1)   A curve bends forward or backward (see Example 4.16). This case is called **saddle-node bifurcation**.
2)   Two curves cross at a point and they are both graphs (of functions depending on a) in a neighborhood. This case is called **transcritical bifurcation**. As we are going to see, the logistic h_a has a transcritical bifurcation at $a = 1$.
3)   Two curves cross at a point where one of them bends either forward or backward. The dynamics of the related DDS may correspond either to the change of the number of equilibria (**pitchfork bifurcation**, see Example 4.16, $a_0 = 1$) or to a **period-doubling** (see h_a, $a_0 = 3$).

For instance (see Fig. 4.10), we can consider the family h_a of logistic maps (without restricting our analysis to the interval $[0, 1]$) as transformations of $\mathbb{R}$ in itself as the parameter a varies, that is the DDS's $\{\mathbb{R}, h_a\}$ with the parameter a in $[0, 4)$:

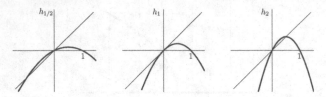

Fig. 4.10 Logistic parabolas: h_a in $\mathbb{R}$, $a = 1/2$, 1, 2

- if $0 < a < 1$, then there are two distinct equilibria: 0 (locally asymptotically stable) and $\alpha_a = (a-1)/a$, (repulsive);
- if $a = 1$, then the two equilibria collapse into a single one: 0, that is stable and attractive from above and unstable and repulsive from below;
- if $1 < a < 3$, then 0 becomes repulsive and another α_a appears: it belongs to $(0, 1)$ and is locally asymptotically stable; therefore $a = 1$ is a value of (transcritical) bifurcation;
- if $a = 3$, then $\alpha_a = 2/3$ is locally asymptotically stable: in fact $h_3'(\alpha_3) = -1$ and the property follows from Remark 3.66. It is easy to verify by numerical simulations that α_3 attracts the orbits much more slowly than α_a with $1 < a < 3$;
- if $a > 3$, then α_a is repulsive too ($h'(\alpha_a) = 2 - a < -1$) and a 2 periodic orbit appears and stays stable and attractive as long as $3 < a < 1 + \sqrt{6}$; so $a = 3$ is a bifurcation value (period-doubling).

To depict the case $1 < a < 3$, we consider $a = 2$.

The graphs of the iterations of h_2 suggest that $\lim_k h_2^k(x) = \alpha_2 = 1/2$ if $0 < x < 1$ (see Fig. 4.11). Notice that the tenth iteration has a graph with two nearly vertical pieces. The proof of the claims related to the periodic orbit

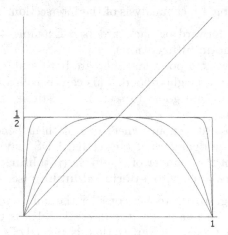

Fig. 4.11 Some iterations of h_2: h_2, h_2^2, h_2^5, h_2^{10}

requires some further analysis. By

$$h_a^2(x) = a^2 x (1-x)(1 - ax(1-x))$$

and recalling that $\alpha = 0$ and α_a are equilibria for $\{[0,1], h_a\}$ if $a > 1$, the quest for fixed points of h_a^2 reduces to solving

$$ax^2 - (1+a)x + \frac{1+a}{a} = 0 \qquad x \in [0,1], \quad 1 < a < 4.$$

The equation of degree two in x has two distinct solutions β_a and γ_a both different from 0 and α_a if and only if $a > 3$ (see Exercise 4.5):

$$\beta_a = \frac{a+1+\sqrt{(a+1)(a-3)}}{2a} \qquad \gamma_a = \frac{a+1-\sqrt{(a+1)(a-3)}}{2a}.$$

Then β_a and γ_a belong to the 2 periodic orbit, indeed $h_a^2(x) = x$ is a degree-four equation, so it cannot have any other solution in addition to 0, α_a, β_a and γ_a. In particular $\{\beta_a, \gamma_a\}$ is the unique 2 periodic orbit. To analyze the stability of the 2 periodic orbit $\{\beta_a, \gamma_a\}$ we notice that $h_a'(x) = a(1-2x)$ implies:

$$h_a'(\beta_a) = -1 - \sqrt{(a+1)(a-3)} \qquad h_a'(\gamma_a) = -1 + \sqrt{(a+1)(a-3)}$$
$$h_a'(\beta_a)\, h_a'(\gamma_a) = 1 - (a+1)(a-3).$$

As $|h_a'(\beta_a)\, h_a'(\gamma_a)| < 1$ if and only if $-2 < -(a+1)(a-3) < 0$, we find that, if $a > 3$, this inequality is equivalent to $a^2 - 2a - 5 < 0$, that is $a < 1 + \sqrt{6}$. From the previous analysis and Theorem 3.69 we deduce that if $a > 3$, then a 2 periodic orbit appears, and it is stable as long as $a < 1 + \sqrt{6}$ while becomes unstable as $a > 1 + \sqrt{6}$.

Thus $a = 3$ is a bifurcation value for h_a, and its bifurcation is of the type period-doubling. We emphasize that the branch of equilibrium points goes on as $a > 3$, but there are more stable equilibria. By further increasing the parameter value a we meet new period-doubling bifurcations, corresponding to the appearance period 4 orbits, period 8 orbits and so on: the period-doubling cascade (increasing powers of 2) reflects the beginning of Sharkovskii's ordering. It could be proved that as a increases it reaches a certain value such that the 2 periodic orbit becomes unstable and repulsive, and this takes place together with the appearance of a stable and attractive 4 periodic orbit. Proceeding in the analysis of the logistic, as the parameter a increases, we find orbits with every period of the kind 2^n, $n \in \mathbb{N}$, that is along the period-doubling cascade. Precisely they all appear long before the value 4 is reached, as it will be proved in the next sections.

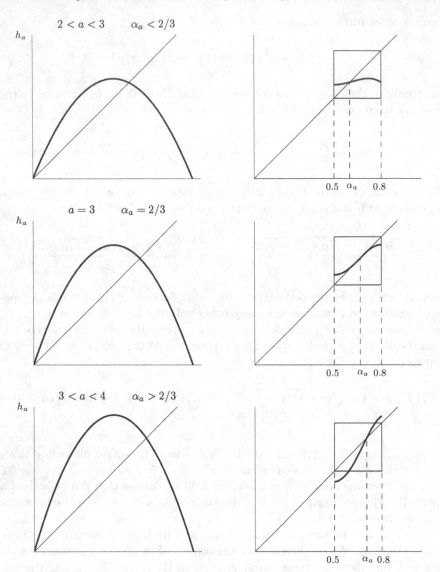

Fig. 4.12 Graphs of h_a, h_a^2 close to the equilibrium α_a in three examples: $a \in (2, 3)$; $a = 3$ and $a \in (3, 4)$

Due to this we expect that the dynamics of h_a must be extremely complicated when a varies from 3 to 4. To the aim of clarifying the whole structure, we exploit a result due to Fatou[2], whose proof is omitted.

---

[2] Pierre Joseph Louis Fatou, 1878-1929.

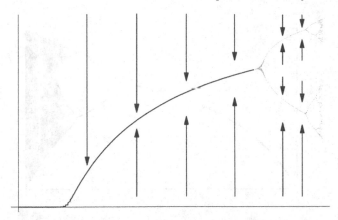

Fig. 4.13 Qualitative bifurcation diagram for $\{\mathbb{R}, h_a\}$

Theorem 4.18 (Fatou). *If the DDS $\{\mathbb{R}, ax^2 + bx + c\}$ with $a \neq 0$ has an attractive periodic orbit, then the critical value $-b/2a$ (zero of the derivative) belongs to the attraction basin of such orbit.*

Corollary 4.19. *A quadratic polynomial can have at most a single attractive and stable periodic orbit.*

Summarizing the previous analysis: the orbits in the dynamics of a quadratic polynomial can be copious but only one can be stable. Actually, in the period-doubling cascade of the logistic dynamics, every new orbit is stable at its appearance and at the same time the former stable orbit becomes unstable.

The unique critical point of h_a is $1/2$, for every value of $a \neq 0$. Therefore, if h_a has an attractive s periodic orbit ($s = 1, 2 \ldots$), then the trajectory starting from $1/2$ is attracted by the unique stable periodic orbit. This is valuable knowledge for drawing a bifurcation diagram that is more precise than the one sketched before, and more informative when a varies on the interval $[3, 4]$.

One way to exploit the consequences of Fatou's Theorem to implement a numerical simulation is as follows: we represent the values of the parameter a on the x-axis and on the y-axis the values $h_a^k(1/2)$ of the iteration with initial datum $1/2$ with large k, for instance k between 100 and 300. Obviously we can only select a finite number of values for a: we can fix an integer N and make a uniform partition of the interval under inspection (for simplicity of the exposition we focus the simulation only on the relevant interval $[2, 4]$):

$$2, \; 2 + \frac{1}{N}, \; 2 + \frac{2}{N}, \; \ldots, \; 4 - \frac{1}{N}, \; 4.$$

Obviously, as N grows, we have to require additional memory allocation and patience in waiting the end of computations.

If for a certain value of a there is an attractive orbit, then iterates $h_a^k(1/2)$ are attracted by this orbit; with an optimistic attitude toward numerical ex-

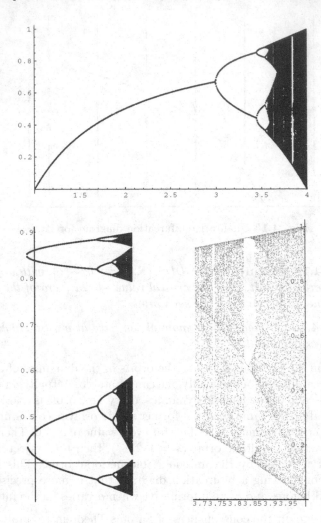

Fig. 4.14 Bifurcation diagram for the logistic map h_a, $a \in [1, 4]$, with zooming on some details

periments, that could be rigorously justified by additional analysis, we hope that for $k > 100$ the computed values are so close to the attractor to be numerically and graphically undistinguishable from it. Plotting (see Fig. 4.14) every iteration from 100 to 300, we can identify the attractive orbits whose period is less than 200. If for a value of a there is no attractive periodic orbit, we expect a "chaotic" cloud of points on the vertical line $x = a$.

The program may require very long time to be implemented, but the resulting graph is surprisingly rich in information: we find again the stability of α_a up to $a = 3$, thereafter we can unscramble the various bifurcations with transi-

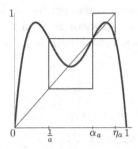 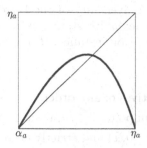 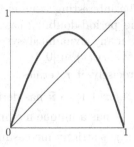

Fig. 4.15 Self-similarity of details in the graph of h_a^2 with the graph of h_b

tion toward stable orbits with double period, obtaining also sharp estimates of corresponding values of bifurcation for the parameter.

At a first glance, there is only a confused cloud of points at the end (the right-hand portion of $[2, 4]$), actually there is a nontrivial but hierarchical and highly structured order.

Two fading bands, one around $a = 3.74$ the other around $a = 3.83$ suggest the presence of attractive orbits of period 5 and 3. Zooming in (see Fig. 4.14 b), we observe a structure similar to the one of the main diagram close to $a = 3$. This kind of self-similarity of the set with some of its details has an analytic explanation, though we limit our discussion to heuristic argument: if we fix a value of a such that $2 < a \leq 4$, then referring to Fig. 4.15 the qualitative graph of h_a^2 in the square box with vertices $(1/a, 1/a)$ and $((a - 1)/a, (a - 1)/a)$ and in the box with vertices[3] $((a - 1)/a, (a - 1)/a)$ and (η_a, η_a) is qualitatively alike to the graph of h_b for a suitable b in $[0, 4]$ (at most up to a reflection), hence (on this subject see Sect. 4.5) the dynamics of the corresponding DDS's are alike.

The period-doubling cascade presents an additional surprising regularity: **the ratio r_n of distances of consecutive bifurcation values a_n**

$$r_n = \frac{a_n - a_{n-1}}{a_{n+1} - a_n}$$

tends, as $n \to +\infty$, to the **Feigenbaum constant**. This constant, approximated with thirteen decimal digits, is:

$$\mathcal{F} = 4.6692016091029\ldots$$

Thus we can estimate[4] which is the value of the parameter a after that the sequence of period-doubling cascade is complete, hence starting from which value of a there are 2^n periodic orbits for any n.

---

[3] $\eta_a = \left(a + \sqrt{a^2 - 4}\right)/2a$ solves the equation $h_a^2(\eta_a) = \alpha_a$.

[4] After setting $p_n = a_n - a_{n-1}$, the limit $\lim_n p_n/p_{n+1} = \mathcal{F} > 1$ entails (see (21) in Appendix A) that the series of step length is convergent: $\sum_{n=1}^{+\infty} p_n < +\infty$.

The interesting issue is the universality of the Feigenbaum constant: indeed the period-doubling cascade exhibiting such asymptotic behavior is typical of many dynamical systems (for instance, $f_a(x) = a \sin(\pi x)$ or $g_a(x) = ax^2 \sin(\pi x)$ in $[0,1]$).

Precisely, if f verifies:

- $f : [0,1] \to \mathbb{R}$ has derivatives of any order;
- f has a unique maximum point $\widetilde{x} \in (0,1)$ such that $f''(\widetilde{x}) < 0$;
- f is strictly increasing in $[0, \widetilde{x})$ and strictly decreasing in $(\widetilde{x}, 1]$;
- f has Schwarzian derivative $(\mathcal{D}f)(x) < 0$ for any $x \in [0,1]$ with $x \neq \widetilde{x}$ (see Definition 4.11);

then, as a varies in $[0, 1/f(\widetilde{x})]$, the bifurcation values a_k corresponding to period-doubling for the family of DDS's $\{I, af\}$ are located in such a way that their differences $a_k - a_{k-1}$ form an asymptotically geometric progression and:

$$\lim_k \frac{a_k - a_{k-1}}{a_{k+1} - a_k} = \mathcal{F}.$$

In contrast to the universality of $\mathcal{F}$, the value of $a_\infty = \lim_n a_n$ does depend on I and f. We collect in a table the approximated numerical values of some bifurcation values of the logistic dynamics: the a_k's corresponding to the appearance of the 2^k cycle, their limit a_∞ and a_ω which denotes the value of the parameter corresponding to the appearance of the 3 cycle.

Bifurcation values of the logistic h_a	Period of the orbits (stable if $a_n < a \leq a_{n+1}$)
$a_1 = 3$	2
$a_2 = 1 + \sqrt{6} = 3.449489\ldots$	4
$a_3 = 3.544090\ldots$	8
$a_4 = 3.564407\ldots$	16
$a_5 = 3.568759\ldots$	32
$a_6 = 3.569692\ldots$	64
$a_7 = 3.569891\ldots$	2^7
$\vdots$	$\vdots$
a_n	2^n
$\vdots$	$\vdots$
$a_\infty = 3.5699456\ldots$	
$a_\omega = 1 + \sqrt{8} = 3.828427\ldots$	3

Exercise 4.4. Plot a bifurcation diagram of $\{\mathbb{R}, f_a\}$ with $f_a(x) = a(x - x^3)$.

Exercise 4.5. Determine the points in the 2 periodic trajectory of the DDS $\{[0,1], h_a\}$ with $3 < a \leq 4$.

Exercise 4.6. Prove that the 2 periodic trajectory $\{\beta_a, \gamma_a\}$ of the logistic dynamics $\{[0,1], h_a\}$ is stable and attractive if $3 < a < 1 + \sqrt{6}$, repulsive if $a > 1 + \sqrt{6}$.

Exercise 4.7. Prove the stability of the 2 cycle $\{\beta_a, \gamma_a\}$ of h_a if $a = a_2 = 1 + \sqrt{6}$. (Hint: first prove that, if $\{I, f\}$ has a periodic orbit $\{\beta, \gamma\}$ and we set $g = f^2$, then

$$g'(\beta) = g'(\gamma) = -1 \qquad \text{and} \qquad \begin{array}{l} 2g'''(\beta) + 3(g''(\beta))^2 > 0 \\ 2g'''(\gamma) + 3(g''(\gamma))^2 > 0 \end{array}$$

entails that $\{\beta, \gamma\}$ is attractive and stable. Second, apply this property to $f = h_{a_2}$, where $a_2 = 1 + \sqrt{6}$, and deduce the stability of the 2 cycle.

Exercise 4.8. Determine by numerical simulations (that is without using the explicit formulas of Exercise 4.2) the 2 cycles of logistic dynamics h_a for $a = 3.1$, $a = 3.2$ and $a = 3.3$. Prove that for these three values of the parameter there exists a unique 2 cycle. Prove directly its stability by using Theorem 3.69.
Draw some iterations with the graphical method for finding the fixed points of h_a^2, with these three values.

Exercise 4.9. Determine by numerical simulations the 4 cycle for h_a if $a = 3.5$.
Then prove that it is stable and attractive. Verify graphically (drawing the cobweb) that it is a 4 cycle.
Plot h_a and its intersections with the identity (it is the fastest way to find the cycle).
We emphasize the convenience of starting the iterations from the critical value: $X_0 = 1/2$. For the numerical evaluation of the intersections Newton's method is recommended.

Exercise 4.10. 1) For the DDS $\{[0,1], h_a\}$, the value α_a is a stable and attractive equilibrium if and only if $1 < a \le 3$. Determine the unique value b_0 of a in $(1,3]$ such that α_a is superattractive (that is $h'(\alpha_a) = 0$).
2) For the DDS $\{[0,1], h_a\}$ the 2 periodic orbit $\{\gamma_a, \beta_a\}$ is stable and attractive if and only if $3 = a_1 < a \le a_2 = 1 + \sqrt{6}$. Determine the unique value b_1 of $a \in (a_1, a_2)$ such that the orbit $\{\gamma_a, \beta_a\}$ is superattractive (that is $(h_a^2)'(\gamma_a) = (h_a^2)'(\beta_a) = 0$). Referring to the bifurcation values a_k corresponding to the period-doubling cascade for the logistic h_a, it is possible to prove that for any $k \in \mathbb{N}$ there exists b_k such that $a_k < b_k < a_{k+1}$ and for $a = b_k$ the logistic h_a has a 2^k superattractive cycle, moreover the value $1/2$ belongs to this cycle.

Exercise 4.11. As a varies in a neighborhood of $a_\omega = 1 + \sqrt{8}$, plot h_a and $(h_a)^3$, study graphically and numerically the solutions of $(h_a)^3(x) = x$. Then deduce that for $a < a_\omega$ there are no 3 periodic orbits and for $a \ge a_\omega$ there are 3 periodic orbits. Evaluate numerically a 3 cycle for h_a corresponding to the value $a = 3,84$. Prove that it is stable and attractive.

Exercise 4.12. Verify that the family of DDS's $\{\mathbb{R}, f_a\}$, with $f_a(x) = ax(1 - x^2)$ and $a \in \mathbb{R}$, has a pitchfork bifurcation.

4.4 Chaos and fractal sets

In this section we introduce some notions that prove useful in the description of the logistic model for values of the parameter that lead to a dynamics where every equilibrium and every periodic orbit is repulsive. Even in very different contexts these notions proved as paradigm of surprising generality in the study of nonlinear dynamics.

We start with the analysis of a simple example, but not as simple as it could appear at a first look.

Example 4.20. The **tent map** T is defined by

$$T : [0,1] \to [0,1] \qquad T(x) = \begin{cases} 2x & x \in [0,1/2] \\ 2-2x & x \in [1/2,1] \end{cases}$$

We study the corresponding dynamical system $\{[0,1], T\}$. The reader is invited to carefully consider the behavior of the iterate functions of T and the related dynamics, before reading this section.

As shown by the graphs in Fig. 4.16, the image of T^k covers the interval $[0,1]$ 2^k times. In particular, after setting $X_{k+1} = T(X_k)$ and choosing $X_0 = 1/7$, we obtain the trajectory X_k with values

$$\frac{1}{7}, \ \frac{2}{7}, \ \frac{4}{7}, \ \frac{6}{7}, \ \frac{2}{7}, \ \frac{4}{7}, \ \frac{6}{7}, \ \cdots$$

that is $1/7$ belongs to the basin of the 3 cycle $\{2/7, 4/7, 6/7\}$. By Sharkowskii's Theorem, the existence of a 3 cycle implies the existence of cycles of any integer period. Then the dynamics associated to T is absolutely nontrivial; rather, as it will be clarified in the sequel, it shows a very sensitive dependence on initial conditions, due to the huge number of oscillations in the iterations of T. We remark that T^k has 2^{k-1} peaks and consequently (Intermediate Value Theorem) 2^k intersections with the graph of the identity (the trivial ones, $x = 0$ and $x = 2/3$, are included): thus, after the elimination of the two fixed points of T, the remaining $2^k - 2$ points necessarily belong to periodic orbits of T, whose period must divide k. In particular they are all k cycles if k is prime: summarizing, T has two equilibria; due to the analysis of T^2, T has two equilibria and one 2 cycle; due to the analysis of T^3, T has two equilibria and two 3 cycles and so on.

Example 4.21. Phase-doubling map $\mathfrak{D}$.

$$\mathfrak{D} : [0,2\pi) \to [0,2\pi) \qquad \mathfrak{D}(\theta) = 2\theta \,(\mathrm{mod}2\pi)$$

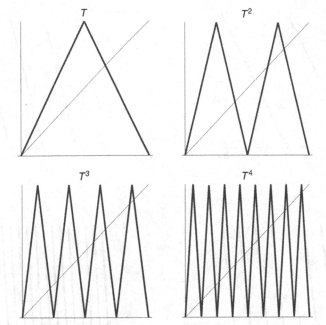

Fig. 4.16 Graphs of T, T^2, T^3, T^4 and of the identity map

The meticulous reader may object to the lack of continuity for $\mathfrak{D}$. Nevertheless, besides the existence of relevant phenomena described by discontinuous laws, we remark that $\mathfrak{D}$ actually has a geometric and kinematic interpretation of great relevance, corresponding to the description of a type of continuous motion: if we read θ as the position (expressed by the angular coordinate) of the hour hand of a clock, or as the position of a body in motion on a circular trajectory, then any position is described by several different values of θ, though the function $\mathfrak{D}$ (that doubles the angle) or "phase", continuously transforms the corresponding points on the unit circle S:[5]

$$\mathfrak{D} : S \to S$$

where $S = \big\{ (x,y) \in \mathbb{R}^2 : x = \cos\theta, y = \sin\theta, \theta \in [0, 2\pi) \big\}$. It is straightforward to verify that also $\mathfrak{D}$ has at least a 3 periodic orbit (it suffices to notice that the graph of $\mathfrak{D}^3$ has intersections with the graph of the identity at points off the graph of $\mathfrak{D}$).

Moreover, the dynamics of $\mathfrak{D}$ has orbits of any integer period. This property has to be directly tested since Theorem 4.2 does not apply to continuous functions from S to S: for instance $f(\vartheta) = \vartheta + \frac{2\pi}{k}$ (as a map from S to S) has k periodic points only, and no other trajectory.   □

---

[5] With a self-explaining abuse of notation, we denote a different (but strictly related to $\mathfrak{D}$) function by the same symbol.

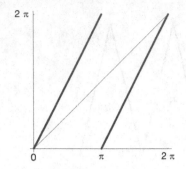
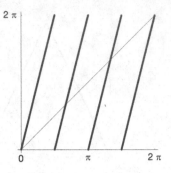
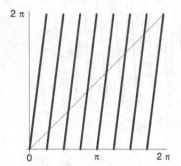
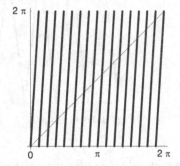

Fig. 4.17 Graphs of $\mathfrak{D}$, $\mathfrak{D}^2$, $\mathfrak{D}^3$, $\mathfrak{D}^4$ and the identity map. Notice that $\mathfrak{D}^k$ has a graph made with 2^k ascending segments

Several definitions of chaotic dynamics are available. Anyway, all of them try to condense the qualitative properties which are shown by the DDS's $\{I, f\}$ that (as in the case of the tent map T in $[0, 1]$) are generated by a function f whose iterations strongly twist the topology of I producing an extreme sensitivity to changes in the initial conditions and thus many periodic orbits that are dense in I.

Definition 4.22. *A DDS $\{I, f\}$ is **chaotic** (we also say that it has a **chaotic dynamics**) if:*

- *the periodic orbits are dense (if we take into account every integer period: 1 is included), that is any interval $(a, b) \subseteq I$ contains at least a point belonging to a periodic orbit;*
- *f is **topologically transitive**, that is, for any $x, y \in I$ and for any $\varepsilon > 0$ there exist $z \in I$ and $k \in \mathbb{N}$ such that*

$$|z - x| < \varepsilon, \qquad \left| f^k (z) - y \right| < \varepsilon;$$

- $\{I, f\}$ *exhibits a **sensitive dependence on initial conditions**, that is there exists $\delta > 0$ such that, for any $x \in I$ and $\varepsilon > 0$ there exist $z \in I$ and $k \in \mathbb{N}$ such that*

$$|x - z| < \varepsilon \qquad \qquad \left| f^k(x) - f^k(z) \right| > \delta.$$

Remark 4.23. The only presence of periodic orbits arbitrarily close to every point of I does not necessarily correspond to a chaotic dynamics (for instance, the linear dynamics $f(x) = -x$ is very simple), nevertheless the density of periodic orbits, together with the other two conditions is a quintessential ingredient of the chaotic dynamics (just ponder the simultaneous presence of orbits of every period whenever the period 3 comes in the phase diagram of a continuous DDS on a real interval).

Remark 4.24. *Topological transitivity of f* is equivalent to the property: "for any pair U, V of nonempty open intervals contained in I there exist $z \in U$ and $k \in \mathbb{N}$ such that $f^k(z) \in V$". This means that, given any two nonempty open intervals, in both intervals there is at least one point whose trajectory crosses the other interval, no matter how "small" the intervals are chosen.

Remark 4.25. The "*sensitive dependence on initial conditions*" has this meaning: if the iterations of a function with such property model the long-term behavior of a (economic, demographical, meteorological, etc.) system, then big differences are expected between the actual value and the one predicted by the model, no matter how small the errors in the measurement of the initial conditions are. Since every experimental measure is stricken by errors, this is an inescapable fact to consider, for instance limiting the predictive use of the model to few iterations.

Nevertheless we cannot a priori exclude the existence of attractive sets (with possibly more complex geometry than equilibria or periodic orbits) whose presence renders some kind of long-term forecasting substantial.

Remark 4.26. Definition 4.22 concerning **chaotic dynamics** is a mixture of chaos and order, in the sense that it excludes any simultaneously simple and ordered description of the dynamics, whereas it requires precise geometric and topological properties of the trajectories: for instance there are periodic orbits arbitrarily close to every point. On the other hand, if we read the DDS $\{I, f\}$ as the deformation of an elastic string I described by the map f and the related dynamics is chaotic, then I is folded up in a more and more complicated manner as the number of iterations f increases.

If I is a nontrivial interval and f is continuous, then the definition of chaotic dynamics can be simplified as specified by the next statement, whose proof is omitted.

Theorem 4.27. *Assume $I \subset \mathbb{R}$ is a nontrivial interval and $f : I \to I$ is a continuous function. If f is topologically transitive and the points belonging* to periodic orbits of $\{I, f\}$ are dense in I, then $\{I, f\}$ exhibits a sensitive dependence on initial conditions too, hence f has a chaotic dynamics in I.

We test the definition of chaotic dynamical system on explicit examples: the dynamics associated to the tent map T and to the phase-doubling map $\mathfrak{D}$ are chaotic in $[0, 1]$ and $[0, 2\pi)$ (or S) respectively.

Theorem 4.28. *The dynamics of $\{[0, 1], T\}$ is chaotic.*

Proof. We have to verify items 1)-2)-3) of Definition 4.22. When performing the analysis, it is useful to look at the graphs of T and its iterations (see Fig. 4.17). On the other hand we emphasize that, in performing computer evaluation of iterated functions, it is worthwhile to store all computed iterated function at every step, to avoid useless repetition of computations.

(1) *Density of periodic orbits.* If $k \in \mathbb{N}$, denoting by I_h the intervals $[h2^{-k}, (h + 1)2^{-k})$, $0 \leq h \leq 2^k - 1$, then T^k restricted to I_h is a one-to-one map with values in $[0, 1]$ (monotone strictly increasing if h is even, strictly decreasing if h is odd); hence $\{[0, 1], T\}$ has exactly 2^k periodic points of period s, $1 \leq s \leq k$, one for each interval I_h, as one can deduce by the Intermediate Value Theorem on $T^k(x) - x$ in each interval I_h.

(2) *T is topologically transitive.* Given $x, y \in [0, 1]$ and $0 < \varepsilon < 1/2$, choose $k \in \mathbb{N}$ such that $2^k \varepsilon > 2$. Then the image of $(x, x + \varepsilon)$ under T^k is exactly $[0, 1]$, so $(x, x + \varepsilon)$ contains points whose image through T^k coincides with any $y \in [0, 1]$. Obviously in the previous argument $(x, x + \varepsilon)$ has to be substituted with $(x - \varepsilon, x)$ if $x + \varepsilon > 1$.

(3) *The sensitive dependence on initial conditions.* It follows by (1) and (2) together with the previous theorem, nevertheless its direct proof is straightforward: it suffices to choose $\delta = 1/2$ and repeat the argument in the proof of (2) to achieve the claim. □

Remark 4.29. Every periodic orbit of $\{[0, 1], T\}$ (whose existence has been proved when verifying item (1)) is repulsive. In fact $|T'(\beta)| = 2 > 1$ at every point β that belongs to these periodic orbits. Moreover 0 (where the derivative does not exists) is repulsive.

Theorem 4.30. *The dynamics of $\{[0, 2\pi), \mathfrak{D}\}$ is chaotic.*

Proof. Items (2) and (3) can be proved as for $\{[0, 1], T\}$ without exploiting Theorem 4.27. About item (1) we notice that, even if $\mathfrak{D}$ lacks continuity, we can still use the same argument of the previous proof since $\mathfrak{D}^k(x) - x$ is a sign-changing continuous function in the interior of each I_h. □

Remark 4.31. All equilibria and periodic orbits of $\{[0, 2\pi), \mathfrak{D}\}$ are repulsive.

If we take time to think, the intricacy of the dynamics associated to T and $\mathfrak{D}$ is not so surprising. It is enough to reflect on the kneading procedures made by a pastry chef to prepare puff pastry: first he squashes and stretches the dough in one direction, then he folds it up; that is, omitting the direction which un-

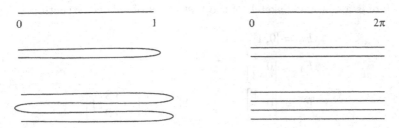

Fig. 4.18 Puff pastry: the first actions of the pastry chef are interpreted as first iterations of T and $\mathfrak{D}$

dergoes no stretch, he performs the transformation T; then he repeats these two steps many times (k), namely he iterates the procedure performing the transformation T^k. The fact that nearby points at the beginning can be very distant at the end, comes as no surprise; moreover (at the start) small portions may cover the whole puff pastry with their image under iteration.

On the other hand, the iterated functions of $\mathfrak{D}$ correspond to stretching, cutting and overlapping (instead of folding), by repeating this cycle many times: again one ends up with a remarkable reshuffling and there is no hope to preserve that any pair of distinct points remain close.

Now we study some basic examples of fractal sets in $\mathbb{R}$ that show up in the phase diagram of several discrete dynamic systems.

We can "informally" define a **fractal set** as a subset of $\mathbb{R}$ fulfilling these properties:

- it has a **fine structure**, namely its details are defined at arbitrarily small scales;
- it is not an elementary geometrical object: it does not coincide neither locally, nor globally with a finite union of intervals or isolated points;
- the "dimension" is noninteger[6];
- often it exhibits properties of **self-similarity** (or approximates these properties in a statistical manner): it can be split in proper subsets, each one with one-to-one correspondence with the whole set, through easy geometrical transformation;
- often a simple definition is available based on recursive repetition of elementary transformations.

Example 4.32. We consider the first archetypal example of fractal set: the **Cantor set** $\mathcal{C}$ (or **middle-third Cantor set**) that is obtained by removing the intermediate open segment $(1/3, 2/3)$ from the interval $[0, 1]$ and repeating this removal on all segments that are left: the residual is the Cantor

---

[6] For a formal definition of dimension see Appendix F.

set. Here is a graphical sketch of the first steps of this construction:

$$E_0 = [0, 1]$$
$$E_1 = \left[0, \frac{1}{3}\right] \cup \left[\frac{2}{3}, 1\right]$$
$$E_2 = \left[0, \frac{1}{9}\right] \cup \left[\frac{2}{9}, \frac{1}{3}\right] \cup \left[\frac{2}{3}, \frac{7}{9}\right] \cup \left[\frac{8}{9}, 1\right]$$

$$\dots$$

Every E_k is the union of 2^k, $k \in \mathbb{N}$, intervals of length 3^{-k} and we can set

$$\boxed{\mathcal{C} = \bigcap_{k=0}^{\infty} E_k}$$

We observe that:
- $\mathcal{C}$ is self-similar: its intersection with anyone of the intervals at step n of the previous construction corresponds to a 3^{-n} scaling of $\mathcal{C}$, up to a translation;
- $\mathcal{C}$ has a fine structure: we cannot exactly draw or imagine it;
- its recursive definition is extremely simple, although its topology is anything but trivial: it is an infinite, uncountable and **totally disconnected** set (namely it is a set containing no nontrivial interval), it has null 1-dimensional measure, but the right dimension for "measuring" it has a value in $(0, 1)$ (see Appendix F).

Example 4.33. Consider the discrete dynamical system $\{\mathbb{R}, g\}$ with $g = 3T/2$. The map g folds $\mathbb{R} \setminus \{1/2\}$ and transforms it in two copies of $(-\infty, 3/2)$. The map g^3 has fixed points that are not fixed points of g, so g has 3 periodic orbits and, by Sharkovskii's Theorem, orbits of any integer period too.   □

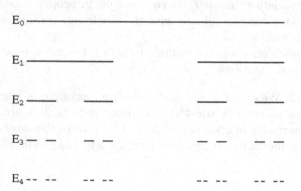

Fig. 4.19 The first four steps of the construction of the Cantor set

Theorem 4.34. *If $g = 3T/2$ and C is the Cantor set, then $g(C) = C$, that is C is invariant for $\{\mathbb{R}, g\}$.*

We omit the proof. We only remark that the invariant set C is *repulsive for the dynamics of g*; indeed:

- if $x < 0$, then $g^k(x) = 3^k x \to -\infty$ if $k \to +\infty$;
- if $x > 1$, then $g(x) < 0$ and $g^k(x) = g^{k-1}(g(x)) = 3^{k-1}g(x) \to -\infty$ if $k \to +\infty$;
- if $x \in [0,1] \setminus C$, then there exists h such that for $x \in [0,1] \setminus E_h$:

$$g^h(x) > 1 \; ; \; g^{h+1}(x) < 0 \; ; \; g^k(x) = g^{k-h}(g^h(x)) \to -\infty \text{ per } k \to +\infty$$

here the sets E_h are the ones that have been introduced in the construction of the Cantor set.

It is possible to show that the trajectories contained in C are extremely erratic, as specified by the next theorem.

Theorem 4.35. *The dynamics of $\{C, 3T/2\}$ is chaotic.*

Example 4.36. If $a > 4$, then the logistic map h_a produces a dynamics in $\mathbb{R}$ that is alike to the one of the map $g = 3T/2$, studied earlier.

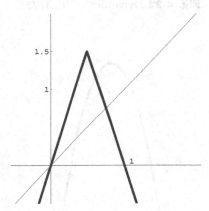

Fig. 4.20 Graph of $g(x) = 3T(x)/2$; two repulsive fixed points: 0, $3/4$

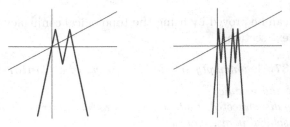

Fig. 4.21 Graphs of g^2 and g^3 where $g(x) = 3T(x)/2$

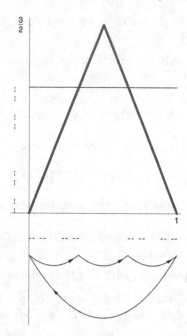

Fig. 4.22 Dynamics of $\{\mathcal{C}, 3T/2\}$

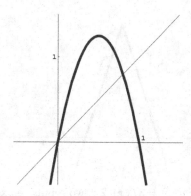

Fig. 4.23 h_a with $a > 4$

This property can be proved by using the topological conjugacy that is introduced in the next section.   □

Definition 4.37. *A nonempty set* $E \subseteq \mathbb{R}$ *is called a **Cantor-like set** if*

- *E is closed and bounded;*
- *E is totally disconnected, that is it contains no nontrivial interval;*
- *E has no isolated point, that is*
  if $p \in E$, *then* $\forall r > 0, (E \setminus \{p\}) \cap (p - r, p - r) \neq \emptyset.$

Example 4.38. C_t: t-middle **Cantor set**, $0 < t < 1$.

Analogously to the construction of the Cantor set and starting again from the closed interval $[0, 1]$, we apply an iterative procedure consisting in removing every open interval of length t times the length of the residual interval where it is contained and centered at the same residual interval.

The ultimate residual is nonempty, since at least the endpoints of every removed open interval (there are infinitely many) belong to the set.

All t-middle Cantor sets (and in particular $C = C_{1/3}$) are Cantor-like set. $\square$

We list some properties of the discrete dynamical system $\{\mathbb{R}, h_a\}$, with $a > 4$, without proofs. Given $a > 4$:

- the set

$$R = \{x \in [0, 1] : \quad h_a^k(x) \in [0, 1] \quad \forall k \in \mathbb{N}\}$$

  (which is invariant with respect to h_a by definition) is a Cantor-like set;
- R is repulsive: $\lim_k h_a^k(x) = -\infty$, $\forall x \in \mathbb{R} \backslash R$;
- the dynamics of h_a restricted to R is chaotic.

We conclude this section by stating without proof an important result that clarifies how the existence of orbits of period 3 entails, besides the orbits of any other integer period, also a dynamics that deeply reshuffles some subsets of the domain.

Theorem 4.39 (T.Y. Li & J.A. Yorke, 1975). *Given a nonempty interval I and a continuous function $f : I \to I$, if the discrete dynamical system $\{I, f\}$ has a periodic orbit of least period 3, then there exists an uncountable subset $E \subseteq I$ that contains no periodic point and such that:*

for any $x, y \in E$ with $x \neq y$

$$\max_k \lim \left| f^k(x) - f^k(y) \right| > 0 \qquad \min_k \lim \left| f^k(x) - f^k(y) \right| = 0;$$

for any $x \in E$ and $y \in I$ with y periodic,

$$\max_k \lim \left| f^k(x) - f^k(y) \right| > 0.$$

The statement of Theorem 4.39 is informally summarized as follows "period 3 implies chaos". This statement is correct for scalar DDS's and only in contexts where the density of periodic orbits in the whole I is not required by the definition of chaos.

Exercise 4.13. Consider the seconds hand of a clock, that moves with uniform unit angular speed, namely it goes around in 60 units of time (seconds).

Describe the DDS that generates the trajectories of the seconds hand at time intervals of one second, starting from the various initial positions.

Establish whether this DDS is topologically transitive, whether the periodic orbits are dense and whether there is sensitive dependence on initial conditions.

Exercise 4.14. Consider a clock with two hands, the hour hand and the minute hand, that move haltingly, every hour and every minute respectively. Describe the dynamics $X_k = \varphi_k - \psi_k$ where φ_k are the angular position of the hour hand and ψ_k of the minute hand respectively (referring to angles in $[0, 2\pi)$ measured in radians).

Exercise 4.15. Consider the DDS $\{\mathbb{R}, g\}$ where $g(x) = 10x$.
Show that this linear DDS is not chaotic, nevertheless it exhibits sensitive dependence on initial conditions. So this last property is not peculiar of nonlinear dynamics but may appear also in the linear ones.

Exercise 4.16. Find a dense set $E \subset [0, 1]$ such that $\lim_k T^k (X_0) = 0$ for any $X_0 \in E$ (precisely, there exists $k_0 = k_0 (X_0)$ such that $T^k(X_0) = 0$ for any $k > k_0$), where T is the tent map.

Exercise 4.17. Describe a dense set $E \subset [0, 1]$ such that $\lim_k \mathfrak{D}^k (X_0) = 0$ for any $X_0 \in E$ (precisely, there exists $k_0 = k_0 (X_0)$ such that $\mathfrak{D}^k(X_0) = 0$ for any $k > k_0$), where $\mathfrak{D}$ is the phase-doubling map.

Exercise 4.18. Using the characterization of self-similar fractals set and Theorem F.1 (see Appendix F), determine the Hausdorff dimension of the Cantor set and, in general, the Hausdorff dimension of a t-middle Cantor set.

Exercise 4.19. Set $A = \mathcal{C}_{1/2}$. Determine the Hausdorff dimension of A and $A \times A$.

Exercise 4.20. The square **Sierpinski gasket** Q is built as follows: one starts from a compact square subdivided in nine identical squares, then removes the ones that are not adjacent to the vertices, then iterates the procedure. Compute its Hausdorff dimension.

Exercise 4.21. Modify the construction of the previous exercise by subdividing in sixteen identical squares and evaluate the Hausdorff dimension of the set W thus obtained.

4.5 Topological conjugacy of discrete dynamical systems

The direct analysis of a DDS may prove to be strikingly hard, nevertheless sometimes it is possible to study another system that is much easier to analyze and has phase diagram qualitatively identical to the one under exam. In this case the two DDS's are called topologically conjugate.
For instance, in this section we show that the tent map in $[0, 1]$ has a topologically conjugate dynamics with the logistic parabola h_4 in $[0, 1]$; incidentally this property allows to prove that the logistic parabola corresponding to parameter 4 has a chaotic dynamics.
The underlying idea is quite simple: given the DDS $\{I, f\}$, if we change suitably the coordinates in the domain I and in the image $f(I)$ and change coherently the function f, then the qualitative features of the phase diagram should not be different.

Definition 4.40. *Let $I, J \subset \mathbb{R}$ be two intervals. Then two functions $f : I \to I$ and $g : J \to J$ are called **topologically conjugate** if there exists a continuous function $\varphi : I \to J$ which is invertible, has a continuous inverse φ^{-1} and fulfills the identity*

$$g = \varphi \circ f \circ \varphi^{-1}. \tag{4.2}$$

*When it exists, then φ is called **topological conjugacy** of f and g. Analogously the DDS's $\{I, f\}$ and $\{J, g\}$ are called **topologically conjugate** if f and g are topologically conjugate.*

Notice that (4.2) is equivalent to anyone of the subsequent identities

$$g \circ \varphi = \varphi \circ f, \qquad f = \varphi^{-1} \circ g \circ \varphi, \qquad \varphi^{-1} \circ g = f \circ \varphi^{-1},$$

as it can be easily verified by first performing suitable compositions with φ and φ^{-1} and then simplifying.

Thanks to this equivalence, to verify the possible topological conjugacy on the examples one can choose in each case the most convenient among the three. The topological conjugacy relationship between f and g is represented in this diagram

$$
\begin{array}{ccc}
I & \xrightarrow{f} & I \\
\varphi^{-1} \uparrow & & \downarrow \varphi \\
J & \xrightarrow{g} & J
\end{array}
$$

Remark 4.41. If $\varphi : I \to J$ is continuous, invertible, with continuous inverse in the interval I, then φ is strictly monotone and:

1) $U \subset I$ is a closed set in I if and only if $\varphi(U)$ is a closed set in J;
2) the sequence $\{X_k\}$ is convergent in I if and only if the sequence $\{\varphi(X_k)\}$ is convergent in J;
3) $A \subset I$ is dense in I $\Leftrightarrow$ $\varphi(A)$ is dense in J.

Theorem 4.42. *Assume $I, J \subset \mathbb{R}$ are two intervals, $f : I \to I, g : J \to J$ two functions and $\varphi : I \to J$ is a topological conjugacy between f and g. Then:*

(i) $\varphi \circ f^k = g^k \circ \varphi$, for any $k \in \mathbb{N}$, that is also f^k and g^k are topologically conjugate;

(ii)   if the sequence $\{X_k\}$ is a trajectory of $\{I, f\}$, then the sequence $\{\varphi(X_k)\}$ is a trajectory of $\{J, g\}$;

(iii) α is a periodic point s for $\{I, f\}$ if and only if $\varphi(\alpha)$ is a periodic point s for $\{J, g\}$;

(iv)   if $A \subset I$ is an attractor and $B \subset I$ is its attraction basin, then $\varphi(A)$ is an attractor and $\varphi(B)$ is its attraction basin;

(v)   the periodic orbits of $\{I, f\}$ are dense in I if and only if the periodic orbits of $\{J, g\}$ are dense in J;

(vi) $\{I, f\}$ is topologically transitive if and only if $\{J, g\}$ is topologically transitive;

(vii) $\{I, f\}$ is chaotic if and only if $\{J, g\}$ is chaotic.

Proof. The first six claims can be easily checked, provided this check is done by following the order in the statement. Here we check only the first one:

$$\varphi \circ f^k = \underbrace{\left(\varphi \circ f \circ \varphi^{-1}\right)\left(\varphi \circ f \circ \varphi^{-1}\right)\cdots\left(\varphi \circ f \circ \varphi^{-1}\right)}_{k \text{ times}}\varphi = g^k \circ \varphi.$$

Item (vii) follows from (v), (vi) and Theorem 4.27.   □

About items (v), (vi) and (vii) we remark that the sensitive dependence on initial conditions alone is not always preserved under topological conjugacy, as it is shown by the next example.

Example 4.43. Consider the DDS's $\{(0,1),x^2\}$ and $\{(1,+\infty),x^2\}$. Then $\varphi(x) = 1/x$ is a topological conjugacy between the two DDS's. But the first one has no sensitive dependence on initial conditions, since 0 attracts all its trajectories, whereas the second has sensitive dependence on initial conditions: $X_0, \widetilde{X}_0 \in (1,+\infty)$ and $\left|X_0 - \widetilde{X}_0\right| = \varepsilon > 0$ implies

$$\left|X_k - \widetilde{X}_k\right| = \left|X_0^{2k} - \widetilde{X}_0^{2k}\right| =$$
$$= \left|X_0 - \widetilde{X}_0\right|\left(X_0^{2k-1} + X_0^{2k-2}\widetilde{X}_0 + \cdots + \widetilde{X}_0^{2k-1}\right) \geq 2k\varepsilon.   □$$

However on the connection between topological conjugacy and sensitive dependence the next theorem provides some useful information.

Theorem 4.44. *The sensitive dependence on initial conditions is preserved under topological conjugacy if the interval I is bounded and closed (and consequently J is bounded and closed too).*

We make explicit some relevant results that are straightforward consequences of the tools introduced in this section.

Theorem 4.45. *The logistic map h_4 in $[0,1]$ is topologically conjugate to the tent map T in $[0,1]$ (T is defined in Example 4.20).*

Proof. The function $\varphi(x) = (\sin(\pi x/2))^2$ is a topological conjugacy between the two DDS's $\{[0,1],h_4\}$ and $\{[0,1],T\}$. In fact $\varphi([0,1]) = [0,1]$, φ is continuous in $[0,1]$, strictly monotone and fulfills

$$h_4(\varphi(x)) = 4\left(\left(\sin\left(\frac{\pi}{2}x\right)\right)^2 - \left(\sin\left(\frac{\pi}{2}x\right)\right)^4\right) = 4\left(\sin\left(\frac{\pi}{2}x\right)\right)^2\left(\cos\left(\frac{\pi}{2}x\right)\right)^2 =$$

$$= (\sin(\pi x))^2 = \begin{cases} (\sin(\pi x))^2 & \text{if } 0 \leq x \leq 1/2 \\ (\sin(\pi x - \pi))^2 & \text{if } 1/2 < x \leq 1 \end{cases} =$$

$$= \varphi(T(x)).   □$$

The example at the beginning of Section 4.2 shows that the dynamics of a DDS associated to a unimodal function like $\{\mathbb{R}, ax^2 + bx + c\}$ with $a < 0$ can show orbits of period 3.

The topological conjugacy and Theorems 4.44 and 4.45 entail that the dynamics of the logistic h_4 in $[0,1]$ produces orbits of period 3: actually this happens even for smaller parmeters, starting from the value a_ω of a (with $a_\omega = 1 + \sqrt{8} = 3.828427...$). Therefore, according to Sharkovskii's Theorem, if $a > a_\omega$ then the logistic map has periodic trajectories of any integer period.

Theorem 4.46. *The logistic map of parameter* 4 *determines a chaotic dynamics in* $[0,1]$ *and the phase diagram of* $\{[0,1], h_4\}$ *qualitatively coincides with the phase diagram of* $\{[0,1], T\}$.

Proof. It is an immediate consequence of Theorems 4.28, 4.42 and 4.45.   □

We emphasize that in general, given two DDS's, it is not easy to verify whether they are topologically conjugate. The acknowledgement of any difference in the dynamics proves that there is no topological conjugacy (see Theorem 4.42 and Remark 4.70). On the other hand, proving the topological conjugacy means to find explicitly the conjugacy map φ, or at least to prove its existence. The techniques in Section 3.9 allow in some particular cases to find the topological conjugacy: Theorem 3.74 provides the explicit formula of the conjugacy map between a nonlinear DDS and a linear DDS, under some restrictive conditions. The topological conjugacy between $\{(0,1/2),\ h_2\}$ and $\{(0,+\infty),\ 2x\}$ is proved in Example 3.77; the reflections after Theorem 3.74 allow to give explicit (non recursive) form to solutions in slightly more general cases through non-bijective transformations (see the analysis of h_4 in the example 3.78). Using the tools introduced in the present section and the notation of Section 3.9, we can rephrase Theorem 3.74, by noticing that $J = \psi(I)$ is an interval, the DDS's $\{I, f\}$ and $\{J, bx\}$ are topologically conjugate, the topological conjugacy is explicitly given by ψ: since $f = \psi^{-1} \circ v \circ \psi$; eventually we notice that the DDS $\{J, bx\}$ is well defined (namely, for any $y \in J$ we obtain $by \in J$); also this point is a consequence of Theorem 3.74, indeed if $y \in J$ then there exists $\overline{x} \in I$ such that $y = \psi(\overline{x})$, therefore, after setting $\widetilde{x} = f(\overline{x})$ we obtain $by = b\psi(\overline{x}) = \psi(\widetilde{x})$ that obviously belongs to J.

Exercise 4.22. Verify that the change of variables $\varphi(x) = x - b/(1-a)$, exploited in the proof of Theorem 2.5 to analyze $\{\mathbb{R}, ax + b\}$ for $a \neq 1$, is actually a topological conjugacy between the affine function $f(x) = ax + b$ and the linear function $g(x) = ax$.

Exercise 4.23. Show an example to prove that two linear DDS cannot be topologically conjugate.

Exercise 4.24. Assume that f and g are topologically conjugate, and $\widetilde{x}$ is a fixed point for f: $f(\widetilde{x}) = \widetilde{x}$. Prove that, if $\widetilde{t}$ is the point corresponding to $\widetilde{x}$ via a topological conjugacy, then $\widetilde{t}$ is a fixed point for g: $g(\widetilde{t}) = \widetilde{t}$ (*the topological conjugacy preserves the fixed points*).

Exercise 4.25. Prove that, if f and g are topologically conjugate through a monotone increasing map φ and there exists an interval H contained in the domain I of f such that $f(x) > x$ for any $x \in H$, then $g(t) > t$ for any $t \in \varphi(H)$.

Exercise 4.26. Show that the dynamical systems $\{[0,1], h_4\}$ and $\{[0,1], g\}$ where $g(x) = \min\{4x^2, 4(x-1)^2\}$ are not topologically conjugate, even though both h_4 and g are unimodal, continuous, surjective and strictly increasing in $(0, 1/2)$, decreasing in $(1/2, 1)$.

Exercise 4.27. Prove that, if f and g are topologically conjugate by φ and f is monotone in an interval H, then g is monotone too (in the same direction) in $\varphi(H)$ (*the topological conjugacy preserves monotonicity*).

Notice that the previous exercises entail a necessary condition for the existence of a topological conjugacy between two maps f and g: *a one-to-one correspondence between all points and/or intervals where f and g coincide with the identity map is mandatory for topological conjugacy.*

Exercise 4.28. Prove that the Sharkovskii Theorem holds for the dynamical system $\{I, f\}$ even if I is an open interval.

Exercise 4.29. Use the Euler and backward Euler numerical schemes (see Example 1.20) to approximate the solution of this Cauchy problem, relative to the ordinary differential equation of the **logistic growth with continuous time**

$$\begin{cases} u' = b(u - u^2) \\ u(0) = u_0, \end{cases}$$

where the unknown u is a differentiable function in $[0, \infty)$ and the prescribed real parameters b and u_0 fulfill $b > 0$, $0 < u_0 < 1$.
Study the behavior of the solutions found by the the two numerical schemes using the theory developed in Chaps. 3 and 4.

4.6 Newton's method

A basic example of DDS is provided by Newton's method for finding the zeroes of functions, not necessarily polynomials. It is a classical tool, nevertheless it is widely used, due to its high numerical speed and robustness. Newton's method is an iterative scheme that achieves very good numerical approximations of the required solution, under mild conditions. In general the best we can get are approximations, since there is no general formula for exact solutions, not even for finding the roots of polynomials of degree five or higher.
Newton's method for finding the zeroes of a differentiable function g consists in the subsequent iterative scheme: first we define the function

$$\boxed{N_g(x) = x - \frac{g(x)}{g'(x)}}$$

whose domain coincides with the domain of g except for the zeroes of g'; then, starting from a value X_0, we generate the sequence $X = \{X_k\}$ recursively,

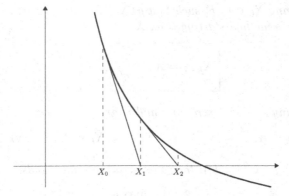

Fig. 4.24 Newton's method

namely we iterate the function N_g:

$$X_1 = N_g(X_0), \quad X_2 = N_g(X_1), \quad \ldots, \quad X_{k+1} = N_g(X_k) = N_g^{k+1}(X_0).$$

If g has at least one zero and the choice of X_0 is suitable, then X is a sequence that rapidly converges to a solution of the equation

$$g(x) = 0, \quad x \in \text{dom}(g).$$

When $X_0 \in \mathbb{R}$ and g is a real-valued function depending on a real variable, the geometrical meaning of the sequence X is as follows: given X_k, we draw the tangent at $(X_k, g(X_k))$ to the graph of g and we label by X_{k+1} the abscissa of its intersection with the real axis. Then we repeat the procedure.

Remark 4.47. The fixed points of N_g are the zeroes of g where the derivative g' does not vanish. Aiming to analyze the stability of a fixed point α of the DDS generated by N_g, if $g \in C^2$ we can compute $(N_g)'(x) = \dfrac{g(x)\,g''(x)}{(g'(x))^2}$ hence, from $g(\alpha) = 0$ and $g'(\alpha) \neq 0$ we deduce $(N_g)'(\alpha) = 0$.

The previous remark together with Theorem 3.57 prove the next statement.

Theorem 4.48. *If $g(\alpha) = 0$ and $g'(\alpha) \neq 0$, then α is locally asymptotically stable for N_g (precisely it is locally superattractive).*

The previous result is useful at the theoretical level, but it gives no hint for numerical computing since it does neither tell how close to α one has to start (the choice of X_0) in order to feel the local attractivity of α, nor it provides any error estimate. The next statement, focused on finding the zeroes of real functions of real variable, is more useful: it is an elementary result that we mention without proving it.

Theorem 4.49. *Let $g : [a, b] \to \mathbb{R}$ be a convex, differentiable function, with $g(a) < 0 < g(b)$. Then g has a unique zero α in (a, b).*

Moreover, for any $X_0 \in [a, b]$ such that $g(X_0) > 0$, the sequence $\{X_k\}$ generated by Newton's method starting from X_0

$$X_{k+1} = X_k - \frac{g(X_k)}{g'(X_k)}$$

is defined for any k in $\mathbb{N}$, decreasing and convergent to α

$$\lim_k X_k = \alpha, \qquad\qquad \alpha < X_{k+1} < X_k < X_0 \qquad \forall k \in \mathbb{N}.$$

If in addiction, $g \in C^2([a, b])$, then this error estimate holds:

$$0 < X_{k+1} - \alpha < \frac{\max g''}{2g'(\alpha)}(X_k - \alpha)^2.$$

In particular, if one starts from X_0 close to α then the convergence is very fast.

When applying Newton's method, one has to pay attention to the conditions in Theorem 4.49. For instance, the statement still holds if one substitutes the convexity condition with the one of concavity, provided one chooses X_0 such that $f(X_0) < 0$, thus obtaining the same error estimate and the increasing monotonicity of X, when starting from X_0 such that $f(X_0) < 0$. Nevertheless, if there are changes of concavity in the interval $[a, b]$, then one can meet unexpected phenomena.

Example 4.50. We consider the function $f : \mathbb{R} \to \mathbb{R}$, $f(x) = \arctan x$. It is well known that there is only one zero (at the origin). Moreover, as it is convex in $(-\infty, 0)$ and concave $(0, +\infty)$, then Theorem 4.49 does not apply, and if we want to use the Newton's method for finding such a zero we have to be careful when choosing X_0. For instance, if $\tilde{x}$ is the unique solution of the equation

$$\arctan x = \frac{2x}{1 + x^2} \qquad x \in (0, +\infty)$$

meaning that the tangent at $(\tilde{x}, \arctan \tilde{x})$ to the graph of arctan intersect the x-axis at $(-\tilde{x}, 0)$, then Newton's method (corresponding to the iterations of $N_f(x) = x - (1 + x^2)\arctan x$) starting from $X_0 = \tilde{x}$ generates a 2 periodic trajectory: $X_1 = -\tilde{x}$, $X_2 = \tilde{x}$, $X_3 = -\tilde{x}$, ..., $X_k = (-1)^k \tilde{x}$. Notice that $(N_f)'(x) = -2x \arctan x < 0$ for any $x \neq 0$ and vanishes at $x = 0$. N_f is odd and strictly decreasing in $\mathbb{R}$. Thus the equation $\{x \in \mathbb{R} : N_f(x) = x\}$ has the unique solution $x = 0$; instead the equation $\left\{x \in \mathbb{R} : (N_f)^2(x) = x\right\}$ has the solutions 0 and $\pm\tilde{x}$ (this last claim follows from the fact that N_f odd and monotone decreasing entails that the fixed points of $(N_f)^2$ coincide with the solution of $\{x \in \mathbb{R} : N_f(x) = -x\}$, namely 0 and $\pm\tilde{x}$ since N_f is convex in $(-\infty, 0)$ and concave in $(0, +\infty)$). By using Algorithm II (see page 96) or an even simpler argument as the sign analysis of the difference

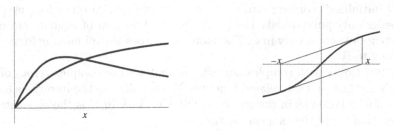

Fig. 4.25 Graphs of $\arctan x$ and $2x/\left(1+x^2\right)$; 2 cycle of $\{\mathbb{R},\, N_{\arctan}\}$

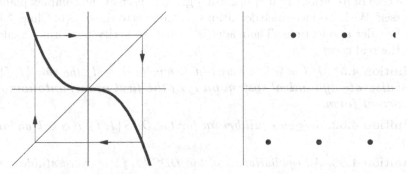

Fig. 4.26 DDS associated to $N_{\arctan}$ and 2 cycle $\{\widetilde{x}, -\widetilde{x}\}$ of $N_{\arctan}$

$\arctan x - 2x/\left(1+x^2\right)$ (see Fig. 4.26), we obtain the complete picture of the phase diagram for the DDS $\{\mathbb{R}, N_f\}$. Summarizing:

- if $|X_0| = \widetilde{x}$, then $X_k = (-1)^k \widetilde{x}$;
- if $|X_0| < \widetilde{x}$, then $\lim_k X_k = 0$, $\operatorname{sign}(X_k) = (-1)^k \operatorname{sign}(X_0)$, $|X_k| \searrow 0$;
- if $|X_0| > \widetilde{x}$, then $\lim_k |X_k| = +\infty$, $\operatorname{sign}(X_k) = (-1)^k \operatorname{sign}(X_0)$, $|X_k| \nearrow +\infty$.

0 is a stable locally attractive equilibrium; $\{\widetilde{x}, -\widetilde{x}\}$ is a repulsive 2 cycle.[7] □

Exercise 4.30. Perform a numerical computation of the zeroes of $f : \mathbb{R} \to \mathbb{R}$, $f(x) = e^x - 3$, up to an error less than 10^{-3}.

4.7 Discrete dynamical systems in the complex plane

The natural framework for analyzing the (discrete dynamical system related to) Newton's method is the complex plane $\mathbb{C}$. In fact the method (when it is

---

[7] The existence of $\widetilde{x} > 1/\sqrt{2}$ follows from the Intermediate Value Theorem together with

$$\frac{2x}{1+x^2}\bigg|_{x=1/\sqrt{2}} = \frac{2}{3}\sqrt{2} > \frac{\pi}{4} = \arctan\frac{1}{\sqrt{2}}.$$

The non existence of other positive values such that $\arctan x = 2x/\left(1+x^2\right)$ (whose existence would generate additional 2 cycles!) follows from the analysis of $(N_f)^2$.

suitably initialized) converges to a zero of the function. Nevertheless, even if we consider only polynomials, the Fundamental Theorem of Algebra ensures the existence of zeroes only in $\mathbb{C}$. Therefore the zeroes we are looking for could not belong to $\mathbb{R}$.

Given a function g with complex domain, we look for the complex zeroes of g: Newton's method and the related function N_g that defines the iterations have identical formulation as in the real case; what we lack in $\mathbb{C}$ is the elementary interpretation using the tangent method.

We formalize the notion of discrete dynamical system (still denoted by DDS) in the case of iterations that transform a given subset of the complex plane $\mathbb{C}$ into itself. We introduce some definitions, analogous to the ones of Chap. 3 but with broader applications. The reader is invited to verify the formal analogy with the real case.

Definition 4.51. *If $I \neq \emptyset$ is a subset of $\mathbb{C}$ and $f : I \to I$, the pair $\{I, f\}$ is called **discrete dynamical system on I, of the first order, autonomous, in normal form**.*

Definition 4.52. *α is an **equilibrium** for the DDS$\{I, f\}$ if $\alpha \in I$ and $\alpha = f(\alpha)$.*

Definition 4.53. *An equilibrium α of the DDS $\{I, f\}$ is called **stable equilibrium** if, $\forall \varepsilon > 0$, there exists $\delta > 0$ such that $|X_0 - \alpha| < \delta$, $X_0 \in I$ and $X_k = f^k(X_0)$ imply*

$$|X_k - \alpha| < \varepsilon \qquad \forall k \in \mathbb{N}.$$

*Vice versa, α is called **unstable equilibrium (or repulsive)** if it is not stable, that is if there exists $\varepsilon_0 > 0$ such that, for any $\delta > 0$, we can find $X_0 \in I$ and $k > 0$ such that*

$$|X_0 - \alpha| < \delta \qquad |X_k - \alpha| > \varepsilon_0.$$

For a comparison with definitions of Sections 3.1 and 3.5, we notice that the set $\{z \in \mathbb{C} : |z - \alpha| < \delta\}$ is a disk, whereas the set $\{x \in \mathbb{R} : |x - \alpha| < \delta\}$ is an interval.

Definition 4.54. *An equilibrium α of a DDS $\{I, f\}$ is called a **locally attractive equilibrium** if there exists $\eta > 0$ such that, for any initial value $X_0 \in I \cap \{z \in \mathbb{C} : |z - \alpha| < \eta\}$, setting $X_k = f^k(X_0)$, we obtain*

$$\lim_k X_k = \alpha.$$

Definition 4.55. *An equilibrium α of the DDS $\{I, f\}$ is called **locally asymptotically stable equilibrium** if it is stable and locally attractive.*

For $\{I, f\}$, with I disk in the complex plane and $f \in C^1(I)$, the stability criterion of Theorem 3.57 still holds, provided that we substitute the modulus of a real number with the modulus of a complex number.

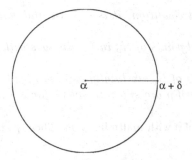

Fig. 4.27 Disk centered at α with radius δ in $\mathbb{C}$

Theorem 4.56. *If α is an equilibrium for the DDS $\{I, f\}$, with $I \subset \mathbb{C}$, f in $C^1(I)$, then*

$$
\begin{aligned}
|f'(\alpha)| < 1 \quad &\Rightarrow \quad \alpha \text{ is locally asymptotically stable} \\
|f'(\alpha)| > 1 \quad &\Rightarrow \quad \alpha \text{ is unstable}
\end{aligned}
$$

Proof. The argument to deal with the complex case is formally identical to the one for the real case, but Lagrange's Theorem does not hold for vector-valued or complex-valued functions. However we can use this identity:

$$
f(z) - f(w) = \int_w^z f'(u)\, du = \int_0^1 f'(w + t(z - w))(z - w)\, dt
$$

(this identity holds whenever the segment joining z and w is contained in I) hence

$$
|f(z) - f(w)| \leq (\max |f'|)\, |z - w|
$$

and $\max |f'|$ can be estimated in a suitable disk centered at α such that $\max |f'| < 1$. □

In the light of Theorem 4.56 it is natural to keep the terminology of Definition 3.59 (hyperbolic, superattractive, neutral) in the complex framework too. We reconsider Newton's method for finding the roots of a polynomial in the complex plane from the viewpoint of complex dynamical systems.

Theorem 4.57. *Let p be a nonconstant polynomial of one complex variable. We set*

$$
N_p(z) = z - \frac{p(z)}{p'(z)}
$$

where the possible simplification of common factors of p and p' is understood. Then:

1) N_p *is defined and differentiable on the set I of all points in the complex plane except the zeroes of p' that are not zeroes of p. In particular, with-*

out any additional condition, it is defined and continuous at every zero of p.

2) *The set of the fixed points of N_p in I coincides with the set of the complex roots of p.*
3) *All the fixed points of N_p are locally asymptotically stable for $\{I, N_p\}$.*
4) *The simple zeroes of p are superattractive for the DDS $\{I, N_p\}$.*

Proof. Let α be a root of p with multiplicity m. Then $p(z) = (z - \alpha)^m q(z)$ with $q(\alpha) \neq 0$ and

$$N_p(z) = z - \frac{p(z)}{p'(z)} = z - \frac{(z - \alpha) q(z)}{(z - \alpha) q'(z) + m q(z)}$$

$$N_p(\alpha) = \alpha.$$

Vice versa, if $N_p(\alpha) = \alpha$, then $\alpha - p(\alpha)/p'(\alpha) = \alpha$ that is[8] $p(\alpha) = 0$. Eventually

$$(N_p)'(z) = 1 - \frac{(p'(z))^2 - p(z) p''(z)}{(p'(z))^2} = \frac{p(z) p''(z)}{(p'(z))^2}.$$

Thus, if α is a simple zero of p, namely $p(\alpha) = 0$ and $p'(\alpha) \neq 0$, then $(N_p)'(\alpha) = 0$ that is α is stable and superattractive.

If α is not a simple zero, namely $p(\alpha) = p'(\alpha) = 0$, then there exists $m \geq 2$ such that $p(z) = (z - \alpha)^m q(z)$ and $q(\alpha) \neq 0$; hence

$$(N_p)'(z) =$$

$$\frac{(z - \alpha)^m q(z) \left(m(m-1)(z-\alpha)^{m-2} q(z) + 2m(z-\alpha)^{m-1} q'(z) + (z-\alpha)^m q''(z) \right)}{\left(m(z-\alpha)^{m-1} q(z) + (z-\alpha)^m q'(z) \right)^2}$$

$$= \frac{q(z) \left(m(m-1) q(z) + 2m(z-\alpha) q'(z) + (z-\alpha)^2 q''(z) \right)}{(m q(z) + (z-\alpha) q'(z))^2}$$

$$(N_p)'(\alpha) = \frac{m(m-1) q(\alpha)^2}{m^2 q(\alpha)^2} = \frac{m-1}{m} < 1$$

that is α is attractive and stable.   □

Since a non constant differentiable function of complex variable can have only isolated zeroes and all of finite integer order, one can prove the more general result below by exactly the same proof of Theorem 4.57.

Theorem 4.58. *Assume $f \in C^1(\Omega)$, $\Omega \subset \mathbb{C}$ is a connected open set and f is non constant. We define*

$$N_f(z) = z - \frac{f(z)}{f'(z)}$$

where, as usual, the possible simplification of common factors of f and f' is understood. Then:

---

[8] In the expression $p(\alpha)/p'(\alpha)$ it is always understood the possible simplification.

1) N_f is defined and differentiable in the set $I = \Omega \backslash \{z \in \Omega : f'(z) = 0 \neq f(z)\}$.

2) N_f is defined and continuous at the zeroes of f.

3) The set of the fixed points of N_f in Ω coincides with the set of the zeroes of f.

4) All the fixed points of N_f are locally asymptotically stable for $\{I, N_f\}$.

5) Every simple zero of f is superattractive for $\{I, N_f\}$.

We emphasize the fact that local asymptotic stability ensures the convergence to the root we want to approximate only if the initial value X_0 is chosen "sufficiently close" to the target. How much close is a delicate question in general, requiring an accurate case by case analysis (about this issue see Theorem 4.49 and Example 4.50).

Example 4.59. Let $p(z) = az + b$ be given with a and b in $\mathbb{C}$, $a \neq 0$. The unique root $(-b/a)$ of p is globally asymptotically stable in $\mathbb{C}$ for $N_p(z) = -b/a$ (all trajectories are eventually constant: $X_k = -b/a$ as $k \geq 1$) and its attraction basin is the whole $\mathbb{C}$. □

In case of nonlinear functions f the phase diagram of N_f is certainly more complex.

Before performing a systematic analysis, the reader is invited to study graphically and by computer simulations the iterations of N_f in $\mathbb{R}$ generated by $f(x) = x^2 \pm 1$, $x \in \mathbb{R}$.

One can easily realize that, even in the simple case when p is a polynomial (hence it has a finite number of roots and they all are locally asymptotically stable for N_p), the domain of N_p (namely $\mathbb{C}$ except the roots of p' that are not roots of p) may be different from the union of the attraction basins of the roots of p.

Now we study the case of quadratic polynomials:

$$p(z) = az^2 + bz + c \qquad a, b, c, \in \mathbb{C}, \quad a \neq 0. \qquad (4.3)$$

The associated function of Newton's method is

$$N_p(z) = z - \frac{az^2 + bz + c}{2az + b} = \frac{az^2 - c}{2az + b}. \qquad (4.4)$$

Lemma 4.60. If p, N_p are defined by (4.3)-(4.4), and $a \neq 0$, then N_p is topologically conjugate to

$$N_q(z) = \frac{z^2 + D}{2z}$$

through the change of variables $\varphi(z) = 2az + b$, where $q(z) = z^2 - D$ and $D = b^2 - 4ac$ is the discriminant of p.

Proof. It suffices to verify $\varphi(N_p(z)) = N_q(\varphi(z))$. □

Thanks to Theorem 4.42 and Lemma 4.60, the analysis of the dynamics of N_p is known as soon as we know the dynamics of N_q, with $q(z) = z^2 - D$, as the constant D varies in $\mathbb{C}$.

We start with the case $D \in \mathbb{R}$, limiting the discussion to real coefficients a, b, c and to the dynamics in $\mathbb{R}$, for simplicity.

Theorem 4.61. *Assume $p(x) = ax^2 + bx + c$ with $a, b, c, x \in \mathbb{R}$ and $a \neq 0$. Then:*

- *If $D = 0$, then $q(x) = x^2$, $N_q(x) = x/2$ and 0 (the double root of q) is the unique fixed point that is globally asymptotically stable for $\{\mathbb{R}, N_q\}$. Analogously $\{\mathbb{R}\setminus\{0\}, N_p\}$ has a unique fixed point that is globally asymptotically stable: $\widetilde{x} = \varphi^{-1}(0) = -b/2a$ (double root of p).*
- *If $D > 0$, then $q(x) = x^2 - D$ has simple real roots $\pm\sqrt{D}$ that coincide with the fixed points of $N_q(x) = (x^2 + D)/2x$ and are locally asymptotically stable and superattractive for $\{\mathbb{R}\setminus\{0\}, N_q\}$. Moreover $+\sqrt{D}$ has the attraction basin $(0, +\infty)$, whereas $-\sqrt{D}$ has the attraction basin $(-\infty, 0)$. Analogously $\{\mathbb{R}\setminus\{-b/2a\}, N_p\}$ has two locally asymptotically stable and superattractive fixed points:*

$$\widetilde{x} = \varphi^{-1}\left(\pm\sqrt{D}\right) = \left(-b \pm \sqrt{b^2 - 4ac}\right)/2a$$

  whose attraction basins are $(-b/2a, +\infty)$ and $(-\infty, -b/2a)$ respectively.
- *If $D < 0$, then $q(x) = x^2 - D$ has imaginary roots $\pm i\sqrt{-D}$ that are the locally asymptotically stable fixed points for $\{\mathbb{C}\setminus\{0\}, N_q\}$ with $N_q(x) = (x^2 + D)/2x$, but the dynamics of $\{\mathbb{R}\setminus\{0\}, N_q\}$ is chaotic. Analogously the dynamics of N_p is chaotic and has no fixed points.*

Proof. Case $D = 0$ is trivial.

If $D > 0$, $0 < X_0 < \sqrt{D}$ and $X_k = (N_q)^k(X_0)$, then $X_0 < \sqrt{D} < X_1$ and $\sqrt{D} < X_{k+1} < X_k$ for $k \geq 1$. The monotonicity of X and Theorem 3.25 imply $\lim_k X_k = \sqrt{D}$.

If $D > 0$ and $\sqrt{D} \leq X_0$, then $\sqrt{D} < X_{k+1} < X_k$ for $k \in \mathbb{N}$ and, by the same monotonicity argument, $\lim_k X_k = \sqrt{D}$.

The case $D > 0$ and $X_0 < 0$ is analogous ($\lim_k X_k = -\sqrt{D}$).

Now we tackle the nontrivial case $D < 0$. To avoid technicalities, we study only the case $D = -1$, namely $q(x) = x^2 + 1$ (this is sufficient since N_{x^2+c} and N_{x^2+1} are topologically conjugate if c is positive through the change of coordinates $\psi(x) = \sqrt{c}\,x$: indeed $\psi \circ N_{x^2+1} = N_{x^2+c} \circ \psi$). The numerical and graphical experiment evokes chaotic dynamics. Actually this is the case, as it can be proved by expressing formally the subsequent argument: we extend the definition of $N_{x^2+1}(x) = (x^2 - 1)/(2x)$ with value ∞ at $x = 0$ and at $x = \infty$. In this way, N_{x^2+1} transforms $\mathbb{R} \cup \{\infty\}$ in itself and is topologically conjugate through $\tau : [0, 2\pi) \to \mathbb{R} \cup \{+\infty\}$, $\tau(x) = \cot(x/2) = 1/tg(x/2)$ to the phase-doubling map $\mathfrak{D} : S \to S$, $\mathfrak{D}(\theta) = 2\theta$ (modulus 2π) that has a chaotic dynamics (Example 4.21 and Theorem 4.30).

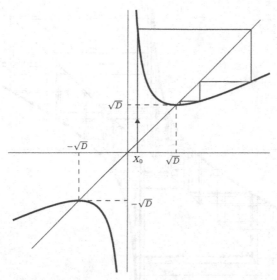

Fig. 4.28 $D > 0$, dynamics of $N_q(x) = (x^2 + D)/2x$, $q(x) = x^2 - D$

The topological conjugacy between N_{x^2+1} and $\mathfrak{D}$ is verified by.

$$N_{x^2+1}(\tau(x)) = \frac{(\cot(x/2))^2 - 1}{2\cot(x/2)} = \frac{(\cos(x/2))^2 - (\sin(x/2))^2}{2\sin(x/2)\cos(x/2)} =$$

$$= \frac{\cos x}{\sin x} = \cot x = \tau(\mathfrak{D}(x)).$$

The claim about the dynamics of N_p follows from the properties of the dynamics of N_q by Lemma 4.60. $\qquad\square$

The previous result is only a particular case of the next theorem where the coefficients a, b, c of the polynomial p are complex numbers, with $a \neq 0$.

Theorem 4.62. *Assume $a, b, c \in \mathbb{C}$ and $a \neq 0$.*
If the polynomial $p(z) = az^2 + bz + c$ has two simple roots z_1, z_2, then the DDS

$$\{\mathbb{C} \setminus \{(z_1 + z_2)/2\}, N_p\} \qquad with \qquad N_p(z) = \frac{az^2 - c}{2az + b}$$

is chaotic on the line r orthogonal to the segment with endpoints z_1, z_2 and crossing it at the middle-point $(z_1 + z_2)/2$. Moreover this line r split the two open half-planes that are the attraction basins of z_1 and z_2 (stable and super-attractive).
If the polynomial $p(z) = az^2 + bz + c$ has a double root, then every point in $\mathbb{C}$ belongs to the attraction basin of this root that is a globally asymptoti-

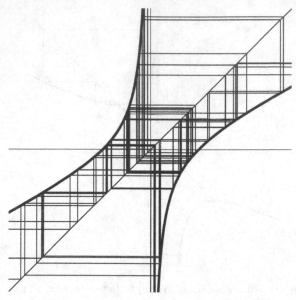

Fig. 4.29 $D < 0$, dynamics of $N_q(x) = (x^2 - 1)/2x$, $q(x) = x^2 + 1$

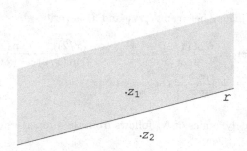

Fig. 4.30 The basin of z_1 is shaded, the one of z_2 is white

cally stable equilibrium (in this case the DDS is defined in the whole complex plane).

Proof. Thanks to Lemma 4.60, in the first case p is topologically conjugate in $\mathbb{C}$ with $z^2 - 1$ through a linear transformation of the complex plane that maps z_1 and z_2 to -1 and $+1$ respectively; in the second case, N_p is topologically conjugate in $\mathbb{C}$ with N_{z^2}.
Thus the claim follows by Theorem 4.42.   □

Remark 4.63. The information concerning the attraction basins is due to Arthur Cayley[9] (1879), whereas the exact description of the dynamics on r is far more recent.

---

[9] Arthur Cayley, 1821-1895.

Until now we kept things intentionally vague when describing the DDS associated to N_p with p polynomial of degree two: though N_p is defined at every $z \in \mathbb{C}$ different from $(z_1 + z_2)/2 = -b/2a$, we remark that there exist infinitely many complex numbers such that, the iterations of N_p do not generate a sequence starting from there. Indeed, if $p(z) = az^2 + bz + c$, then

$$N_p(z) = \frac{az^2 - c}{2az + b} \text{ and,}$$

- if $b^2 - 4ac = 0$, then $N_p(z) = z/2 - b/(4a)$ and $\{\mathbb{C}, N_p\}$ is a well-defined DDS;
- if $b^2 - 4ac \neq 0$, then there are infinitely many points (a set that we label with Z; Z is contained in the axis r of the the segment $[z_1, z_2]$) whose iterations through N_p lead to the forbidden value $-b/2a$ in a finite number of steps; so in this case the accurately defined DDS is $\{\mathbb{C} \backslash Z, N_p\}$.

To prove the claim it is sufficient to study the case when $p(z) = z^2 + 1$, $z_{1,2} = \pm i$ (the general case follows by topological conjugacy): in this simple case $N_p(z) = (z^2 - 1)/2z$ and solving for z the equation $\{z \in \mathbb{C} : w = (z^2 - 1)/2z\}$ we get the two branches of N_p^{-1}: $N_p^{-1}(w) = w \pm \sqrt{w^2 + 1}$. Thus Z is the set of all values that we compute by the iterations of N_p^{-1} starting from 0: $\varphi_{j_1} \circ \varphi_{j_2} \circ \cdots \circ \varphi_{j_k}(0)$, $j_k = 1, 2$, $\varphi_1(z) = z + \sqrt{z^2 + 1}$, $\varphi_2(z) = z - \sqrt{z^2 + 1}$. So it is straightforward that Z is infinite and contained in $\mathbb{R}$ (φ_1 is strictly monotone).

Example 4.64. Assume $p(z) = z^3 - 1$. Then $N_p(z) = (2z^3 + 1)/3z^2$, whose locally attractive fixed points are 1, $e^{2\pi i/3}$, $e^{4\pi i/3}$.

We advise the reader to try a numerical experiment: choose a mesh of points in the square region of the complex plane with vertices $2(\pm 1 \pm i)$, then compute the values of the first 60 iterations of N_p, starting from each point of the mesh; thanks to the superattractivity property, if $|X_{60} - 1| < 1/4$ it is reasonable to presume that X_0 is in the attraction basin of 1 and decide to color it in gray, if $|X_{60} - e^{2\pi i/3}| < 1/4$ it is reasonable to presume that X_0 is in the attraction basin of $e^{2\pi i/3}$ and decide to color it in black, if $|X_{60} - e^{4\pi i/3}| < 1/4$ it is reasonable to presume that X_0 is in the attraction basin of $e^{4\pi i/3}$ and decide to color it in white.

If it is allowed by the computational resources without making too long the computation time, substitute 60 with 100 and/or refine the mesh.

We make some comment on how the computations were made and the diagrams were plotted in order to print Figs. 4.31-4.32.

Fig. 4.31 represents the attraction basins of the three complex cubic roots of 1 with respect to the dynamics of N_{z^3-1}. Starting from every point of a 200×200 mesh in the square $\{z \in \mathbb{C} : |\text{Re}(z)| \leq 2, |\text{Im}(z)| \leq 2\}$ we computed the iterations of N_{z^3-1} until we reached a value whose distance from one of the cubic roots $(1, \exp 2\pi i/3, \exp 4\pi i/3)$ is less than 0.5. As this proximity criterion was fulfilled we stop the computation and attributed a conventional color to the starting point: gray around 1, black around $\exp(2\pi i/3)$, white around $\exp(4\pi i/3)$. $\square$

Fig. 4.31 Dynamics of Newton's method relative to $f(z) = z^3 - 1$: basins of 1, $e^{2\pi i/3}$, $e^{4\pi i/3}$ denoted by different gray levels

Figure 4.32 shows with different gray hues the attraction basins of the New-ton's method $N_{z^n - 1}$ when n varies from 1 to 12. The algoritm is more refined than in the previous plot and exploits the computational routines built in the software Mathematica©: at every point of a mesh in $\{z \in \mathbb{C} : |\mathrm{Re}(z)| \leq 1.3\,, |\mathrm{Im}(z)| \leq 1.3\}$ one computes 35 iterations (recall that every zero is simple, hence superattractive) then attributes a value to the argument of the result-ing complex number (that turns out to be numerically indiscernible from the argument of the corresponding attractor).

We examine comparatively the various case in Fig. 4.32.

If $n = 1$, then $z = 1$ is globally asymptotically stable and the dynamics is very simple (recall that $N_{z-1}(z) = 1$). Every trajectory is eventually constant.

If $n = 2$, we find again the results obtained by the analytical study of N_{z^2-1}. We recall that the boundary of the two basins, though an elementary geomet-rical object (it is a straight line), is swept by a chaotic dynamics and contains an infinite set such that any trajectory starting from one of its points is de-fined only for finitely many steps.

Passing to $n \geq 3$ the situation ramifies a lot: all basins share common bound-aries (common to each one of them); since $n > 2$ these boundaries are neces-sarily topologically complicate items, exhibiting many properties of symmetry and self-similarity, but it is not an elementary geometrical object. Figure 4.32 suggests a richness of properties for the boundaries whose details cannot be graphically visualized.

We observe that the boundary separating the attraction basins is not an ele-mentary curve: here we meet a display of chaotic behavior or at least of quite unpredictable evolution, of a new kind; the competition between several at-

Fig. 4.32 Attraction basins in $\mathbb{C}$ of the n-tuple complex roots of unity for the dynamics associated to Newton's method N_{z^n-1} with $n = 1, \ldots, 12$

tractors establishes fractal regions with dimension strictly bigger than 1 where **sensitive dependence on initial conditions** appears: an initial condition X_0 placed on the boundary of an attraction basin automatically is on the boundary of every other basin[10]. Moreover, the ramification of this boundary makes any prediction on the asymptotic dynamics for a generic initial condition uncertain, but when we know that the initial condition is in the interior of one basin.

In the case of degree-three polynomials p, the dynamics of N_p in $\mathbb{C}$ is extremely complex and at the same time rich of topologically and aesthetically interesting structures.

We conclude with the qualitative study of some other examples of complex dynamics. We limit the discussion to polynomials, because there is no problem about the domain: they are transformations of $\mathbb{C}$ in itself.

Example 4.65. Assume $f(z) = az$, $a \in \mathbb{C}$. The multiplication by a complex number a corresponds to a rotation of the complex plane by an angle equal to the argument θ of a composed with a dilation by $|a|$ of a:

$$a = |a|(\cos\theta + i\sin\theta) = |a|e^{i\theta} \qquad \theta \in [0, 2\pi).$$

---

[10] We recall that z belongs to the boundary of the set $E \subset \mathbb{C}$ if for any $r > 0$ there exist $w \in E$ and $u \in \mathbb{C}\backslash E$ such that $|z - w| < r$ and $|z - u| < r$. Therefore z on the boundary of all basins means that for any $r > 0$ there exist $w_1, \ldots, w_n$ such that $|z - w_j| < r$ and every w_j is in the basin of the j-th complex root of 1.

$$a = (8/7)\,(\cos \pi/6 + i \sin \pi/6) \qquad\qquad a = (7/8)\,(\cos \pi/6 + i \sin \pi/6)$$

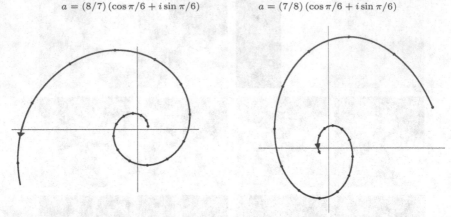

Fig. 4.33 Spiralling trajectories of the DDS $\{\mathbb{C}, az\}$: convergent if $|a| < 1$, divergent if $|a| > 1$

Therefore all the trajectories of $\{\mathbb{C}, az\}$ are of the kind $X_k = a^k X_0$. The next figure displays the first steps of some trajectories of $\{\mathbb{C}, az\}$ with initial point $X_0 = 1$.

Then:

- if $|a| < 1$, then 0 is the unique equilibrium of $\{\mathbb{C}, az\}$ and it is globally asymptotically stable too;
- if $|a| > 1$, then 0 is still the unique equilibrium, but now it is repulsive;
- if $a = 1$, then every point in $\mathbb{C}$ is an equilibrium and every trajectory is stationary;
- if $|a| = 1$ but $a \neq 1$, then 0 is the unique equilibrium again and it is stable but not attractive: the trajectories wind around the origin; they are all periodic if the argument of a is a divisor of 2π, otherwise there are no periodic orbits.

We observe that if we restrict the domain of f to the unit circle $S = \{z \in \mathbb{C} : |z| = 1\}$, then the discrete dynamical system $\{S, az\}$, with $|a| = 1$ and $\arg a/2\pi \notin \mathbb{Q}$, is topologically transitive and has sensitive dependence on initial conditions. $\qquad\qquad\square$

The dynamics generated by quadratic polynomials in $\mathbb{C}$ is much more interesting.

Example 4.66. The DDS $\{\mathbb{C}, z^2\}$ has two equilibria, 0 and 1.
0 is locally asymptotically stable and its basin is $\{z \in \mathbb{C} : |z| < 1\}$.
1 is repulsive. All points z with unit modulus and different from 1 are initial values of trajectories that go around the circle of radius 1. All points $z \in \mathbb{C}$

such that $|z| > 1$ are initial values of trajectories that diverge to the point at infinity:

$$X_{k+1} = X_k^2, \qquad |X_0| > 1 \quad \Rightarrow \quad \lim_k |X_k| = +\infty$$

As in the previous example, every quadratic polynomial $az^2 + bz + c$ exhibits this alternative: initial data whose trajectories are divergent and initial data whose trajectories remain bounded. □

Exercise 4.31. Given the polynomial $q(z) = z^2 + c$, with $c, z \in \mathbb{C}$, prove that, if $|z| > |c| + 1$, then $\lim_k |q^k(z)| = +\infty$; hence for any trajectory of $\{\mathbb{C}, z^2 + c\}$ there is this alternative: either it is divergent or it is contained in the disk $\{z \in \mathbb{C} : |z| \le |c| + 1\}$.

The previous exercise may lead to presume that the dynamics of the DDS $\{\mathbb{C}, z^2 + c\}$ is quite simple, but this conjecture is far from the truth. Actually if $c \ne 0$, then the boundary between the two sets defined by initial data related to either diverging or bounded trajectories has a quite complex structure that is called **Julia set**. The name refers to the French mathematician Gaston Julia (1893-1978) who studied the properties of this boundary far before the modern computer graphics allowed an easy visualization by numerical simulations.

The set of initial points of trajectories of the DDS $\{\mathbb{C}, z^2 + c\}$ that stay bounded is called **filled Julia set** and is denoted by K_c. So the Julia set is the boundary of the filled Julia set.

The next general result (that subsumes Theorem 4.18) proves very useful when one implements numerical simulations .

Theorem 4.67. *If a complex polynomial p has an attractive periodic orbit then the basin of this orbit contains at least a zero of p'.*

When looking for periodic orbits of $q(z) = z^2 + c$, the previous statement suggests to perform the iteration starting from $X_0 = 0$; moreover it ensures that there exists at most one stable periodic orbit. We briefly mention the dynamical systems of higher order in the complex plane.

Definition 4.68. *For any subset I of $\mathbb{C}$ and any continuous function*

$$F : \underbrace{I \times I \times \cdots \times I}_{n \text{ terms}} \to I,$$

*the pair $\{I^n, F\}$ is called **discrete dynamical system in I, of order n, autonomous, in normal form**.*

Given a DDS of order n, a system of infinitely many equations of this type is associated to it

$$X_{k+n} = F(X_{k+n-1}, X_{k+n-2}, \ldots, X_k) \qquad\qquad k \in \mathbb{N}. \qquad (4.5)$$

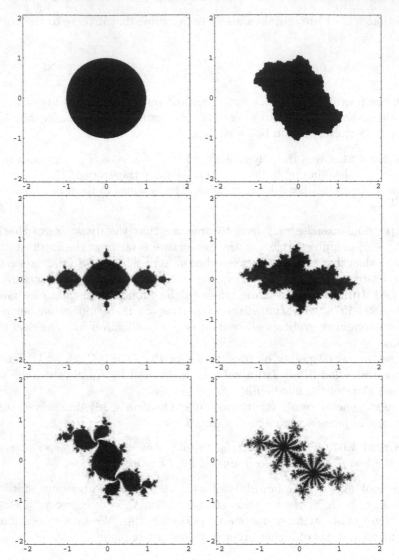

Fig. 4.34. Filled Julia sets K_c, $c = 0$; $c = 0,01 + 0,4i$; $c = -1$; $c = -0,7 + 0,2i$; $c = -0,09 + 0,7i$; $c = -0,41 + 0,59i$

Every n-tuple of initial conditions $\{X_0, X_1, \ldots, X_{n-1}\}$ uniquely determines the sequence solving (4.5), that is called trajectory of the DDS associated to n-tuple of initial conditions.

Example 4.69. let $F : \mathbb{C} \times \mathbb{C} \to \mathbb{C}$ be defined by $F(z, w) = z + w$. Then:

$$X_{k+2} = X_{k+1} + X_k$$

that is the recursion that, with the initial conditions $X_0 = 0$ and $X_1 = 1$ produces the Fibonacci numbers and, starting from a generic pair of complex numbers X_0 and X_1 produces the sequence

$$\frac{1}{2\sqrt{5}}\left(2X_1 - \left(1 - \sqrt{5}\right)X_0\right)\left(\frac{1+\sqrt{5}}{2}\right)^n + \frac{1}{2\sqrt{5}}\left(2X_1 + \left(1 + \sqrt{5}\right)X_0\right)\left(\frac{1-\sqrt{5}}{2}\right)^n.$$

4.8 Summary exercises

Exercise 4.32. If α is an equilibrium for $\{I, f\}$ with $f \in C^1(I)$, the number $f'(\alpha)$ is called *multiplier of the equilibrium*.
Prove that a topological conjugacy $C^1(I)$ with C^1 inverse preserves the multiplier of any equilibrium of the dynamical system.

Exercise 4.33. If $\{\alpha_1, \ldots, \alpha_s\}$ is an s periodic orbit of the DDS $\{I, f\}$ with $f \in C^1(I)$, the number $(f^s(\alpha_1))' = \prod_{j=1}^{s} f'(\alpha_j)$ is called *multiplier of the orbit*.
Prove that a $C^1(I)$ topological conjugacy with C^1 inverse preserves the multiplier of any periodic orbit of the dynamical system.

Remark 4.70. The properties seen in the two previous exercises hold also for DDS's having as domain I a subset of the complex plane. In the real case the two previous exercises provide a refinement of item (iv) in Theorem 4.42 about equilibria and periodic points.
We emphasize that these properties can be exploited also to disprove a property: for instance to prove that there are no C^1 topological conjugations with C^1 inverse between $\{[0, 1/2], x - x^2\}$ and a DDS associated to a linear function in an interval $[a, b]$. Indeed a should be an equilibrium and the topological conjugacy should preserve the multiplier 1 in the equilibrium a, therefore the linear function would necessarily be the identity map that has infinitely many equilibria, contrarily to the other DDS: a contradiction with Theorem 4.42.

5

Discrete dynamical systems: one-step vector equations

In this chapter we consider the discrete time evolution of vector-valued, instead of scalar, quantities. We confine the analysis to the study of one-step linear or affine laws describing the changes of these quantities.
Some applications to demographic models as well as to genetics are presented. A good understanding of this chapter requires some basic results of linear algebra that, for completeness, are listed in Appendix D.

5.1 Definitions and notation

Let us consider the Leslie model (see Example 1.16) for a population $\mathbf{X}$ divided into separated age groups of amplitude equal to the time step. For simplicity we consider only three age groups, in increasing order of seniority. A, B and C denote the three corresponding scalar sequences

$$A = \{A_k\} \qquad B = \{B_k\} \qquad C = \{C_k\}$$

where A_k, B_k and C_k denote the numerical consistency of each age group at time k. Let $\mathbf{X}$ denote the three-dimensional vector $\mathbf{X} = \begin{bmatrix} A\ B\ C \end{bmatrix}^T$.
Let a, b, c and α, β, γ be the fertility and mortality rates, respectively within classes A, B, C. Assuming for simplicity $\gamma = 1$ (i.e. individuals of class C have life expectancy not exceeding the time-step), the system that describes the evolution of the population is:

$$\begin{cases} A_{k+1} = aA_k + bB_k + cC_k \\ B_{k+1} = (1 - \alpha)\,A_k \\ C_{k+1} = (1 - \beta)\,B_k\,. \end{cases}$$

In vector notation, we obtain the recursive law

$$\boxed{\mathbf{X}_{k+1} = \mathbb{M}\,\mathbf{X}_k}$$

(5.1)

E. Salinelli, F. Tomarelli: *Discrete Dynamical Models*.
UNITEXT – La Matematica per il 3+2 76
DOI 10.1007/978-3-319-02291-8_5, © Springer International Publishing Switzerland 2014

where

$$\mathbf{X}_k = \begin{bmatrix} A_k \\ B_k \\ C_k \end{bmatrix} \qquad \mathbb{M} = \begin{bmatrix} a & b & c \\ 1-\alpha & 0 & 0 \\ 0 & 1-\beta & 0 \end{bmatrix}.$$

It is straightforward to express (prove it by induction) the general term of (5.1) in closed form

$$\boxed{\mathbf{X}_k = \mathbb{M}^k \mathbf{X}_0 \qquad \forall k \in \mathbb{N}}$$

$$\overbrace{\phantom{\mathbb{M} \cdot \mathbb{M} \cdots \mathbb{M}}}^{k \text{ times}}$$

where $\mathbb{M}^k = \mathbb{M} \cdot \mathbb{M} \cdots \mathbb{M}$ and $\mathbb{M} \cdot \mathbb{M}$ is the matrix product. However, despite the explicit (non-recursive) knowledge of the sequence $\mathbf{X}$, a computational problem still remains: in fact, matrix multiplication is a long operation even if performed with automatic calculation procedures, and has to be computed for large values of k. For instance, knowing that at the time 50 we have $\mathbf{X}_{50} = \mathbb{M}^{50}\mathbf{X}_0$, does not give much information on the division into age groups of the population $\mathbf{X}_{50}$, unless one actually evaluates $\mathbb{M}^{50}$. In practice the size of $\mathbb{M}$ is much greater than three, a fact that substantially increases the computational complexity of the problem.

A bit of linear algebra can, however, provide a lot of qualitative information on $\mathbb{M}^k \mathbf{X}_0$, even without performing the exact calculation, or reduce the computational complexity if the matrix has a particular structure (as indeed is the case in the Leslie model).

We recall two important definitions from linear algebra (we refer to Appendix D for more details).

Definition 5.1. *Let $\mathbb{M}$ be a square matrix of order n. We call **eigenvalues** of $\mathbb{M}$ the roots (in the complex plane $\mathbb{C}$) of its **characteristic polynomial** $\mathcal{P}(\lambda) = \det(\mathbb{M} - \lambda\mathbb{I})$:*

$$\{\lambda \in \mathbb{C} : \ \mathcal{P}(\lambda) = 0\}.$$

According to the Fundamental Theorem of Algebra, a matrix $\mathbb{M}$ of order n has exactly n eigenvalues in the complex plane, provided that they are counted with the same multiplicity of the roots of $\mathcal{P}(\lambda)$.

Definition 5.2. *A vector $\mathbf{V} \in \mathbb{C}^n \setminus \{\mathbf{0}\}$ is an **eigenvector** of the square matrix $\mathbb{M}$ of order n if there exists a complex number λ such that*

$$\mathbb{M}\mathbf{V} = \lambda\mathbf{V}.$$

In this case λ is an eigenvalue of $\mathbb{M}$ and we say that λ is the eigenvalue associated to the eigenvector $\mathbf{V}$.

If the $n \times n$ matrix $\mathbb{M}$ has n linearly independent[1] eigenvectors $\mathbf{V}^1, \ldots, \mathbf{V}^n$, then the $\mathbf{V}^1, \ldots, \mathbf{V}^n$ form a basis of $\mathbb{R}^n$; in this case each vector $\mathbf{W} \in \mathbb{R}^n$ can be represented in a unique way through this basis as

$$\mathbf{W} = c_1 \mathbf{V}^1 + c_2 \mathbf{V}^2 + \cdots + c_n \mathbf{V}^n. \tag{5.2}$$

Hence, in such case a simple evaluation[2] of $\mathbb{M}^k \mathbf{W}$ is available:

$$\boxed{\mathbb{M}^k \mathbf{W} = \mathbb{M}^k \left(c_1 \mathbf{V}^1 + c_2 \mathbf{V}^2 + \cdots + c_n \mathbf{V}^n \right) = c_1 \lambda_1^k \mathbf{V}^1 + \cdots + c_n \lambda_n^k \mathbf{V}^n}$$

$$\tag{5.3}$$

where λ_j is the eigenvalue associated to $\mathbf{V}^j$, $j = 1, 2, \ldots, n$.

If there are not n linearly independent eigenvectors of $\mathbb{M}$ the explicit representation of $\mathbb{M}^k \mathbf{W}$ is more technical (see Remark 5.9).

Analogously to the scalar case, we will call **orbit** or **trajectory** any sequence of vectors $\{\mathbf{X}_k\}$ that solves system (5.1).

Exercise 5.1. Determine the explicit solution of the vector DDS $\mathbf{X}_{k+1} = \mathbb{M} \mathbf{X}_k$ with the initial condition $\mathbf{X}_0$, in the various cases:

1) $\mathbb{M} = \begin{bmatrix} 1 & 3 \\ 0 & 2 \end{bmatrix}$, $\mathbf{X}_0 = \begin{bmatrix} 1 & 2 \end{bmatrix}^T$; 2) $\mathbb{M} = \begin{bmatrix} 1 & 0 & 0 \\ 2 & -1 & 0 \\ 3 & -2 & -3 \end{bmatrix}$, $\mathbf{X}_0 = \begin{bmatrix} 4 & -1 & 0 \end{bmatrix}^T$;

3) $\mathbb{M} = \begin{bmatrix} 1 & 0 & -1 \\ 0 & -1 & 0 \\ -1 & 0 & -1 \end{bmatrix}$, $\mathbf{X}_0 = \begin{bmatrix} \sqrt{2} - 1 & 3 & -1 \end{bmatrix}^T$.

5.2 Applications to genetics

Many traits of the individuals in a biological species are determined by genes inherited from their parents. Assume that a particular gene $\mathcal{G}$ has only two forms called **alleles** (allele G and allele g), since each individual inherits an allele of this gene from each parent, he may have in its chromosomal inheritance four types of pairs of alleles of the gene $\mathcal{G}$: (G, G), (g, g), (G, g), (g, G). The first two types are called **homozygous**, the latter **heterozygous**.

Let us consider, for the sake of simplicity, the case in which the order of the alleles has no influence, i.e. we consider indistinguishable (G, g) and (g, G). So, we have three types of individuals with regard to the characteristics determined by gene $\mathcal{G}$: two homozygous (G, G), (g, g) and one heterozygous (G, g).

---

[1] This is not always true. However, this hypothesis is verified in many cases; for instance, if $\mathbb{M}$ is a symmetric matrix: $\mathbb{M}_{ij} = \mathbb{M}_{ji}$, $\forall i, j$, or if all the eigenvalues of $\mathbb{M}$ are simple.

[2] Recall that if λ is an eigenvalue of a square matrix $\mathbb{S}$ with eigenvector $\mathbf{V} \neq \mathbf{0}$, then λ^k is an eigenvalue of $\mathbb{S}^k$ with eigenvector $\mathbf{V}$.

Hardy-Weinberg law

Suppose that a parent population has a trait determined by two allelomorphic genes G and g which are present in the population with initial proportions respectively p and q with $p, q \in [0, 1]$ and $p + q = 1$ and that

1)   *there are no mutations from G to g or vice-versa;*
2)   *none of the individuals (G, G), (g, g) or (G, g) has an advantage over the other and the mating occurs randomly;*
3)   *the number of individuals in the population is very large;*
4)   *there are no immigrations of genes G or g as a result of interbreeding with neighboring populations.*

Then the proportions of the two alleles remain unchanged in subsequent generations.

Furthermore, the proportions of individuals respectively G homozygous, g homozygous and heterozygous are p^2, q^2 and $2pq$, starting from the first generation.

For instance, if G determines the skin color and is recessive[3], while g prevents the formation of pigment (albinism gene), assuming the percentages of the two alleles are respectively 99.9% and 0.1% (that is $p = 0.999$ and $q = 0.001$) in generation 0, then these percentages remain unaltered in subsequent generations. Moreover, starting from generation 1 there is a fraction $q^2 = 0.000001$ equal to 0.0001% (an individual in a million) of homozygous albino individuals and there is a fraction $2pq = 0.001998$ that is the 0.1998% of heterozygous albino individuals (in this situation nearly two out of a thousand individuals are albinos).

Remark 5.3. Notice that also the inverse problem has a practical interest: one observes the phenotypic traits in the population, then deduces the frequency of each of the two gene alleles by exploiting the Hardy-Weinberg law.

Remark 5.4. Assumptions 1)-4) in the Hardy-Weinberg law correspond to stability (absence of evolution); if one or more among them are not fulfilled, evolutionary selection emerges:

not 1)   mutations $g \to G$ or $G \to g$;
not 2)   advantage of individuals with a certain genetic composition (best fitness) and consequent natural selection;
not 3)   numerical limitation of the population;
not 4)   migration of genes from neighboring populations.

Proof of the Hardy-Weinberg law. It is not necessary to know P_0, that is p at time 0 (Q_0, i.e. q at time 0, can be deduced from $Q_0 = 1 - P_0$), nor the number of males or females.

---

[3] If an heterozygous individual (G, g) presents the phenotypic traits corresponding to the gene G, then G is called **dominant**, if it has those related to gene g, then G is called **recessive**.

We write a first-order scalar DDS, that is we express P_{k+1} as function of P_k. Let M and F respectively be the (unknown) number of males and females in the generation k.
Set, at the generation $k + 1$:

> A the fraction of (G, G) homozygous
> a the fraction of (g, g) homozygous
> b the fraction of (G, g) heterozygous
> N the total number of individuals.

An individual (G, G) receives two alleles G from the parents of generation k (one from each parent) and there are $(P_k M) (P_k F)$ possible ways to receive them from the possible pairs of parents MF. So, the fraction of homozygous (G, G) is

$$A = \frac{(P_k M) (P_k F)}{MF} = P_k^2 .$$

Correspondingly, there are $P_k^2 N$ homozygous (G, G) individuals in the generation $k + 1$ and they have a total of $2P_k^2 N$ alleles G.
Similarly, in the generation $k + 1$, there is a portion $2P_k (1 - P_k)$ of heterozygous individuals, i.e. there are $2P_k (1 - P_k) N$ heterozygous individuals, and they have a total of $2P_k (1 - P_k) N$ alleles G and $2P_k (1 - P_k) N$ alleles g. The population has $\left(2P_k^2 + 2P_k (1 - P_k)\right) N = 2P_k N$ alleles G (homozygous (g, g) have none), while the total number of alleles is $2N$. Therefore, the portion P_{k+1} of alleles G in the generation $k + 1$ is

$$P_{k+1} = \frac{2N P_k}{2N} = P_k$$

namely P_k is constant with respect to k: $P_k = P_0 = p$ for any k, and by substitution:

- G homozygous individuals in $k + 1$-st generation are $p^2 N$, i.e. the portion p^2 of the total population;
- heterozygous individuals in $k + 1$-st generation are $2pqN$, i.e. the portion $2pq$ of the total population;
- g homozygous individuals in $k + 1$-st generation are $q^2 N$, i.e. the portion q^2 of the total population.   □

The Hardy-Weinberg law explains why the recessive alleles do not disappear from the population.
Now, in contrast with the previous assumptions, we assume that condition 2 fails: for instance, the homozygous (g, g) are infertile or die before reaching sexual maturity and g is recessive. Then G has a *selective advantage* and g is a *lethal gene* (although recessive).

Selection principle
If the (g, g) homozygous do not reproduce, while heterozygous and (G, G) homozygous reproduce normally, then (in absence of mutations) the fraction of

g alleles in the k-th generation is given by

$$Q_k = \frac{Q_0}{1 + kQ_0}$$

where Q_0 is the initial fraction.
So the lethal allele g tends (albeit slowly) to extinction.

Proof. As in the previous case let P_k^2, $2P_kQ_k$ and $(1 - P_k)^2$ be the population proportions of (G, G), (G, g), (g, g) in the $(k + 1)$-st generation, which consist of N individuals. We are interested in the portion of recessive alleles g that are present in the fertile part of the population, namely

$$\left(P_k^2 + 2P_k (1 - P_k)\right) N = P_k (1 + Q_k) N.$$

There are $2P_k (1 + Q_k) N$ alleles. But the g allele is present only in heterozygous individuals, that is, they are $2P_kQ_kN$. Therefore the portion of g alleles in the population of the $(k + 1)$-th generation which can breed is:

$$Q_{k+1} = \frac{2P_kQ_kN}{2P_k (1 + Q_k) N} = \frac{Q_k}{1 + Q_k}.$$

We have a non-linear first order DDS $Q_{k+1} = f (Q_k)$ where f is a Möbius function (see Section 2.7 and in particular Example 2.69). By substituting $W_k = Q_k^{-1}$ we obtain the explicit solution

$$Q_k = \frac{Q_0}{1 + kQ_0}.$$ □

It is appropriate to make few comments on the selection principle. We have already observed that the fraction of g alleles tends to zero, but slowly: if, for instance, the initial population presents only 1% of these alleles, i.e. $Q_0 = 0.01$, then $Q_1 = 1/101 = 0.0099009901$ i.e. Q_1 is not very far from Q_0; in order to halve the initial fraction ($Q_k = Q_0/2$) one has to solve $\frac{Q_0}{2} = \frac{Q_0}{1 + kQ_0}$. Therefore $k = 1/Q_0 = 100$, i.e. 100 generation are necessary (about 2500 years for a human population).

The selection principle is active in nature when the (g, g) homozygous are infertile while heterozygous and (G, G) homozygous are not.

Over the years several attempt have been planned to apply the selection principle to the human species in the case of non-lethal alleles that do not prevent the reproduction, but correspond to physical or mental characteristics which are not desired in certain cultural contexts, trying to simulate the phenomenon described above, through with social prescriptions (*negative eugenics*): prohibition on having children or even suppression of offspring, with the intent of eliminating the unwanted traits from the population (if allele g is recessive such social "stretching" obviously could be exercised only on homozygous (g, g) before there was the possibility to determine the genetic makeup. Regardless of ethical considerations, and of the subjectivity which makes some specific traits undesirable, the quantitative considerations above illustrate the reason for the total ineffectiveness of negative eugenics on the human population: the slow effect makes sure the variation of "tastes" about the desired traits occurs before the effects are significant. We must also emphasize that in the long term the hypothesis of absence of mutations is very weak. Moreover, the biological variety is always an asset and, in particular the genetic diversity in a species must be preserved, because sometimes seemingly insignificant (and relatively uncommon) genes turn out to be surprisingly combined with greater resistance to certain diseases.

Eventually, we analyze the effects of a *mutation without selective advantage*: assume that none of the two alleles is dominant and that the chance of breeding of an individual does not depend on the alleles.
Let P_k and Q_k be respectively the fraction of G allele and g allele in the population at the k-th generation before the mutation. In absence of mutation, we would have $P_{k+1} = P_k$ and $Q_{k+1} = Q_k$ according to the Hardy-Weinberg law. Conversely, we assume that a fraction $s \in (0, 1)$ of G alleles mutates in g alleles before reproduction. Repeating the argument of the proof of Hardy-Weinberg, we obtain:

$$P_{k+1} = (1 - s)\, P_k \qquad \forall k$$
$$P_k = (1 - s)^k P_0$$

In summary, we got the subsequent principle.

Mutation principle
If

1) *the fraction of G alleles that turn to in g is $s \in (0, 1)$ and there is no mutation of g into G;*
2) *none of the individuals (G, G), (g, g) or (G, g) has a selective advantage with respect to the others and the matings are random;*
3) *the number of individuals in the population is very large;*
4) *there is no immigration of genes G and g as a result of interbreeding with contiguous populations;*

then the fraction of G alleles at the k-th generation is

$$P_k = (1 - s)^k P_0 \ .$$

We observe that P_k tends to zero but not as fast as one might expect, because s is in general small. The changes are more rapid if the generations have short time duration, as in the case of some bacteria.

Remark 5.5. In the three examples above the evolution of the vector $\begin{bmatrix} P_k \ Q_k \end{bmatrix}^T$ is derived from the study of the scalar P_k, by exploiting the simple and explicit relationship between P_k and Q_k: $P_k + Q_k = 1$.

Example 5.6 (Sex-linked genes). The visible traits (called *phenotypes*) of the individuals are determined by genetic characteristics (called *genotypes*). Many phenotypes are determined by pairs of genes, called *alleles*. If an allele a is dominant while the other is recessive, then an individual has the recessive trait α if and only if both alleles of the corresponding pair in his genetic makeup are recessive, i.e. they are (α, α).

For some characters, women have a pair of alleles, while men have only one (inherited from the mother): these characters are called *sex-linked traits*. Color blindness and hemophilia are examples (which involve disadvantages, although not lethal) of these characters related to sex. Therefore, if a male inherits the a allele, he will show the dominant phenotype (a), if he inherits the α allele, he will show the recessive phenotype (α). Because alleles that match color blindness or hemophilia are both recessive, we will see later how it is much more likely that these phenotypes (sex-linked traits) occur in men than in women.

From a pair consisting of a woman of genotype (a, α) and a man (a) (both unaffected by the disease, because a is dominant), any son will inherit with a probability of 50% allele a or allele α by the mother, and therefore he will have a 50% probability of being genotype (α), that is, to be affected by the disease (phenotype α). A daughter will have an equal chance with one of the two genotypes (a, a) or (a, α) and in any case she will be not affected by the disease (in the second case, she will be an healthy carrier).

We want to study the evolution of the character over several generations. Let P_k be the proportion of a alleles and $Q_k = 1 - P_k$ the proportion of α alleles in the female population of generation k.

Similarly, let $\mathcal{P}_k$ and $\mathcal{Q}_k$ be the corresponding fraction (with $\mathcal{P}_k + \mathcal{Q}_k = 1$) in the male population of the k-th generation. Let u, v, w be respectively the fractions in the female population at the $(k+1)$-st generation of dominant homozygous (a, a), heterozygous (a, α) and recessive homozygous (α, α). We deduce

$$\begin{cases} u = P_k \mathcal{P}_k \\ v = P_k \mathcal{Q}_k + \mathcal{P}_k Q_k \\ w = Q_k \mathcal{Q}_k \end{cases}.$$

If there are D women in the $(k+1)$-th generation, they will have a total of $2D$ alleles of the type under consideration. The a alleles are $2uD$ (due to homozygous dominant) plus vD (due to heterozygous), namely

$$(2u + v) D = D (2P_k \mathcal{P}_k + P_k \mathcal{Q}_k + \mathcal{P}_k Q_k) =$$

$$\boxed{\text{from} \quad P_k + Q_k = 1 = \mathcal{P}_k + \mathcal{Q}_k}$$

$$= D (P_k \mathcal{P}_k + P_k \mathcal{Q}_k + \mathcal{P}_k P_k + \mathcal{P}_k Q_k) =$$
$$= D (P_k + \mathcal{P}_k).$$

The fraction P_{k+1} of a alleles in the female population at the $(k+1)$-th generation is

$$P_{k+1} = \frac{D\,(P_k + \mathcal{P}_k)}{2D} = \frac{1}{2}\,(P_k + \mathcal{P}_k).$$

Coming to the males, who have only one allele, the proportion of each group that owns the allele a (respectively α) to the k-th generation coincides with the proportion of a alleles (respectively α) in the same generation, i.e. $\mathcal{P}_k$ (respectively $\mathcal{Q}_k$).

To switch to the fraction of men with dominant allele in the generation $k+1$, it is necessary to calculate the probability that a male inherits the dominant allele a. Because it inherits from a woman of generation k, we have

$$\mathcal{P}_{k+1} = P_k \qquad\qquad \mathcal{Q}_{k+1} = Q_k.$$

Summing up, the portion of a allele (which is dominant) between men and women verifies the one-step vector discrete dynamical system

$$\begin{cases} P_{k+1} = \dfrac{1}{2}\,(P_k + \mathcal{P}_k) \\[2mm] \mathcal{P}_{k+1} = P_k. \end{cases}$$

1° method of analysis of Example 5.6

We reduce ourselves to a two-steps scalar discrete dynamical system by replacing $k+1$ with k in the first equation and using the second one:

$$P_{k+2} = \frac{1}{2}\,(P_{k+1} + \mathcal{P}_{k+1}) = \frac{1}{2}P_{k+1} + \frac{1}{2}P_k$$

that is

$$\boxed{2P_{k+2} - P_{k+1} - P_k = 0}$$

The characteristic equation $2\lambda^2 - \lambda - 1 = 0$ admits the solutions $\lambda_1 = 1$ and $\lambda_2 = -1/2$; then

$$P_k = c_1 + c_2 \left(-\frac{1}{2}\right)^k \qquad \forall k \in \mathbb{N}.$$

By imposing the initial conditions P_0 and P_1, we obtain

$$\begin{cases} P_0 = c_1 + c_2 \\[1mm] P_1 = c_1 - \dfrac{1}{2}c_2 \end{cases} \Rightarrow \begin{cases} c_1 = \dfrac{P_0 + 2P_1}{3} \\[2mm] c_2 = \dfrac{2}{3}\,(P_0 - P_1) \end{cases}$$

We observe that

$$\lim_k P_k = c_1 = \frac{P_0 + 2P_1}{3}.$$

Because the current population is the result of a sequence of many previous generations, we can assume k is large, and then $P_k = c_1$ is the actual portion of the dominant alleles in the current female population (in this framework an "initial datum" is meaningless).

Since also $\mathcal{P}_k = P_{k-1} = c_1 + c_2 \left(-1/2 \right)^{k-1}$ is close to c_1, we can assume $\mathcal{P}_k = c_1$ is the same value of the portion of a alleles in the male population. Therefore, the portion of dominant a allele (respectively recessive α allele) stabilizes at a value $c_1 = r$ (respectively $1 - r$) both for the male and for the female populations. It follows that in each generation the portions of dominant homozygous, heterozygous and recessive homozygous in the female population are, respectively,

$$u = r^2 \qquad v = 2r \left(1 - r \right) \qquad w = \left(1 - r \right)^2.$$

For a particular type of color blindness, the portion of α allele, and correspondingly the proportion of males who have this vision problem, is about $1 - r = 0.01$. Instead, the fraction of women with the same disease is $\left(1 - r \right)^2 = \left(0.01 \right)^2 = 0.0001$.

2° method of analysis of Example 5.6

If we set

$$\mathbf{X} = \begin{bmatrix} P \\ \mathcal{P} \end{bmatrix} \qquad \mathbb{M} = \begin{bmatrix} 1/2 & 1/2 \\ 1 & 0 \end{bmatrix}$$

the vector representing the portion of a dominant allele of the female and male populations is a sequence of values that verifies the vector discrete dynamical system

$$\mathbf{X}_{k+1} = \mathbb{M} \, \mathbf{X}_k$$

whose solution is

$$\mathbf{X}_k = \mathbb{M}^k \, \mathbf{X}_0$$

with $\mathbf{X}_0 = \begin{bmatrix} P_0 & \mathcal{P}_0 \end{bmatrix}^T$. The characteristic equation $\det \left(\mathbb{M} - \lambda I \right) = 0$, namely

$$2\lambda^2 - \lambda - 1 = 0$$

admits the solutions

$$\lambda_1 = 1 \qquad \lambda_2 = -1/2.$$

An eigenvector $\mathbf{V}^1$ associated to λ_1 is a solution of the linear homogeneous system $\left(\mathbb{M} - \lambda_1 I \right) \mathbf{V}^1 = \mathbf{0}$ i.e., in components

$$\begin{cases} -\dfrac{1}{2}V_{11} + \dfrac{1}{2}V_{12} = 0 \\[2mm] V_{11} - V_{12} = 0 \end{cases} \qquad \Rightarrow \qquad V_{11} = V_{12}.$$

We choose $\mathbf{V}^1 = \begin{bmatrix} 1 & 1 \end{bmatrix}^T$.

Similarly, an eigenvector $\mathbf{V}^2$ associated to the eigenvalue λ_2 solves the linear homogeneous system $(\mathbb{M} - \lambda_2 \mathbb{I}) \, \mathbf{V}^2 = \mathbf{0}$ i.e., in components

$$
\begin{cases}
V_{11} + \dfrac{1}{2} V_{12} = 0 \\[2ex]
V_{11} + \dfrac{1}{2} V_{12} = 0
\end{cases}
\qquad \Rightarrow \qquad V_{11} = -\dfrac{1}{2} V_{12}.
$$

We choose $\mathbf{V}^2 = \begin{bmatrix} -1 & 2 \end{bmatrix}^T$.

In order to calculate $\mathbb{M}^k \mathbf{X}_0$, we first write $\mathbf{X}_0$ as a linear combination of $\mathbf{V}^1$ and $\mathbf{V}^2$:

$$
\mathbf{X}_0 = c_1 \mathbf{V}^1 + c_2 \mathbf{V}^2 \qquad c_1, c_2 \in \mathbb{R}
$$

(the constants c_1 and c_2 exist and are unique). It follows (see (5.3))

$$
\mathbf{X}_k = \mathbb{M}^k \mathbf{X}_0 = c_1 1^k \mathbf{V}^1 + c_2 \left(-\frac{1}{2} \right)^k \mathbf{V}^2.
$$

Since $\lim\limits_{k} (-1/2)^k = 0$, we get

$$
\lim\limits_{k} \mathbf{X}_k = c_1 \mathbf{V}^1 = c_1 \begin{bmatrix} 1 \\ 1 \end{bmatrix}.
$$

Since $\mathbf{X}_k = \begin{bmatrix} P_k & \mathcal{P}_k \end{bmatrix}^T$, one gets back to the conclusion

$$
\lim\limits_{k} P_k = \lim\limits_{k} \mathcal{P}_k = c_1 = \frac{1}{3} \left(P_0 + 2 P_1 \right)
$$

that is, after many generations, the two portions are stabilized and equal. Genetic selection does not completely eliminate recessive diseases that are not lethal before reaching the reproductive age. We have already observed that biological diversity is a good thing for the genetic heritage of a species because sometimes to the same recessive genotype is accompanied by more than one character of the phenotype, only some of which have biological disadvantages. A classic example in this respect is given by the resistance to malaria associated to thalassemia or Mediterranean Anemia.

Remark 5.7. The fact of having two methods of analysis (multi-step scalar or one-step vector) is of a general nature.

More precisely: a linear homogeneous scalar equation of order n

$$
X_{k+n} + a_{n-1} X_{k+n-1} + a_{n-2} X_{k+n-2} + \cdots + a_0 X_k = 0 \qquad a_0 \neq 0
$$

with given initial conditions $X_0, X_1, \ldots, X_{n-1}$, is equivalent, as can be verified by simple substitution, to the one-step linear system

$$
\mathbf{V}_{k+1} = \mathbb{F} \mathbf{V}_k
$$

with initial datum $\mathbf{V}_0 = \begin{bmatrix} X_0 \; X_1 \; \cdots \; X_{n-1} \end{bmatrix}^T$, where

$$\mathbb{F} = \begin{bmatrix} 0 & 1 & 0 & 0 & \cdots & 0 \\ 0 & 0 & 1 & 0 & \cdots & 0 \\ 0 & 0 & 0 & 1 & \cdots & 0 \\ \vdots & \vdots & \vdots & \vdots & \ddots & \vdots \\ 0 & 0 & 0 & 0 & \cdots & 1 \\ -a_0 & -a_1 & -a_2 & -a_3 & \cdots & -a_{n-1} \end{bmatrix} \qquad \mathbf{V}_k = \begin{bmatrix} X_k \; X_{k+1} \; \cdots \; X_{k+n-1} \end{bmatrix}^T .$$

Obviously, the characteristic polynomial is the same (up to the sign!) and hence it is indicated with the same name. $\mathbb{F}$ is called elementary Frobenius form.

Vice versa, each vector DDS of dimension n, $\mathbf{V}_{k+1} = \mathbb{M}\mathbf{V}_k$, with $\det \mathbb{M} \neq 0$ and all eigenvalues of the matrix $\mathbb{M}$ with geometric multiplicity 1 (it is not excluded that some algebraic multiplicity is greater than 1) can be transformed into an "n-step linear homogeneous equation"

$$Z_{k+n} = -\sum_{j=1}^{n} b_{n-j} Z_{k+n-j} \tag{5.4}$$

with initial conditions

$$Z_0 = W_0 \quad Z_1 = W_1 \quad \ldots \quad Z_{n-1} = W_{n-1} .$$

The polynomial

$$\mathcal{P}(\lambda) = (-1)^n \left(\lambda^n + \sum_{j=1}^{n} b_{n-j} \lambda^{n-j} \right)$$

is the characteristic polynomial of $\mathbb{M}$ evaluated at λ and $\mathbf{W} = \mathbb{U}\mathbf{V}_0$ where

$$\mathbb{U}\mathbb{M}\mathbb{U}^{-1} = \begin{bmatrix} 0 & 1 & 0 & 0 & \cdots & 0 \\ 0 & 0 & 1 & 0 & \cdots & 0 \\ 0 & 0 & 0 & 1 & \cdots & 0 \\ \vdots & \vdots & \vdots & \vdots & \ddots & \vdots \\ 0 & 0 & 0 & 0 & \cdots & 1 \\ -b_0 & -b_1 & -b_2 & -b_3 & \cdots & -b_{n-1} \end{bmatrix} .$$

Notice that (5.4) is an n-step equation because $b_0 = (-1)^n \det \mathbb{M} \neq 0$.

In case of eigenvalues with geometric multiplicity greater than 1 the vector DDS $\mathbf{V}_{k+1} = \mathbb{M}\mathbf{V}_k$ is not equivalent to a single equation with n step (because each matrix in the elementary form of Frobenius, has only eigenvalues with geometric multiplicity exactly equal to 1). However, the vector system decouples in more equations each of which is equivalent to a single multi-step equation (the sum of the number of steps of the various equations is n).

One may be wondering what is the best method: actually they coincide. However, the one-step structure (vector equation) is more natural in many problems and in high dimension the direct treatment of the vector system is simpler in order to obtain qualitative information on the solutions.

In any case, if the size of the vector n is large, then the roots of the characteristic polynomial cannot be calculated exactly but can only be estimated.

In general, however, it is not true that any one-step linear vector system $\mathbf{V}_{k+1} = \mathbb{M}\mathbf{V}_k$ can be reduced to a scalar linear equation of order n: if $n > 1$, then $\mathbb{I}_n$ is not similar to a matrix in elementary Frobenius form.

5.3 Stability of linear vector discrete dynamical systems

Let's go back to examine the case of **one-step linear homogeneous vector discrete dynamical system** (of dimension n):

$$\mathbf{X}_{k+1} = \mathbb{M}\,\mathbf{X}_k \qquad \forall k \in \mathbb{N} \tag{5.5}$$

denoting with $\lambda_j \in \mathbb{C}$, $j = 1, 2, \ldots, n$, the eigenvalues of $\mathbb{M}$.

It is natural to define **equilibrium** of (5.5) any solution $\mathbf{V} \in \mathbb{R}^n$ of $\mathbf{V} = \mathbb{M}\mathbf{V}$, namely the eigenvectors of $\mathbb{M}$ associated to the eigenvalue 1 and, obviously, the origin of $\mathbb{R}^n$, which is always an equilibrium because $\mathbf{0} = \mathbb{M}\mathbf{0}$ for all $\mathbb{M}$.

The definitions of stability and attractivity of an equilibrium are adaptations to the vector case of Definitions 3.36 - 3.39 introduced in the scalar case: the only difference consists in replacing the modulus (wherever it is present) with the Euclidean norm. We remark that in the linear vector-valued case local attractivity coincides with global attractivity, in particular an equilibrium of the DDS (5.5) is globally asymptotically stable if and only if it is locally asymptotically stable.

A solution $\mathbf{X}$ of (5.5) is called a **periodic orbit** of (least) period $s \in \mathbb{N}$ of the DDS (5.5) if $\mathbf{X}_{k+s} = \mathbf{X}_k$ $\forall k$ and $\mathbf{X}_0$, $\mathbf{X}_1$, $\mathbf{X}_{s-1}$ are distinct vectors.

Theorem 5.8. *If $|\lambda_j| < 1$ $\forall$ λ_j eigenvalue of $\mathbb{M}$, then $\mathbf{0}$ is the unique equilibrium of the DDS and all solutions of (5.5) for any initial condition $\mathbf{X}_0$ satisfy*

$$\lim_k \mathbf{X}_k = \mathbf{0}\,.$$

*In this case, the **equilibrium** $\mathbf{0}$ is said to be **stable** and **attractive**. If there exists an eigenvalue λ_j s.t. $|\lambda_j| > 1$, then $\mathbf{0}$ is not attractive nor stable: the dynamics exhibits some orbits diverging in norm.*

Proof. If there is a basis of eigenvectors $\mathbf{V}^j$ of M, then, for each $\mathbf{X}_0$, there exist (and are unique) n constants $c_1, \ldots, c_n$ such that

$$\mathbf{X}_0 = \sum_{j=1}^{n} c_j \mathbf{V}^j$$

$$\mathbf{X}_k = M^k \mathbf{X}_0 = M^k \left(\sum_{j=1}^{n} c_j \mathbf{V}^j \right) = \sum_{j=1}^{n} c_j M^k \mathbf{V}^j = \sum_{j=1}^{n} c_j \lambda_j^k \mathbf{V}^j. \tag{5.6}$$

So

$$\lim_k \|\mathbf{X}_k\|_{\mathbb{R}^n} = \lim_k \left\| \sum_{j=1}^{n} c_j \lambda_j^k \mathbf{V}^j \right\| \leq \quad \boxed{\text{triangle inequality}}$$

$$\leq \lim_k \sum_{j=1}^{n} \left\| c_j \lambda_j^k \mathbf{V}^j \right\| = \quad \boxed{\text{homogeneity of the norm}}$$

$$= \lim_k \sum_{j=1}^{n} \left(|c_j| \left\| \mathbf{V}^j \right\| \left| \lambda_j^k \right| \right) = \quad \boxed{\text{limit of a sum}}$$

$$= \sum_{j=1}^{n} |c_j| \left\| \mathbf{V}^j \right\| \lim_k |\lambda_j|^k = 0.$$

If there is no basis of eigenvectors, the proof is more technical, but the result still holds. It is sufficient to consider a basis of $\mathbb{R}^n$ associated to the canonical Jordan form (see Appendix D): this basis consists of generalized eigenvectors. Explicitly, they are solutions of $(M - \lambda_j \mathbb{I}_n)^{m_j} \mathbf{W} = \mathbf{0}$ where m_j is the algebraic multiplicity of λ_j. M can be decomposed as $M = \mathbb{S} + \mathbb{T}$, where $\mathbb{S}$ is diagonal (the terms on the principal diagonal are the eigenvalues) and $\mathbb{T}$ is nilpotent, therefore $\mathbb{S}\mathbb{T} = \mathbb{T}\mathbb{S}$, and the k-th power of M can be easily expressed as:

$$M^k = \mathbb{S}^k + k\mathbb{S}^{k-1}\mathbb{T} + \frac{k(k-1)}{2}\mathbb{S}^{k-2}\mathbb{T}^2 + \cdots + k\mathbb{S}\mathbb{T}^{k-1} + \mathbb{T}^k. \tag{5.7}$$

Notice that in (5.7), for all k, at most the first n summands[4] may be different from $\mathbb{O}$.

So, the solution $\mathbf{X}_k$ is expressed as a finite linear combination of the generalized eigenvectors $\mathbf{W}^j$ with coefficients that depend on the representation of $\mathbf{X}_0$ in the Jordan basis, times $k^h \lambda_j^{k-h}$ with $0 \leq h \leq n$. Thus, also in this case, $\lim_{k \to +\infty} \|\mathbf{X}_k\| = 0$ holds for each initial datum $\mathbf{X}_0$, thanks to the implication $|\lambda| < 1 \quad \Rightarrow \quad \lim_k k^n \lambda^k = 0$. $\qquad \square$

Remark 5.9. We make the argument of the previous proof explicit for a non diagonalizable matrix M. Starting from (5.7) it is possible to prove that, if $\mathbf{W}^{j,i,r}$ denote the generalized eigenvectors of M in the Jordan basis, where

$\quad j = 1, \ldots, J$;

$\qquad J$ is the number of distinct eigenvalues of M;

---

[4] In fact: $\mathbb{T}^m = \mathbb{O}$ if m is the the maximum size of a Jordan sub-block, and $m \leq n$.

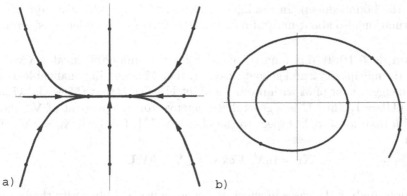

Fig. 5.1 Illustration of Theorem 5.8: a) $\begin{bmatrix} 0.5 & 0 \\ 0 & 0.3 \end{bmatrix}$; b) $\begin{bmatrix} 0 & -0.5 \\ 0.5 & 0 \end{bmatrix}$

$i = 1, \ldots, J_j$;

J_j is the number of the Jordan sub-blocks associated to the eigenvalue λ_j;

$r = 1, \ldots, s_i$;

s_i is the dimension of the i-th Jordan sub-block associated to λ_j;

$m_j = \sum\limits_{i=1}^{J_j} s_i$ is the algebraic multiplicity of λ_j,

then, by

$$\mathbf{X}_0 = \sum_{j=1}^{J} \sum_{i=1}^{J_j} \sum_{r=1}^{s_i} c_{j,i,r} \mathbf{W}^{j,i,r}$$

and

$$\begin{cases} \mathbb{M}\mathbf{W}^{j,i,r} = \lambda_j \mathbf{W}^{j,i,r} + \mathbf{W}^{j,i,r+1} & r < s_i \\ \mathbb{M}\mathbf{W}^{j,i,s_i} = \lambda_j \mathbf{W}^{j,i,s_i}. \end{cases}$$

So we have e formula for the general term defined by $\mathbf{X}_k = \mathbb{M}^k \mathbf{X}_0$:

$$\mathbf{X}_k = \sum_{j=1}^{J} \sum_{i=1}^{J_j} \sum_{r=1}^{s_i} \sum_{h=1}^{\min(s_i,k)} \binom{k}{h-1} \lambda_j^{k-h+1} c_{j,i,r-h+1} \mathbf{W}^{j,i,r}.$$

This expression reduces to (5.6) when (diagonalizable case) $m_j = 1$ for all j.

Without developing in detail the qualitative analysis of the orbits corresponding to the solutions of the linear homogeneous DDS $\mathbf{X}_{k+1} = \mathbb{M}\mathbf{X}_k$, we can say that the natural modes of the solutions (integer powers of the eigenvalue, of which vector linear combinations generate the general solution) have a qualitative behavior that depends only on the position of the corresponding eigenvalue in the complex plane (in particular, we have convergence to zero when $|\lambda| < 1$, divergence when $|\lambda| > 1$, while $|\lambda| = 1$ and multiplicity 1 ensure the boundedness). In the light of Remark 5.7, Fig. 2.7 gives qualitative information also about natural modes of the vector case under consideration.

Example 5.10. If M is a matrix of order 2 with a unique eigenvalue λ with algebraic multiplicity 2 and geometric one 1, then M is not diagonalizable. Let V be an eigenvector of M with norm 1 and let U be such that $(M - \lambda \mathbb{I}_n) U = V$ and $\|U\| = 1$. Then U is a generalized eigenvector independent of V, and M is in Jordan form with respect to the basis $\{V, U\}$. Lastly, if $X_0 = aV + bU$, then

$$X_k = \left(a\lambda^k + bk\lambda^{k-1}\right) V + b\lambda^k U. \qquad \square$$

We now analyze the cases in which the eigenvalues λ_j only verify the inequality $|\lambda_j| \leq 1$. We observe that the equilibria are characterized by the condition $X = MX$ and so they are all the eigenvectors of any eigenvalue $\lambda = 1$. We limit ourselves, for the sake of simplicity, to the case $n = 2$ with real M. In what follows

$$X_0 = c_1 V^1 + c_2 V^2$$

is fixed, where V^1 and V^2 are generalized eigenvectors (in particular, at least V^2 is always an eigenvector) of M corresponding to the basis associated to the canonical Jordan form.

1)   If $\lambda_1 = 1$, $|\lambda_2| < 1$, then the whole space generated by V^1 consists of equilibria and the solution is

$$X_k = c_1 V^1 + c_2 \lambda_2^k V^2.$$

Since $\lim_k \lambda_2^k = 0$, we have

$$\lim_k X_k = c_1 V^1$$

where c_1 depends only on the initial value. 0 is an equilibrium which is stable but not attractive because the solutions may have as a limit any multiple of V^1, and all these multiples are equilibria.
This is the case of Example 5.6.

2)   If $\lambda_1 = -1$, $|\lambda_2| < 1$, reasoning as above, from $\lambda_2^k \to 0$ it follows

$$X_k \sim c_1 \left(-1\right)^k V^1.$$

The orbit of X_k approximates a 2 periodic orbit that depends on X_0 through c_1.

3)   If $\lambda_1 = 1$, $\lambda_2 = -1$, we have

$$X_k = c_1 V^1 + c_2 \left(-1\right)^k V^2$$

namely, an equilibrium if $c_2 = 0$, a periodic orbit of period 2 which depends on X_0 through c_1 and c_2 if $c_2 \neq 0$.

4)   If $\lambda_1 = a + ib$ and $\lambda_2 = a - ib$, where a and b are real numbers such that $a^2 + b^2 = 1$, then there exists $\theta \in [0, 2\pi)$ such that

$$M = \begin{bmatrix} a & -b \\ b & a \end{bmatrix} = \begin{bmatrix} \cos\theta & -\sin\theta \\ \sin\theta & \cos\theta \end{bmatrix}.$$

M is then a **rotation matrix** (i.e. left multiplying a vector $\mathbf{V} \in \mathbb{R}^2$ by M corresponds to a positive turn by an angle of θ radians).

Therefore, if $\theta = 2\pi k/h$ where $k, h \in \mathbb{Z}$ have no common factors, then $\mathbf{X}_k$ has a periodic behavior of period $|h|$, i.e.

$$\mathbf{X}_{k+l|h|} = \mathbf{X}_k \qquad \forall l \in \mathbb{N}.$$

Furthermore, $\|\mathbf{X}_k\| = \|\mathbf{X}_0\|$, $\forall k \in \mathbb{N}$.

If there are no such h, k there is no periodicity, but all $\mathbf{X}_k$ belong to the same circle: $\|\mathbf{X}_k\| = \|\mathbf{X}_0\|$, $\forall k \in \mathbb{N}$. Moreover, one could prove that in this case every orbit is dense in this circle.

5)   If $\lambda_1 = \lambda_2 = 1$ and there exist two linearly independent eigenvectors $\mathbf{V}^1$ and $\mathbf{V}^2$, then

$$\mathbf{X}_k = c_1 \mathbf{V}^1 + c_2 \mathbf{V}^2 = \mathbf{X}_0 \qquad \forall k \in \mathbb{N}.$$

6)   If $\lambda_1 = \lambda_2 = -1$ and there exist two linearly independent eigenvectors $\mathbf{V}^1$ and $\mathbf{V}^2$, then

$$\mathbf{X}_k = c_1 (-1)^k \mathbf{V}^1 + c_2 (-1)^k \mathbf{V}^2 = (-1)^k \mathbf{X}_0 \qquad \forall k \in \mathbb{N}.$$

7)   If $\lambda_1 = \lambda_2 = 1$ and

$$M = \begin{bmatrix} 1 & 0 \\ a & 1 \end{bmatrix} \qquad a \neq 0$$

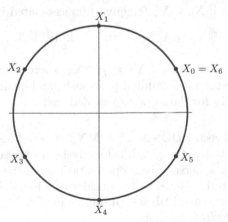

Fig. 5.2 Example of an orbit (case 4) : $\theta = \pi/3$, $X_0 = \begin{bmatrix} \cos(\pi/6) & \sin(\pi/6) \end{bmatrix}^T$

then

$$M^k = \begin{bmatrix} 1 & 0 \\ ka & 1 \end{bmatrix}$$

and, by $X_0 = c_1 V^1 + c_2 V^2$, one gets the possible solution:

$$X_k = M^k X_0 = c_1 V^1 + c_2 V^2 + ka \begin{bmatrix} 0 \\ c_1 V_1^1 + c_2 V_1^2 \end{bmatrix} =$$
$$= c_1 V^1 + c_2 V^2 + c_1 a k V^2 .$$

8) If $\lambda_1 = \lambda_2 = -1$ and

$$M = \begin{bmatrix} -1 & 0 \\ a & -1 \end{bmatrix} = (-1) \begin{bmatrix} 1 & 0 \\ -a & 1 \end{bmatrix} \qquad a \neq 0$$

taking into account that

$$M^k = (-1)^k \begin{bmatrix} 1 & 0 \\ -ka & 1 \end{bmatrix}.$$

by $X_0 = c_1 V^1 + c_2 V^2$ one obtains the following solution, that can be unbounded:

$$X_k = M^k X_0 = (-1)^k \left(c_1 V^1 + c_2 V^2 - ka \begin{bmatrix} 0 \\ c_1 V_1^1 + c_2 V_1^2 \end{bmatrix} \right) =$$
$$= (-1)^k \left(c_1 V^1 + c_2 V^2 - c_1 a k V^2 \right) .$$

The conclusions of the last two cases follow by the following decomposition $M = S + T$ where S is diagonal and T is nilpotent. As $ST = TS$ and $T^2 = O$, it is easy to prove (see Exercise 5.7) that

$$M^k = S^k + k T S^{k-1}.$$

Otherwise, if there exists λ_j such that $|\lambda_j| > 1$, then there are unbounded solutions. For instance, if $X_0 = V_j$ (eigenvectors associated to λ_j), one deduces

$$M^k X_0 = M^k V_j = \lambda_j^k X_0 \quad \text{and} \quad \lim_k \|M^k X_0\| = +\infty.$$

To find out more, one can study $Y_k = \mu^{-k} X_k$ where $\mu = \max_j |\lambda_j| > 1$. In this case, Y_k falls in the cases studied previously, and qualitative information on the behavior of X_k for "large" k can be deduced.

For two-dimensional vector DDS $X_{k+1} = M X_k$, where M is a matrix of order 2, it is still possible to use the graphical analysis to represent the orbits.
It is possible to give a perspective representations in the three-dimensional space (t, X_1, X_2) as in Fig. 5.4a, or to consider (see Fig. 5.4b) the projections on the (X_1, X_2) plane, called **phase plane**, as in Fig. 5.3 which summarizes the results of the previous analysis.

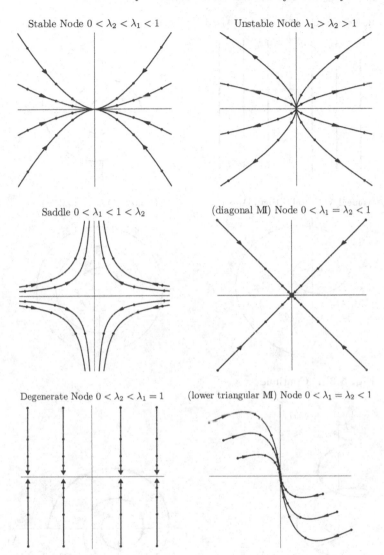

Fig. 5.3a. Orbits of linear two-dimensional DDS's in the phase plane: λ_1, λ_2 are eigenvalue of $\mathbb{M}$. The horizontal and vertical axes are oriented respectively as the corresponding generalized eigenvectors $\mathbf{V}^1$ and $\mathbf{V}^2$. Notice that in the 8-th and 9-th graph (on the next page) the convergence or divergence of trajectories proceeds clockwise. However, since the eigenvalue has geometric multiplicity 2, it is reasonable to replace $\mathbf{V}^1$ with $\mathbf{V}^2$ in the graphical representation: in this way the curves are oriented counterclockwise

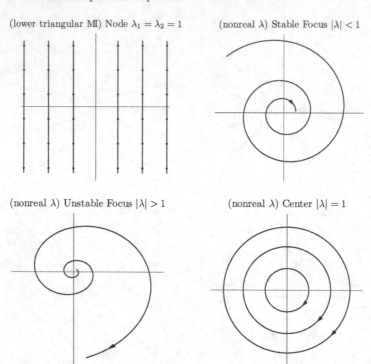

(lower triangular M) Node $\lambda_1 = \lambda_2 = 1$ (nonreal λ) Stable Focus $|\lambda| < 1$

(nonreal λ) Unstable Focus $|\lambda| > 1$ (nonreal λ) Center $|\lambda| = 1$

Fig. 5.3b. Continue

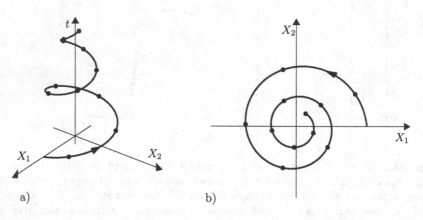

a)   b)

Fig. 5.4. a) graph of an orbit; b) an orbit in the phase plane

Exercise 5.2. Determine the solution of the vector DDS $\mathbf{X}_{k+1} = \mathbb{M}\mathbf{X}_k$ with

$$1)\ \mathbb{M} = \begin{bmatrix} 1 & 1 \\ 0 & 1 \end{bmatrix} \qquad 2)\ \mathbb{M} = \begin{bmatrix} 1 & 0 \\ 0 & 1 \end{bmatrix} \qquad 3)\ \mathbb{M} = \begin{bmatrix} 1 & 1 \\ 1 & 1 \end{bmatrix}.$$

5.4 Strictly positive matrices and the Perron–Frobenius Theorem

In this section we prove a theoretical result on the algebraic properties of matrices whose elements are all positive. This result will be useful for the demographic applications of the next section.

A **linear positive discrete dynamical system** is a system

$$\mathbf{X}_{k+1} = \mathbb{M}\mathbf{X}_k \qquad k \in \mathbb{N} \tag{5.8}$$

in which all the elements of $\mathbf{X}_k$ are nonnegative for every k. This situation occurs, for example, if one starts from a given $\mathbf{X}_0$ with all nonnegative components and if all the elements of $\mathbb{M}$ are positive, and is typical of many economic, management, demographic and game-theory applications. The properties of non-negativity of the matrix $\mathbb{M}$ have deep and elegant consequences. The main result is the Perron–Frobenius[5] Theorem, which ensures that, if $m_{ij} > 0$ for each i, j, then every orbit of system (5.8) has the same asymptotic behavior. To conduct a thorough analysis, we specify the definition of positivity of a matrix.

Definition 5.11. *A matrix* $\mathbb{M} = [m_{ij}]$ *is called*

$\mathbb{M} \geq \mathbb{O}$, **weakly positive or nonnegative,** *if* $m_{ij} \geq 0$ *for each* i, j;
$\mathbb{M} > \mathbb{O}$, **positive,** *if* $m_{ij} \geq 0$ *for each* i, j *and* $m_{ij} > 0$ *for at least one pair of indices* i, j;
$\mathbb{M} \gg \mathbb{O}$, **strictly positive,** *if* $m_{ij} > 0$ *for each* i, j.[6]

Similarly, a vector $\mathbf{V}$ *is called nonnegative, positive or strictly positive if, respectively,* $V_j \geq 0$ *for each* j, $\mathbf{V}$ *is nonnegative and there exists* j *such that* $V_j > 0$, *for each* j, $V_j > 0$, *and* j.

Theorem 5.12 (Perron–Frobenius). *If* $\mathbb{M} \gg \mathbb{O}$, *then its eigenvalue of maximum modulus (called **dominant eigenvalue**), denoted by* $\lambda_\mathbb{M} > 0$, *is unique, real, greater than zero and simple (algebraically and, therefore, geometrically). Furthermore, there exists a strictly positive eigenvector* $\mathbf{V}^\mathbb{M}$ *(called **dominant eigenvector**) associated to* $\lambda_\mathbb{M}$.

Theorem 5.13. *Consider* $\mathbb{M} > \mathbb{O}$ *and* $h \in \mathbb{N}$ *such that* $\mathbb{M}^h \gg \mathbb{O}$. *Then the conclusions of the Perron–Frobenius Theorem holds for* $\mathbb{M}$.

---

[5] Oskar Perron (1880–1975), Ferdinand Georg Frobenius (1849–1917).

[6] The should not confuse a positive matrix with a positive definite matrix: compare Definition 5.11 with the one of *positive definite* matrix in Appendix D. For instance, $\begin{bmatrix} 2 & -1 \\ -1 & 2 \end{bmatrix}$ is positive definite but it is not positive. Vice versa, $\begin{bmatrix} 1 & 2 \\ 2 & 1 \end{bmatrix}$ is strictly positive but it is not positive definite.

Theorem 5.14. *Let* $\mathbb{M} \geq \mathbb{O}$. *Then there exists an eigenvalue* $\lambda \geq 0$ *that verifies* $|\mu| \leq \lambda$ *for every eigenvalue* μ *of* $\mathbb{M}$ *and such that to* λ *corresponds an eigenvector* $\mathbf{V} > \mathbf{0}$. *The uniqueness of this eigenvalue, denoted by* $\lambda_{\mathbb{M}}$ *and called dominant eigenvalue in this case too, follows.*

We observe that $\mathbf{V}^{\mathbb{M}}$ is uniquely determined (up to positive multiple) when the assumptions of Theorem 5.12 or 5.13 hold. Instead if the assumption of Theorem 5.14 holds, there may be positive linearly independent eigenvectors associated with the dominant eigenvalue $\lambda_{\mathbb{M}}$ (the geometric multiplicity of λ can be greater than 1): just consider the identity matrix in dimension 2.

The three previous results are "optimal", as illustrated by the following examples, and will be proved later.

Example 5.15. If $\mathbb{M} = \begin{bmatrix} 1 & 1 \\ 1 & 1 \end{bmatrix}$ then $\mathbb{M} \gg \mathbb{O}$ and by Theorem 5.12 there exists a positive and simple dominant eigenvalue ($\lambda_{\mathbb{M}} = 2$) with corresponding eigenvector which is strictly positive $\begin{bmatrix} 1 & 1 \end{bmatrix}^T$. $\qquad \square$

Example 5.16. The matrix $\mathbb{A} = \begin{bmatrix} 1 & 1 \\ 1 & 0 \end{bmatrix}$ is not strictly positive, but $\mathbb{A}^2 \gg \mathbb{O}$ and by Theorem 5.13 there exists a positive and simple dominant eigenvalue $\lambda_{\mathbb{M}} = \frac{1}{2}(1 + \sqrt{5})$, with corresponding strictly positive eigenvector $\begin{bmatrix} \frac{1}{2}(1 + \sqrt{5}) & 1 \end{bmatrix}^T$. $\qquad \square$

Example 5.17. Let $\mathbb{B} = \begin{bmatrix} 1 & 0 \\ 0 & 1 \end{bmatrix}, \mathbb{D} = \begin{bmatrix} 0 & 1 \\ 1 & 0 \end{bmatrix}, \mathbb{E} = \begin{bmatrix} 0 & 0 \\ 1 & 0 \end{bmatrix}$. None of the three matrices has a strictly positive power, but Theorem 5.14 applies. $\mathbb{B}$ has the positive dominant eigenvalue $\lambda_{\mathbb{B}} = 1$ which is not simple and has algebraic and geometric multiplicity 2. $\mathbb{D}$ has two eigenvalues with maximum modulus (± 1). $\mathbb{E}$ has dominant eigenvalue $\lambda_{\mathbb{E}} = 0$ having algebraic multiplicity 2 and geometric multiplicity 1. $\qquad \square$

Theorem 5.18. *If* $\mathbb{M} \gg \mathbb{O}$ *or* $\mathbb{M} > \mathbb{O}$, *and there exists* h *such that* $\mathbb{M}^h \gg \mathbb{O}$, *then all the orbits of*

$$\mathbf{X}_{k+1} = \mathbb{M}\mathbf{X}_k$$

satisfy

$$\boxed{(\lambda_{\mathbb{M}})^{-k} \mathbf{X}_k = c_{\mathbb{M}}\mathbf{V}^{\mathbb{M}} + \sigma(k)}$$

(5.9)

Here $\lambda_{\mathbb{M}}$ *denotes the dominant eigenvalue of* $\mathbb{M}$, $\mathbf{V}^{\mathbb{M}} \gg 0$ *is the corresponding dominant eigenvector (see Theorems 5.12 and 5.13),* $c_{\mathbb{M}}$ *is the coefficient of* $\mathbf{X}_0$ *along the component* $\mathbf{V}^{\mathbb{M}}$ *with respect to the Jordan basis of* $\mathbb{M}$ *and* $\sigma(k)$ *is infinitesimal when* $k \to +\infty$.

Proof. Let $\lambda_{\mathbb{M}}$ and $\mathbf{V}^{\mathbb{M}}$ be respectively the simple and positive dominant eigenvalue and the corresponding positive eigenvector of $\mathbb{M}$, whose existence is guaranteed by Theorems 5.12 and 5.13.

If $\mathbf{X}_0 = c_M \mathbf{V}^M + \sum_j c_j \mathbf{V}^j$ where $\mathbf{V}^j$ are generalized eigenvectors that form a basis of $\mathbb{R}^n$ together with $\mathbf{V}^M$, then

$$\mathbf{X}_k = c_M \lambda_M{}^k \mathbf{V}^M + \mathbf{W}_k$$

where the vector $\mathbf{W}_k$ is a finite sum (with coefficients that are independent of k) with terms that grow at most as $k^n \lambda_j^k$, with $|\lambda_j| < |\lambda_M|$.
Thus we can write $\lambda_M{}^{-k} \mathbf{X}_k = c_M \mathbf{V}^M + \sigma(k)$. □

The above theorem reduces the analysis of the long-term behavior of the orbits to the computation of the dominant eigenvalue and eigenvector (and vice versa: from any knowledge of the time series of a quantity generated by a strictly positive linear system, it is possible to obtain some estimate of the dominant eigenvalue and eigenvector). These facts are stated in the following theorem.

Theorem 5.19. *If $\mathbf{X}_0 > 0$ and $\mathbb{M} \gg \mathbb{O}$ (or $\mathbf{X}_0 > 0$, $\mathbb{M} > \mathbb{O}$ and $\exists h$: $\mathbb{M}^h \gg \mathbb{O}$), then the solution $\mathbf{X}$ of the vector DDS $\mathbf{X}_{k+1} = \mathbb{M}\mathbf{X}_k$, $k \in \mathbb{N}$, with initial datum $\mathbf{X}_0$, tends to align itself with the corresponding dominant eigenvector $\mathbf{V}^M \gg \mathbb{O}$ of $\mathbb{M}$:*

$$\lim_k \frac{\mathbf{X}_k}{\|\mathbf{X}_k\|} = \frac{\mathbf{V}^M}{\|\mathbf{V}^M\|}. \tag{5.10}$$

Moreover, the dominant eigenvalue $\mathbf{V}^{M^T} \gg 0$ of $\mathbb{M}^T$ is orthogonal to every eigenvector of $\mathbb{M}$ different from $\mathbf{V}^M$ and to each generalized eigenvector of $\mathbb{M}$ different from $\mathbf{V}^M$; finally, if c_M is the coefficient of $\mathbf{V}^M$ in the expansion of $\mathbf{X}_0$ with respect to the Jordan basis of $\mathbb{M}$, then

$$c_M > 0. \tag{5.11}$$

Proof. It is sufficient to prove

$$\mathbf{V}^{M^T} \perp \mathbf{V} \qquad \forall\, \mathbf{V} \text{ generalized eigenvector of } \mathbb{M} \text{ different from } \mathbf{V}^M. \tag{5.12}$$

In fact (5.12) implies that the space generated by all the generalized eigenvectors of $\mathbb{M}$ different from $\mathbf{V}^M$ intersects $\{\mathbf{X} \in \mathbb{R}^n : \mathbf{X} \geq 0\}$ only at $\mathbf{0}$, so $\mathbf{X}_0 > 0$ implies (5.11). Finally, (5.10) is an immediate consequence of (5.11) and of Theorem 5.18. Therefore, there remains to prove (5.12).
If λ_j is an eigenvalue different from λ_M and $\mathbf{V}^j$ is the corresponding eigenvector, one obtains:

$$\lambda_j \langle \mathbf{V}^j, \mathbf{V}^{M^T} \rangle = \langle \mathbb{M}\mathbf{V}^j, \mathbf{V}^{M^T} \rangle = \langle \mathbf{V}^j, \mathbb{M}^T \mathbf{V}^{M^T} \rangle =$$

$$= \overline{\lambda_M} \langle \mathbf{V}^j, \mathbf{V}^{M^T} \rangle = \lambda_M \langle \mathbf{V}^j, \mathbf{V}^{M^T} \rangle.$$

It follows $\langle \mathbf{V}^j, \mathbf{V}^{M^T} \rangle = 0$ as $\lambda_j \neq \lambda_M$.
If the eigenvectors of $\mathbb{M}$ do not form a basis, consider the Jordan basis that consists of generalized eigenvectors of $\mathbb{M}$. In this case, for each fixed maximal Jordan block $\mathbb{B}$ of dimension r corrisponding to the eigenvalue λ_j (with reference to Theorem D.6

and to Fig. D.2 in Appendix D, notice that to each eigenvalue can correspond several Jordan blocks) we order and label the sub-base for that block $\mathbb{B}$ as follows:

$$\mathbf{V}^{j-r+1},\ \mathbf{V}^{j-r+2},\ \ldots,\ \mathbf{V}^{j-1}, \mathbf{V}^{j} \quad \text{with } \mathbf{V}^{j} \text{ eigenvector,}$$

therefore

$$\mathbf{M}\mathbf{V}^{j} = \lambda_{j}\mathbf{V}, \qquad \mathbf{M}\mathbf{V}^{j-s} = \lambda_{j}\mathbf{V}^{j-s} + \mathbf{V}^{j-s+1}, \quad s = 1,\ldots,r-1. \tag{5.13}$$

One deduces: $\mathbf{V}^{j-r+1},\ \mathbf{V}^{j-r+2},\ \ldots,\ \mathbf{V}^{j-1}$, are all orthogonal to $\mathbf{V}^{\mathbf{M}^{T}}$. In fact we know that $\mathbf{V}^{j}$ is orthogonal to $\mathbf{V}^{\mathbf{M}^{T}}$, hence it is sufficient to proceed by induction proving that $\mathbf{V}^{j-s+1} \perp \mathbf{V}^{\mathbf{M}^{T}}$ implies $\mathbf{V}^{j-s} \perp \mathbf{V}^{\mathbf{M}^{T}}$. This implication is proved as follows: by (5.13) it follows $\lambda_{j}\mathbf{V}^{j-s} = \mathbf{M}\mathbf{V}^{j-s} - \mathbf{V}^{j-s+1}$ and

$$\lambda_{j}\langle \mathbf{V}^{j-s}, \mathbf{V}^{\mathbf{M}^{T}} \rangle = \langle \mathbf{M}\mathbf{V}^{j-s} - \mathbf{V}^{j-s+1}, \mathbf{V}^{\mathbf{M}^{T}} \rangle = \langle \mathbf{M}\mathbf{V}^{j-s}, \mathbf{V}^{\mathbf{M}^{T}} \rangle =$$
$$= \langle \mathbf{V}^{j-s}, \mathbf{M}^{T}\mathbf{V}^{\mathbf{M}^{T}} \rangle = \overline{\lambda_{\mathbf{M}}}\langle \mathbf{V}^{j-s}, \mathbf{V}^{\mathbf{M}^{T}} \rangle = \lambda_{\mathbf{M}}\langle \mathbf{V}^{j-s}, \mathbf{V}^{\mathbf{M}^{T}} \rangle$$

that implies $\langle \mathbf{V}^{j-s}, \mathbf{V}^{\mathbf{M}^{T}} \rangle = 0$ as $\lambda_{j} \neq \lambda_{\mathbf{M}}$.   □

Proof of the Perron–Frobenius Theorem. In the following, the inequalities among vectors are to be understood in the sense of comparison of their difference with $\mathbf{0}$, namely:

$$\mathbf{V} \geq \mathbf{W} \Leftrightarrow \mathbf{V} - \mathbf{W} \geq \mathbf{0};\quad \mathbf{V} > \mathbf{W} \Leftrightarrow \mathbf{V} - \mathbf{W} > \mathbf{0};\quad \mathbf{V} \gg \mathbf{W} \Leftrightarrow \mathbf{V} - \mathbf{W} \gg \mathbf{0}.$$

Set

$$\lambda_{\mathbf{M}} = \max\{\lambda \in \mathbb{R} :\ \exists \mathbf{X} > \mathbf{0} \text{ such that } \mathbf{M}\mathbf{X} \geq \lambda\mathbf{X}\}. \tag{5.14}$$

Notice that the set we compute the maximum of is not empty (it contains $\lambda = 0$), closed (it is always possible to choose $\mathbf{X}$ so that $\|\mathbf{X}\| = 1$) and bounded:

$$(\mathbf{M}\mathbf{X})_{j} = \sum_{i}\mathbf{M}_{ji}X_{i} \leq \left(\sum_{i}\mathbf{M}_{ij}\right)\max_{i}X_{i} \leq \left(\sum_{i,j}\mathbf{M}_{ij}\right)\max_{i}X_{i} \quad\Rightarrow\quad \lambda \leq \sum_{i,j}\mathbf{M}_{ij}.$$

Then, this $\lambda_{\mathbf{M}}$ exists and satisfies

$$0 \leq \lambda_{\mathbf{M}} < +\infty.$$

Up to this point only the hypothesis $\mathbf{M} \geq \mathbb{O}$ has been used. Moreover, if $\mathbf{M} \gg \mathbb{O}$, then $\lambda_{\mathbf{M}} \geq \min_{j}\mathbf{M}_{jj} > 0$. Hence $0 < \lambda_{\mathbf{M}} < +\infty$.
Let $\mathbf{V}^{\mathbf{M}} > \mathbf{0}$ be such that $\mathbf{M}\mathbf{V}^{\mathbf{M}} \geq \lambda_{\mathbf{M}}\mathbf{V}^{\mathbf{M}}$.

(I) We prove that $\lambda_{\mathbf{M}}$ is an eigenvalue and $\mathbf{V}^{\mathbf{M}}$ is the corresponding eigenvector satisfying $\mathbf{V}^{\mathbf{M}} \gg \mathbf{0}$.
Since $\mathbf{M} \gg \mathbb{O}$ it follows $\mathbf{M}\mathbf{X} \gg \mathbf{0}$ for each $\mathbf{X} > \mathbf{0}$, then

$$\mathbf{M}\left(\mathbf{M}\mathbf{V}^{\mathbf{M}} - \lambda_{\mathbf{M}}\mathbf{V}^{\mathbf{M}}\right) \gg \mathbf{0} \quad \text{or} \quad \mathbf{M}\mathbf{V}^{\mathbf{M}} = \lambda_{\mathbf{M}}\mathbf{V}^{\mathbf{M}}.$$

If $\mathbf{M}\mathbf{V}^{\mathbf{M}} > \lambda_{\mathbf{M}}\mathbf{V}^{\mathbf{M}}$, then $\mathbf{Y} = \mathbf{M}\mathbf{V}^{\mathbf{M}} > \mathbf{0}$ and so $\mathbf{M}\mathbf{Y} \gg \lambda_{\mathbf{M}}\mathbf{Y}$. Thus $\lambda_{\mathbf{M}}$ could be increased slightly contradicting its definition as the maximum.

Therefore $\mathbb{M}\mathbf{V}^{\mathrm{M}} = \lambda_{\mathrm{M}}\mathbf{V}^{\mathrm{M}}$.

Moreover, since $\mathbf{V}^{\mathrm{M}} > \mathbf{0}$ implies $\mathbb{M}\mathbf{V}^{\mathrm{M}} \gg \mathbf{0}$, then $\mathbb{M}\mathbf{V}^{\mathrm{M}} = \lambda_{\mathrm{M}}\mathbf{V}^{\mathrm{M}}$ implies $\mathbf{V}^{\mathrm{M}} \gg \mathbf{0}$.

(II) We prove that $|\lambda| < \lambda_{\mathrm{M}}$ for each eigenvector λ of $\mathbb{M}$.

Let $\lambda \neq \lambda_{\mathrm{M}}$ be an eigenvalue of $\mathbb{M}$ and $\mathbf{Y} \neq \mathbf{0}$ the eigenvector that corresponds to λ. We define the vector $\widetilde{\mathbf{Y}}$ of components $\widetilde{Y}_j = |Y_j|$, with $j = 1, \cdots, n$, and we consider the vector $\mathbb{M}\widetilde{\mathbf{Y}}$ of components:

$$\left(\mathbb{M}\widetilde{\mathbf{Y}}\right)_j = m_{j1}|Y_1| + m_{j2}|Y_2| + \cdots + m_{jn}|Y_n| \geq |m_{j1}Y_1 + m_{j2}Y_2 + \cdots + m_{jn}Y_n|$$

i.e.

$$\mathbb{M}\widetilde{\mathbf{Y}} \geq \widetilde{(\mathbb{M}\mathbf{Y})}$$

$$\mathbb{M}\widetilde{\mathbf{Y}} \geq \widetilde{(\lambda\mathbf{Y})} = |\lambda|\,\widetilde{\mathbf{Y}}.$$

By the definition of λ_{M} it follows $|\lambda| \leq \lambda_{\mathrm{M}}$. To prove the strict inequality, we consider the matrix $\mathbb{M}_\delta = \mathbb{M} - \delta\mathbb{I}_n$ where $\delta > 0$ is chosen in such a way that $\mathbb{M}_\delta \gg \mathbb{O}$ (for instance, by setting $\delta = \dfrac{1}{2}\min_{ij}\mathbb{M}_{ij}$).

From the identity $(\mu - \delta)\,\mathbb{I}_n - \mathbb{M}_\delta = \mu\mathbb{I}_n - \mathbb{M}$, for every $\mu \in \mathbb{C}$, it follows that $(\lambda_{\mathrm{M}} - \delta)$ and $(\lambda - \delta)$ are eigenvalues of $\mathbb{M}_\delta$. Furthermore, by the strict positivity of $\mathbb{M}_\delta$, we have

$$|\lambda - \delta| \leq \lambda_{\mathrm{M}} - \delta \tag{5.15}$$

because $\lambda_{\mathrm{M}} - \delta$ is the maximum eigenvalue of $\mathbb{M}_\delta$.

If $|\lambda| = \lambda_{\mathrm{M}}$ and $\lambda \neq \lambda_{\mathrm{M}}$ there is a contradiction with (5.15). We have used the fact (shown in Fig. 5.5) that the subtraction of a positive quantity by a complex number reduces most (at constant modulus) the modulus of the positive real number.

(III) We prove that λ_{M} has geometric multiplicity 1.

Suppose by contradiction that there exists an eigenvector $\mathbf{Y}$ linearly independent from $\mathbf{V}^{\mathrm{M}}$. From the inequality $\mathbf{V}^{\mathrm{M}} \gg \mathbf{0}$ follows the existence of a real constant α such that, by setting $\mathbf{W} = \alpha\mathbf{V}^{\mathrm{M}} + \mathbf{Y}$, one has $\mathbf{W} > \mathbf{0}$ but not $\mathbf{W} \gg \mathbf{0}$. However, by $\mathbb{M} \gg \mathbb{O}$ it follows $\lambda_{\mathrm{M}}\mathbf{W} = \mathbb{M}\mathbf{W} \gg \mathbf{0}$ which is a contradiction.

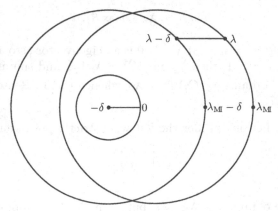

Fig. 5.5 $|\lambda| = \lambda_{\mathrm{M}} \neq \lambda$ and $\delta > 0$ $\Rightarrow$ $|\lambda - \delta| > \lambda_{\mathrm{M}} - \delta$

(IV) We prove that λ_M has algebraic multiplicity 1.

Assume by contradiction that the algebraic multiplicity of λ_M is greater than 1. We have then at least two generalized eigenvectors $\mathbf{V}^M$ and $\mathbf{U}$ associated to λ_M such that

$$(M - \lambda_M I_n) \, \mathbf{U} = \mathbf{V}^M$$
$$(M - \lambda_M I_n) \, \mathbf{V}^M = \mathbf{0}.$$

Let $\mathbf{W}$ be the unique strictly positive eigenvector of M^T corresponding to λ_M. Then

$$\left(M^T - \lambda_M I_n\right) \mathbf{W} = \mathbf{0} \qquad\qquad \mathbf{W}^T \left(M - \lambda_M I_n\right) = \mathbf{0}$$

$$0 = \mathbf{W}^T \left(M - \lambda_M I_n\right) \mathbf{U} = \left\langle \mathbf{W}, \mathbf{V}^M \right\rangle$$

which is a contradiction because $\mathbf{V}^M \gg \mathbf{0}$ e $\mathbf{W} \gg \mathbf{0}$. $\square$

Note that we have implicitly also proved the statement of Theorem 5.14 with a slightly different statement: the eigenvector $\mathbf{V}^M$ is weakly positive. As, by definition, an eigenvector cannot be zero, and Theorem 5.14 follows.

In order to prove Theorem 5.13 it is enough to observe that $A = M^h$ matches the hypotheses of Theorem 5.12 and between the eigenvalues of M and A there exists the relationship $\lambda_M = \sqrt[h]{\lambda_A}$.

The previous proof entails also that definition (5.14), called Collatz-Wielandt formula, provides the value of the dominant eigenvalue:

$$\lambda_M = \max \left\{ \lambda \in \mathbb{R} : \; \exists \mathbf{X} > \mathbf{0} \text{ such that } M\mathbf{X} \geq \lambda\mathbf{X} \right\}.$$

Remark 5.20. In some cases it may be useful to estimate the dominant eigenvalue, without having to resort to computations. The following result is true:

If $M \geq \mathbb{O}$ and $\lambda_M \geq 0$ is its dominant eigenvalue, then, if C_j and R_j are respectively the sums of the elements of the j-th column and j-th row of M, we have

$$\min_j C_j \leq \lambda_M \leq \max_j C_j \tag{5.16}$$

$$\min_j R_j \leq \lambda_M \leq \max_j R_j. \tag{5.17}$$

In fact, by setting $M = [m_{ij}]$, if $\mathbf{V}^M \geq \mathbf{0}$ is an eigenvector associated with λ_M such that $\sum_{j=1}^n V_j^M = 1$, then $\sum_j m_{ij} V_j^M = \lambda_M V_i^M$ and summing also with respect to i, one obtains $\sum_j C_j V_j^M = \lambda_M$ whence (5.16). Reasoning on M^T, one obtains (5.17).

Example 5.21. Let us consider the following strictly positive matrix:

$$M = \begin{bmatrix} 1 & 2 & 3 \\ 2 & 2 & 2 \\ 4 & 1 & 1 \end{bmatrix}.$$

From (5.16) one obtains $5 \leq \lambda_M \leq 7$, but from (5.17) one obtains $6 \leq \lambda_M \leq 6$ i.e. $\lambda_M = 6$ without even writing the characteristic polynomial. $\square$

5.5 Applications to demography

We continue the analysis of the demographic model of Leslie (Example 1.16). If we denote by Y^j the number of individuals aged j ($j = 1, 2, \ldots, n$), φ_j the fertility rate (born for individual) of individuals aged j and σ_j the survival rate of individuals aged between j and $j + 1$, at each discrete time k (with uniform step, equal to the width of the age classes) one has a vector $\mathbf{Y}_k = \begin{bmatrix} Y_k^1 & Y_k^2 & \ldots & Y_k^n \end{bmatrix}^T$ which describes the structure of the population. The evolution of $\mathbf{Y}$ is described by the equation

$$\mathbf{Y}_{k+1} = \mathbb{L}\mathbf{Y}_k$$

where

$$\mathbb{L} = \begin{bmatrix} \varphi_1 & \varphi_2 & \cdots & \varphi_{n-1} & \varphi_n \\ \sigma_1 & 0 & \cdots & 0 & 0 \\ 0 & \sigma_2 & \cdots & 0 & 0 \\ \vdots & \vdots & \ddots & \vdots & \vdots \\ 0 & 0 & \cdots & \sigma_{n-1} & 0 \end{bmatrix}.$$

At time $k + 1$ the number of individuals of the i-th class if given by $Y_{k+1}^i = \sum_{j=1}^n \mathbb{L}_{ij} Y_k^j$ while the total population at the generic time k is $\sum_{i=1}^n Y_k^i$.

If $\sigma_i > 0$, $i = 1, \ldots, n-1$ e $\varphi_j > 0$, $j = 1, \ldots, n$, then the matrix $\mathbb{L}$ although not strictly positive, has a strictly positive power ($\mathbb{L}^n \gg \mathbb{O}$, see Fig. 5.6). Therefore, by Theorem 5.13, $\mathbb{L}$ has a simple dominant eigenvalue $\lambda_\mathbb{L} > 0$ such that $|\lambda| < \lambda_\mathbb{L}$ for any other eigenvalue λ of $\mathbb{L}$. Furthermore, Theorem 5.18 applies to the orbits of the DDS $\mathbf{Y}_{k+1} = \mathbb{L}\mathbf{Y}_k$: in particular, the *asymptotic distribution of the population is described by the dominant eigenvector* (which is strictly positive by the Perron–Frobenius Theorem), that is, the distribution of the various age groups is proportional to the components of the positive dominant eigenvector $\mathbf{V}^\mathbb{L}$.

The dominant eigenvector represents the so-called *stable age distribution* that describes the proportions in which the population tends to be distributed in the various classes (in the absence of perturbations).

Depending on the values of survival and fertility, the dominant eigenvalue describes the perspective growth or extinction of the population:

$\lambda_\mathbb{L} > 1$ population in exponential growth;

$\lambda_\mathbb{L} = 1$ population tending to an equilibrium (strictly positive);

$0 < \lambda_\mathbb{L} < 1$ population exponentially proceeding to the extinction.

If $0 < \lambda_\mathbb{L} < 1$, then $\mathbf{0}$ is a stable and attractive equilibrium.
If $\lambda_\mathbb{L} = 1$, then every positive multiple of the dominant eigenvector $\mathbf{V}^\mathbb{L}$ (associated to $\lambda_\mathbb{L}$) is a stable but not attractive equilibrium (half line of equilibria).

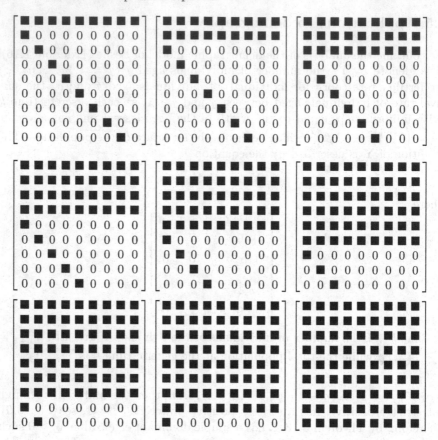

Fig. 5.6 $\mathbb{L}$, $\mathbb{L}^2$, ..., $\mathbb{L}^9$ with $\mathbb{L}$ Leslie matrix of order 9 (black squares indicate strictly positive elements)

If $\lambda_{\mathbb{L}} > 1$, then $\mathbf{0}$ is an unstable and repulsive equilibrium. Observe in Fig. 5.6 how an increase in the power of $\mathbb{L}$ makes the rows $\gg \mathbb{0}$ increase, while the diagonal $\gg \mathbb{0}$ moves downward.

The dominant eigenvalue and eigenvector can be calculated by numerical methods starting from the algebraic definition or by simulating the system for a sufficiently long period: if $\mathbf{X}_0$ has non-zero components $c_{\mathbb{L}}$ along $\mathbf{V}^{\mathbb{L}}$ in the Jordan basis of $\mathbb{L}$, then (thanks to Theorem 5.18) we have:

$$\lambda_{\mathbb{L}} = \lim_k \frac{\|\mathbf{X}_{k+1}\|}{\|\mathbf{X}_k\|} \qquad \mathbf{V}^{\mathbb{L}} = \lim_k \lambda_{\mathbb{L}}^{-k} c_{\mathbb{L}}^{-1} \mathbf{X}_k .$$

Remark 5.22. If $\mathbb{L}$ is a Leslie matrix, then the determinant value is:

$$\det \mathbb{L} = (-1)^{1+n} \varphi_n \sigma_1 \sigma_2 \cdots \sigma_{n-1}.$$

males	1911	females		males	2001	females
0,02	90 e +	0,02		0,17	90 e +	0,50
0,07	85-89	0,08		0,47	85-89	1,04
0,24	80-84	0,26		0,69	80-84	1,27
0,53	75-79	0,54		1,54	75-79	2,37
0,97	70-74	1,00		2,10	70-74	2,73
1,38	65-69	1,39		2,48	65-69	2,89
1,81	60-64	1,89		2,88	60-64	3,14
2,00	55-59	2,03		2,77	55-59	2,90
2,33	50-54	2,43		3,40	50-54	3,47
2,45	45-49	2,55		3,24	45-49	3,27
2,56	40-44	2,74		3,55	40-44	3,53
2,70	35-39	2,95		4,08	35-39	4,00
2,96	30-34	3,32		4,11	30-34	4,01
3,27	25-29	3,79		3,85	25-29	3,75
4,03	20-24	4,40		3,14	20-24	3,02
4,51	15-19	4,84		2,70	15-19	2,56
5,39	10-14	5,23		2,53	10-14	2,39
5,53	5-9	5,33		2,47	5-9	2,33
6,36	0-4	6,12		2,39	0-4	2,26
males	total (thousands)	females		males	total (thousands)	females
17.021,7		17.649,7		28.094,8		29.749,2

Fig. 5.7 Italy's resident population, grouped by sex and age (by ISTAT)

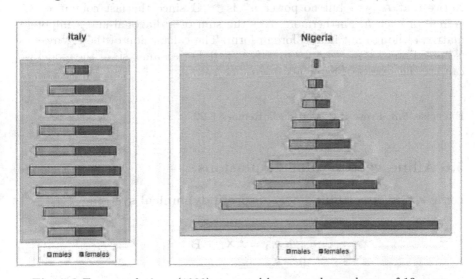

Fig. 5.8 Two populations (1998), grouped by sex and age classes of 10 years

Example 5.23. *A model of student distribution.*
Consider the students who attend a university five-year course. Define:

S^j the number of students enrolled to the j-th year of study, $j = 1, 2, \ldots, 5$
S^6_k the number of graduates in the k-th year

a_j the coefficient corresponding to the fraction of dropouts in the j-th year
p_j the coefficient of transition from year j to year $j + 1$, $j = 1, \ldots, 4$
p_5 the proportion of graduates
$\mathbf{M}_k$ the people registered or transferred from other locations in the k-th year
(it is a vector with 6 components, where the last is zero).

At any time k we have a vector with six components $\mathbf{S}_k$ that describes the
structure of the student population.

The evolution model is described by

$$\mathbf{S}_{k+1} = \mathbb{A}\mathbf{S}_k + \mathbf{M}_k$$

with

$$\mathbb{A} = \begin{bmatrix} 1 - p_1 - a_1 & 0 & 0 & 0 & 0 & 0 \\ p_1 & 1 - p_2 - a_2 & 0 & 0 & 0 & 0 \\ 0 & p_2 & 1 - p_3 - a_3 & 0 & 0 & 0 \\ 0 & 0 & p_3 & 1 - p_4 - a_4 & 0 & 0 \\ 0 & 0 & 0 & p_4 & 1 - p_5 - a_5 & 0 \\ 0 & 0 & 0 & 0 & p_5 & 0 \end{bmatrix}.$$

Notice that $\mathbb{A} \geq \mathbb{O}$ but no power $\mathbb{A}^h$ is $\gg \mathbb{O}$ since the last column of $\mathbb{A}^h$
is null for each h. Furthermore, $\mathbb{A}$ is the sum of a diagonal and a nilpotent
matrix, although not in the Jordan form. The calculation of the eigenvalues
is immediate: $1 - p_1 - a_1$, $1 - p_2 - a_2$, $\ldots$, 0. For an analysis of the model we
can use the results of Section 5.6.   □

Exercise 5.3. Prove the assertion in Remark 5.22.

5.6 Affine vector-valued equations

Let us consider the **affine linear discrete dynamical system**

$$\boxed{\mathbf{X}_{k+1} = \mathbb{A}\mathbf{X}_k + \mathbf{B}}$$

$$(5.18)$$

where $\mathbf{B} \in \mathbb{R}^n$ and the matrix $\mathbb{A}$ of order n are given. We are interested in
finding $\mathbf{X}_k$ explicitly, that is, to give a non-recursive expression and, subsequently, to study the existence of possible equilibria and discuss their stability.
Following the same formal reasoning of the affine scalar case (Section 2.1, Theorem 2.5) it is useful to make a change of variable and reduce to the linear
case: we try to adapt the steps of the proof of Theorem 2.5, paying attention
to the fact that $\mathbb{A}$ is a matrix.

Theorem 5.24. *If all the eigenvalues of $\mathbb{A}$ are different from 0 and 1 then $\mathbb{I}_n - \mathbb{A}$ is invertible and*

$$\boxed{\mathbf{X}_k = \mathbb{A}^k \left(\mathbf{X}_0 - \mathbf{A} \right) + \mathbf{A}} \tag{5.19}$$

where the vector $\mathbf{A} = \left(\mathbb{I}_n - \mathbb{A} \right)^{-1} \mathbf{B}$ is the unique equilibrium of the DDS (5.18).

Notice the strong similarity of (5.19) with formula (2.4) relative to the scalar affine case.

Proof. In the following we will write $\mathbb{I}$ instead $\mathbb{I}_n$. The matrix $\mathbb{I} - \mathbb{A}$ is invertible because 1 is not an eigenvalue of $\mathbb{A}$. The equilibria $\mathbf{A}$ of the DDS (5.18) solve

$$\mathbf{A} = \mathbb{A}\mathbf{A} + \mathbf{B}$$

whose (unique) solution is $\mathbf{A} = \left(\mathbb{I} - \mathbb{A} \right)^{-1} \mathbf{B}$.
Set $\mathbf{Y}_k = \mathbf{X}_k - \mathbf{A}$ $\forall k$. By substituting into equation (5.18), one obtains the DDS:

$$\mathbf{Y}_{k+1} + \mathbf{A} = \mathbb{A} \left(\mathbf{Y}_k + \mathbf{A} \right) + \mathbf{B}$$

equivalent to

$$\mathbf{Y}_{k+1} = \mathbb{A}\mathbf{Y}_k + \mathbb{A}\mathbf{A} - \mathbf{A} + \mathbf{B}$$

that is

$$\mathbf{Y}_{k+1} = \mathbb{A}\mathbf{Y}_k - \left(\mathbb{I} - \mathbb{A} \right) \mathbf{A} + \mathbf{B}.$$

By definition of $\mathbf{A}$, we finally get:

$$\mathbf{Y}_{k+1} = \mathbb{A}\mathbf{Y}_k - \left(\mathbb{I} - \mathbb{A} \right) \left(\mathbb{I} - \mathbb{A} \right)^{-1} \mathbf{B} + \mathbf{B} =$$
$$= \mathbb{A}\mathbf{Y}_k - \mathbf{B} + \mathbf{B} = \mathbb{A}\mathbf{Y}_k. \qquad \square$$

In the general case (that is, if 0 or 1 are eigenvalues of $\mathbb{A}$) $\mathbb{R}^n$ is decomposed as $\mathbb{R}^n = \mathcal{V}_0 \oplus \mathcal{V}_1 \oplus \mathcal{V}_2$ where

$\mathcal{V}_0$ is the generalized eigenspace associated to the eigenvalue 0;
$\mathcal{V}_1$ is the generalized eigenspace associated to the eigenvalue 1;
$\mathcal{V}_2$ is the union of the generalized eigenspaces of all the other eigenvalues.

The matrix $\mathbb{A}$ in the coordinates of the Jordan form is expressed as

$$\mathbb{A} = \begin{bmatrix} \mathbb{A}_0 & \mathbb{O} & \mathbb{O} \\ \mathbb{O} & \mathbb{A}_1 & \mathbb{O} \\ \mathbb{O} & \mathbb{O} & \mathbb{A}_2 \end{bmatrix}$$

where $\mathbb{A}_0$, $\mathbb{A}_1$, $\mathbb{A}_2$ are square matrices with $\mathbb{A}_0$ nilpotent. All vectors $\mathbf{W}$ of $\mathbb{R}^n$ can be decomposed as $\mathbf{W} = \left[\mathbf{W}^0 \ \mathbf{W}^1 \ \mathbf{W}^2 \right]^T$. System (5.18) decouples into three systems.

Each equilibrium $\mathbf{A} = \begin{bmatrix} \mathbf{A}^0 & \mathbf{A}^1 & \mathbf{A}^2 \end{bmatrix}^T$ solves $\mathbf{A} = \mathbb{A}\mathbf{A} + \mathbf{B}$, namely

$$(\mathbb{I} - \mathbb{A}_0)\,\mathbf{A}^0 = \mathbf{B}^0 \qquad (\mathbb{I} - \mathbb{A}_1)\,\mathbf{A}^1 = \mathbf{B}^1 \qquad (\mathbb{I} - \mathbb{A}_2)\,\mathbf{A}^2 = \mathbf{B}^2 \ .$$

Since $(\mathbb{I} - \mathbb{A}_0)$ and $(\mathbb{I} - \mathbb{A}_2)$ are invertible, $\mathbf{A}^0$ and $\mathbf{A}^2$ are uniquely determined

$$\mathbf{A}^0 = (\mathbb{I} - \mathbb{A}_0)^{-1}\,\mathbf{B}^0 \qquad\qquad \mathbf{A}^2 = (\mathbb{I} - \mathbb{A}_2)^{-1}\,\mathbf{B}^2 \ .$$

However $(\mathbb{I} - \mathbb{A}_1)$ is not invertible (it is nilpotent), therefore solutions $\mathbf{A}^1$ may exist or not, and if they exist they are not unique (precisely, they exist if and only if $\mathbf{B}^1 \perp \ker (\mathbb{I} - \mathbb{A}_1)^T$). The conclusions are summarized in the statement below.

Theorem 5.25. *If 0 or 1 (or both) are eigenvalues of $\mathbb{A}$, then system (5.18) admits equilibria $\mathbf{A} = \begin{bmatrix} \mathbf{A}^0 & \mathbf{A}^1 & \mathbf{A}^2 \end{bmatrix}^T$ if and only if the component $\mathbf{B}^1$ of $\mathbf{B}$ in $\mathcal{V}_1$ is orthogonal to the kernel of $(\mathbb{I} - \mathbb{A}_1)^T$.*
If so, the solutions $\mathbf{X}_k = \begin{bmatrix} \mathbf{X}^0 & \mathbf{X}^1 & \mathbf{X}^2 \end{bmatrix}^T$ decomposed according to $\mathcal{V}_0$, $\mathcal{V}_1$, $\mathcal{V}_2$ can be made explicit:

$$\mathbf{X}_k^0 = (\mathbb{A}_0)^k \left(\mathbf{X}_0^0 - \mathbf{A}^0 \right) + \mathbf{A}^0$$

$$\mathbf{X}_k^2 = (\mathbb{A}_2)^k \left(\mathbf{X}_0^2 - \mathbf{A}^2 \right) + \mathbf{A}^2$$

$$\mathbf{X}_k^1 = \begin{bmatrix} \mathbf{X}_k^{\mathcal{U}} & \mathbf{X}_k^{\mathcal{W}} \end{bmatrix}^T$$

where $\mathcal{U}$ is the space associated to the trivial Jordan block and $\mathcal{V}_2$ is the union of the spaces associated to the non trivial Jordan blocks related to the eigenvalue 0:
$\mathbf{X}_k^{\mathcal{U}} = k\mathbf{B}^{\mathcal{U}}$ *for each k, where $\mathbf{B}^{\mathcal{U}}$ is the component of $\mathbf{B}$ in $\mathcal{U}$, and $\mathbf{X}_k^{\mathcal{W}}$ can be expressed in terms of $\mathbf{B}^{\mathcal{W}}$ and $\mathbf{X}_0^{\mathcal{W}}$ in each non trivial Jordan sub-block of dimension $l - j$ corresponding to the eigenvalue 0.*
Hence:

$$\exists \text{ equilibria } \mathbf{A} \quad \Leftrightarrow \quad \begin{cases} (I) \quad 1 \text{ is not an eigenvalue, that is } \mathcal{V}_1 = \mathcal{U} \oplus \mathcal{W} = \{0\} \\[4pt] \text{or} \\[4pt] (II) \quad 1 \text{ is an eigenvalue and } \mathbf{B}_j^1 = 0 \text{ for every } j, \\ \qquad\quad \text{except possibly the last components in} \\ \qquad\quad \text{each nontrivial Jordan block} \\ \qquad\quad \text{of the eigenvalue 1.} \end{cases}$$

In the first case there exists a unique equilibrium

$$\mathbf{A} = (\mathbb{I} - \mathbb{A})^{-1}\mathbf{B} = \begin{bmatrix} (\mathbb{I} - \mathbb{A}^0)^{-1}\mathbf{B}^0 & (\mathbb{I} - \mathbb{A}^2)^{-1}\mathbf{B} \end{bmatrix}^T .$$

In the second case there are infinitely many equilibria:

$$\mathbf{A} = \begin{bmatrix} \mathbf{A}^0 & \mathbf{A}^1 & \mathbf{A}^2 \end{bmatrix}^T$$

$$\mathbf{A}^1 = \begin{bmatrix} \mathbf{A}^U & \mathbf{A}^W \end{bmatrix}^T = \begin{bmatrix} \mathbf{0} & \mathbf{A}^W \end{bmatrix}^T$$

$$\mathbf{A}^W = consists\ of\ vectors\ \begin{bmatrix} 0\ 0\ \ldots\ 0\ \mathbf{B}_l \end{bmatrix}^T\ in\ each\ l\text{-}dimensional\ block.$$

- *If $|\lambda| < 1$ for each eigenvalue, then we are in case I and the equilibrium $\mathbf{A}$ is stable and attractive.*
- *If there exists an eigenvalue λ such that $|\lambda| > 1$, then the possible equilibrium (cases I or II) is instable.*
- *If $\lambda = 1$ is an eigenvalue and there exists an equilibrium, then we are in case II and $\mathbf{A}$ is stable if the algebraic and geometric multiplicities of the eigenvalue 1 coincide (however $\mathbf{A}$ is not attractive).*

It is recommended to compare the discussion on equilibria and their stability with the one concerning m-steps scalar difference equations.

Example 5.26. Let $\mathbb{M}$ be a given matrix of order 9, that in Jordan coordinates corresponding to the basis $\mathbf{e}_1, \mathbf{e}_2, \ldots, \mathbf{e}_9$ (the components not listed are zero) reads

$$\mathbb{M} = \begin{bmatrix} 0 & 0 & & & & & & & \\ 1 & 0 & & & & & & & \\ & & 1 & 0 & & & & & \\ & & 1 & 1 & & & & & \\ & & & & 1 & 0 & & & \\ & & & & 1 & 1 & & & \\ & & & & & & \lambda_1 & 0 & 0 \\ & & & & & & 0 & \lambda_2 & 0 \\ & & & & & & 0 & 0 & \lambda_3 \end{bmatrix}$$

with

$$\mathcal{V}_0 = \mathbf{e}_1 \oplus \mathbf{e}_2 \qquad \mathcal{V}_1 = \mathbf{e}_3 \oplus \mathbf{e}_4 \oplus \mathbf{e}_5 \oplus \mathbf{e}_6 \qquad \mathcal{V}_2 = \mathbf{e}_7 \oplus \mathbf{e}_8 \oplus \mathbf{e}_9 \ .$$

The natural modes are identified by
$$0, \ 1, \ k, \ k^2, \ k^3, \ \lambda_1^k, \ \lambda_2^k, \ \lambda_3^k \ . \qquad \qquad \square$$

Remark 5.27. The model of student distribution considered in Example 5.23 in a stationary state ($\mathbf{M}_k = \mathbf{M}$, $\forall k$, namely the number of registrations and transfers is assumed constant) is in the form (5.10) studied in this section.

We conclude with some general considerations on the positive linear DDS. An affine system $\mathbf{X}_{k+1} = \mathbb{A}\mathbf{X}_k + \mathbf{B}$ may not have equilibria $\mathbf{A}$ that verify $\mathbf{A} \geq \mathbf{0}$ even if $\mathbb{A} \gg \mathbb{O}$ and $\mathbf{B} \gg \mathbf{0}$ (for instance, if $n = 1$, $\mathbb{A} = 2$ and $\mathbf{B} = 1$ then $\mathbf{A} = -1$). It is also relevant to ask when, given nonnegative $\mathbb{A}$ and $\mathbf{B}$, the equilibria are non-negative. This issue is closely connected to the stability. We state a result that partially illustrates this problem.

Theorem 5.28. *Consider the vector-valued DDS* $\mathbf{X}_{k+1} = \mathbb{A}\mathbf{X}_k + \mathbf{B}$, *where* $\mathbb{A} \gg \mathbb{O}, \mathbf{B} \gg \mathbf{0}$.
The following conditions are equivalent:

(i) $|\lambda| < 1$ *for every eigenvalue* λ *of* $\mathbb{A}$;
(ii)   *there exists an equilibrium* $\mathbf{A} \geq \mathbf{0}$ *of the system.*
        In particular, if (i) or (ii) holds then $\mathbf{A}$ *is stable and attractive and* $\mathbf{A} \gg \mathbf{0}$.

Proof. (i) $\Rightarrow$ (ii): If $|\lambda| < 1$ for each eigenvalue λ, then $\mathbb{I} - \mathbb{A}$ is invertible, and by Theorem 5.24 there exists a unique equilibrium $\mathbf{A}$ which is stable and attractive. Moreover, if the dynamical system is initialized with a datum $\mathbf{X}_0 \geq \mathbf{0}$, it turns out $\mathbf{X}_k \gg \mathbf{0}$ for each $k \geq 1$ and it follows $\mathbf{A} \geq \mathbf{0}$. Furthermore, by $\mathbb{A}\mathbf{A} \geq \mathbf{0}, \mathbf{B} \gg \mathbb{O}$ it follows $\mathbf{A} = \mathbb{A}\mathbf{A} + \mathbf{B} \gg \mathbf{0}$.
(ii) $\Rightarrow$ (i): If there exists $\mathbf{A} \geq \mathbf{0}$ such that $(\mathbb{I} - \mathbb{A})\mathbf{A} = \mathbf{B}$, then $\mathbf{A} = \mathbb{A}\mathbf{A} + \mathbf{B} \gg \mathbf{0}$. After observing that $\mathbb{A}^T \gg \mathbb{O}$ and $\mathbb{A}$ and $\mathbb{A}^T$ have the same eigenvalues, we denote by $\lambda_\mathbb{A} > 0$ the dominant eigenvalue of $\mathbb{A}$ (and therefore also of $\mathbb{A}^T$), and $\mathbf{U}$ is a strictly positive dominant eigenvector of $\mathbb{A}^T$ (see Theorem 5.12):

$$\mathbf{U} \gg \mathbf{0} \qquad \mathbb{A}^T\mathbf{U} = \lambda_\mathbb{A}\mathbf{U} \qquad \langle \mathbf{B}, \mathbf{U} \rangle > 0.$$

Hence

$$(1 - \lambda_\mathbb{A})\langle \mathbf{A}, \mathbf{U} \rangle = \left\langle \mathbf{A}, \left(\mathbb{I} - \mathbb{A}^T \right) \mathbf{U} \right\rangle = \langle (\mathbb{I} - \mathbb{A})\mathbf{A}, \mathbf{U} \rangle = \langle \mathbf{B}, \mathbf{U} \rangle > 0$$

and from inequality $\langle \mathbf{A}, \mathbf{U} \rangle > 0$ it follows $\lambda_\mathbb{A} < 1$.
So, $0 < \lambda_\mathbb{A} < 1$ and $0 \leq |\lambda| < \lambda_\mathbb{A} < 1$ for every eigenvalue λ. $\qquad \square$

Exercise 5.4. Given the square matrix $\mathbb{A}$ of order n whose elements are all equal to $1/(2n)$ and a vector $\mathbf{B} \in \mathbb{R}^n, \mathbf{B} \gg \mathbf{0}$, study the DDS $\mathbf{X}_{k+1} = \mathbb{A}\mathbf{X}_k + \mathbf{B}$.

5.7 Nonlinear vector discrete dynamical systems

We briefly address the case of one-step non linear vector systems.

Definition 5.29. *Let* $F : I^n \to I^n$ *be a continuous function. Then the pair* $\{I^n, F\}$ *is called* **vector discrete dynamical system**.

To each vector DDS $\{I^n, F\}$ is associated a system of infinitely many equations $\mathbf{Z}_{k+1} = F(\mathbf{Z}_k), k \in \mathbb{N}$. For every choice of n initial data $\mathbf{Z}_0, \mathbf{Z}_1, \ldots, \mathbf{Z}_{n-1}$ there is a unique sequence $\{\mathbf{Z}_k\}$ that solves this system: this sequence is called **orbit** of the DDS associated to the n-tuple of initial data.
All the definitions introduced in the scalar case about equilibria and their stability can be restated with the obvious changes in the vector case.

Example 5.30. Set $n = 2$, $F(x, y) = \left[f(x, y) \; g(x, y) \right]^T$, $\mathbf{Z}_k = \left[X_k \; Y_k \right]^T$.
Then, $\forall k \in \mathbb{N}$

$$\begin{cases} X_{k+1} = f(X_k, Y_k) \\ Y_{k+1} = g(X_k, Y_k) \, . \end{cases} \qquad \qquad \square$$

As in the scalar case, there are no general methods to find explicit solutions of nonlinear systems. We shall only discuss the asymptotic behavior and stability of equilibria.
The following definition is useful.

Definition 5.31. *Let* $\mathbb{A}$ *be an* $n \times n$ *matrix. The number*

$$\|\mathbb{A}\| = \max \{ \|\mathbb{A}\mathbf{V}\| : \; \mathbf{V} \in \mathbb{R}^n \, , \, \|\mathbf{V}\| = 1 \}$$

is called **norm of** $\mathbb{A}$.

Theorem 5.32. *Let* $n = 2$ *and* $\mathbf{A} = \left[\alpha_1 \; \alpha_2 \right]^T$ *be an equilibrium:* $F(\mathbf{A}) = \mathbf{A}$,
that is in components

$$\alpha_1 = f(\alpha_1, \alpha_2) \qquad \qquad \alpha_2 = g(\alpha_1, \alpha_2)$$

where $F : I^2 \to I^2$ *is a differentiable function with continuous partial derivatives and*

$$\mathbb{J}(\mathbf{A}) = \begin{bmatrix} \dfrac{\partial f}{\partial x}(\alpha_1, \alpha_2) & \dfrac{\partial f}{\partial y}(\alpha_1, \alpha_2) \\[2ex] \dfrac{\partial g}{\partial x}(\alpha_1, \alpha_2) & \dfrac{\partial g}{\partial y}(\alpha_1, \alpha_2) \end{bmatrix} \, .$$

If all the eigenvalues of $\mathbb{J}$ *satisfy* $|\lambda| < 1$, *then* α *is a stable and locally attractive equilibrium (i.e. if* $\left[X_0 \; Y_0 \right]^T$ *is "close" to* $\left[\alpha_1 \; \alpha_2 \right]^T$, *the orbit associated to this initial datum converges to* $\mathbf{A}$).
If there is an eigenvalue λ *such that* $|\lambda| > 1$, *then* α *is iunstable.*

Example 5.33 (Lotka-Volterra (predator-prey) model). .
Let us consider two species that interact in an environment: *preys*, described by the sequence X (X_0 is the number of individuals of the population at time 0, X_1 at time 1, ...). In the absence of predators X_k would evolve according to the linear law

$$X_{k+1} = aX_k \, .$$

If we consider the presence of a predator Y which feeds mainly on X it is appropriate to amend the law that describes the dynamics as follows

$$X_{k+1} = aX_k - bX_kY_k \, .$$

The evolution of predators (for which the social competition is not neglected) is described by

$$Y_{k+1} = cY_k - dY_k^2 + eX_kY_k$$

which takes into account the "encounters" with preys.

In the quadrant $x > 0$, $y > 0$ there are four regions determined by the intersection of the straight lines of equation $y = a/b$ and $y = (e/d)\,x + (c-1)\,/d$ in which X_k and Y_k are monotone. All the orbits "rotate" clockwise around the equilibrium α whose stability depends on the value of the parameters. According to the variation of the positive parameters a, b, c, d, e, with $a > 1$ and $c > 1$, the system displays different behaviors.

Let us consider a numerical example: if

$$a = 2 \quad b = 1 \quad c = 1.5 \quad d = 2 \quad e = 1$$

then

$$\begin{cases} X_{k+1} = 2X_k - X_k Y_k \\ Y_{k+1} = 1.5 Y_k - 2Y_k^2 + X_k Y_k \end{cases}$$

namely

$$F\left(x, y\right) = \begin{bmatrix} 2x - xy \\ \dfrac{3}{2}y - 2y^2 + xy \end{bmatrix}.$$

The equations satisfied by the equilibria are

$$\begin{cases} \alpha_1 = \alpha_1 \alpha_2 \\ 0.5\alpha_2 + \alpha_1\alpha_2 = 2\alpha_2^2 \end{cases}.$$

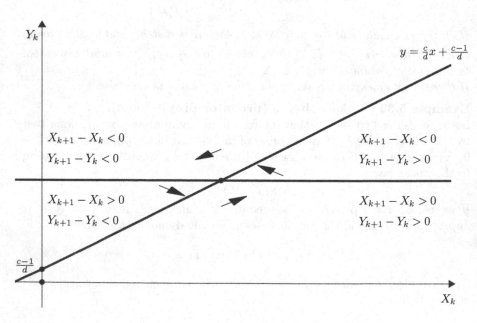

Fig. 5.9 Phase plane of the Lotka-Volterra model

There are three equilibria: $(0,0)$, $(0,1/4)$, and $(3/2,1)$. The significant part of the plane is the quadrant $x > 0$, $y > 0$, to which only one equilibrium belongs: $(3/2,1)$. Since

$$\mathbb{J}\left(\frac{3}{2},1\right) = \begin{bmatrix} 1 & -3/2 \\ 1 & -1 \end{bmatrix} \qquad \lambda_{1,2}(\mathbb{J}) = \pm\frac{\sqrt{2}}{2}\,i$$

by Theorem 5.32 the equilibrium $(3/2,1)$ is stable and locally attractive. In the quadrant $x > 0$, $y > 0$ there are four regions determined by the intersection of the straight lines of equation

$$y = 2 \qquad\qquad y = \frac{1}{2}x + \frac{1}{4}$$

where X_k and Y_k have unique monotonicity. All the orbits "rotate" counterclockwise around the equilibrium α whose stability depends on the parameters values.

To discuss the stability of the equilibrium $(0,0)$ we observe that

$$\mathbb{J}(0,0) = \begin{bmatrix} a & 0 \\ 0 & c \end{bmatrix} = \begin{bmatrix} 2 & 0 \\ 0 & 3/2 \end{bmatrix}$$

hence $(0,0)$ is stable if and only if $|a| < 1$ and $|c| < 1$.
Verify that $(0,1/4)$ is instable. $\qquad\qquad\qquad\qquad\qquad\qquad\qquad\qquad\square$

5.8 Numerical schemes for solving linear problems

Often one is faced with the problem of calculating the solutions of a system of linear algebraic equations (see Appendix D): for instance, numerical schemes for the solution of linear differential equations consist in the solution of such systems with large size (number of equations). In Example 1.21 we met a finite difference scheme that allows to approximate the solution of a partial differential equation by solving a system of linear algebraic equations. From a theoretical standpoint, the Theorem of Rouche-Capelli and Cramer's rule give a complete answer to the problem of solving such algebraic system. However, these methods are inapplicable to the actual determination of the solutions if the size of the system is not very small or if the associated matrix is not diagonal. Moreover, in the finite difference schemes that approximate differential equations the size is very large and the associated matrix is not trivial. It is therefore important to identify efficient numerical schemes for the exact or approximate solution of linear algebraic systems of large size, at least in the case of matrices with special structure (as in the case of the discretization of the differential equations).

We start by considering an important example of *sparse matrix* that is, with many null elements.

Example 5.34. The $N \times N$ tridiagonal matrix (with zero entries outside the three main diagonals)

$$
\mathbb{T} = \begin{bmatrix}
2 & -1 & & & & & \\
-1 & 2 & -1 & & & & \\
& -1 & 2 & \ddots & & & \\
& & \ddots & \ddots & \ddots & & \\
& & & \ddots & \ddots & -1 & \\
& & & & -1 & 2 & -1 \\
& & & & & -1 & 2
\end{bmatrix}
\tag{5.20}
$$

has eigenvalues [7] $\lambda_k (\mathbb{T}) = 4 \left(\sin \dfrac{k\pi}{2(N+1)} \right)^2$ per $1 \le k \le N$, associated to the eigenvectors $\mathbf{U}_k (\mathbb{T})$ of components $(U_k)_j = \sin (jk\pi/(N+1))$.

It follows that, for each N, the matrix $\mathbb{T}$ is positive definite and its eigenvalues $\lambda_k (\mathbb{T})$ satisfy $0 < \lambda_k (\mathbb{T}) \le 4$ for each k; moreover, $\lambda_{\min} (\mathbb{T}) = \lambda_1 (\mathbb{T}) \sim \left(\dfrac{\pi}{N+1} \right)^2$. □

Example 5.35 (Eigenvalue problem for the operator $-d^2/dx^2$ in an interval). The search for constants λ which correspond to non-identically zero functions $u = u(x)$ that solve

$$
\begin{cases}
-u''(x) = \lambda u(x) & 0 < x < l \\
u(0) = u(l) = 0
\end{cases}
\tag{5.21}
$$

is called *eigenvalue problem*.

Problem (5.21) has infinitely many solutions given by the pairs $\lambda_k^l = \left(\dfrac{k\pi}{l} \right)^2$ and $u_k(x) = C_k \sin \dfrac{k\pi x}{l}$ with $k = 1, 2, \ldots$ and arbitrary constants C_k.

Approximating the second derivative with centered finite differences:

$$
u''(x_k) \sim \frac{u_{k-1} - 2u_k + u_{k-1}}{h^2} \quad \text{where}
$$

$$
h = l/(N+1) \; ; \quad x_k = kh \; ; \quad u_k \sim u(x_k) \; ; \qquad 1 \le k \le N
$$

one obtains the following problem (numerical approximation of problem (5.21)):

Determine the constants λ for which there exists a nonzero vector

$\mathbf{U} = \begin{bmatrix} U_1 \ U_2 \ \cdots \ U_N \end{bmatrix}^T$ *that solves the linear system $N \times N$:*

$$
\frac{1}{h^2} \mathbb{T} \mathbf{U} = \lambda \mathbf{U} .
$$

---

[7] One can use the identities:

$$
\left(\sin \tfrac{\alpha}{2} \right)^2 = \frac{1 - \cos \alpha}{2} \; ; \quad \sin(\alpha + \beta) = \sin \alpha \cos \beta + \sin \beta \cos \alpha .
$$

This last system consists in the determination of the eigenvalues of the matrix $h^{-2}\mathbb{T}$ where $\mathbb{T}$ is the tridiagonal matrix (5.20): therefore, the eigenvalues of $h^{-2}\mathbb{T}$ are

$$\lambda_{h,k} = h^{-2}\lambda_k(\mathbb{T}) = \frac{4}{h^2}\left(\sin\frac{k\pi}{2(N+1)}\right)^2 \qquad 1 \le k \le N.$$

Note that, for fixed l and k, if $h \to 0$, then $N \to +\infty$ and $\lambda_{h,k} = \nearrow \lambda_k^l$, $\forall k$. $\square$

Example 5.36. (Euler method to solve the heat equation) According to Example 1.21 and our notation, an approximated solution of problem

$$\begin{cases} \dfrac{\partial v}{\partial t} = \dfrac{\partial^2 v}{\partial x^2} & x \in (0,a), \ 0 < t < T \\ v(x,0) = f(x) & x \in [0,a] \\ v(0,t) = v(a,t) = 0 \ \ t > 0 \end{cases} \tag{5.22}$$

can be obtained using the *Euler method*, that is, the numerical scheme

$$\begin{cases} W_{j,k+1} = \alpha W_{j-1,k} + (1-2\alpha)W_{j,k} + \alpha W_{j+1,k} & \alpha = h/s^2 \\ W_{0,k+1} = W_{N,k+1} = 0. \end{cases} \tag{5.23}$$

In terms of discrete dynamical systems the numerical scheme (5.23) produces an orbit of the vector-valued DDS $\mathbf{W}_{k+1} = \mathbb{B}\mathbf{W}_k$ from $\mathbf{W}_0$, sampling of the initial datum f, with $\mathbb{B} = \mathbb{I}_N - \alpha\mathbb{T}$, where $\mathbb{T}$ is the tridiagonal matrix (5.20), $\alpha = h/s^2$, h is the step of the time discretization, s the spatial one.

By performing numerical experiments with a computer, it is easy to verify that, if the **numerical stability condition** $0 < \alpha \le 1/2$ holds, then the Euler method has a good behavior, that is, the obtained numerical values approximate the exact solution. Vice versa, if the **numerical instability condition** $\alpha > 1/2$ holds, then the numerical values obtained oscillate uncontrollably instead of approximating the solution.

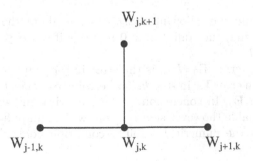

Fig. 5.10 Finite difference discretization for Euler method (5.17)

The goal is to avoid any amplifications of the errors introduced in the evaluation (with finite arithmetic precision) of the initial datum.

$$\mathbb{B} = \begin{bmatrix} 1-2\alpha & \alpha & & & & \\ \alpha & 1-2\alpha & \alpha & & & \\ & \alpha & 1-2\alpha & \ddots & & \\ & & \ddots & \ddots & \ddots & \\ & & & \ddots & \ddots & \alpha \\ & & & \alpha & 1-2\alpha & \alpha \\ & & & & \alpha & 1-2\alpha \end{bmatrix}.$$

The condition $\alpha \leq 1/2$ (i.e. $h \leq s^2/2$) should therefore be set to obtain significant results; However, it is very costly in practice: spatially doubling the grid points of the discretization involves a quadrupling of such points in time, then the total number of points and the relative time of calculation increases eight times. For this reason the backward Euler method, which is stable without any condition on the numerical value of $\alpha = h/s^2$ is preferred, even if such a method requires the solution of a linear algebraic system with non-diagonal matrix, while in the case of the Euler method (5.23) the algebraic system is already solved.

To understand the meaning of the condition of numerical stability for the explicit Euler method, according to Example 5.36, we observe that matrix $\mathbb{T}$ is diagonalizable. So also $\mathbb{B} = \mathbb{I}_N - \alpha\mathbb{T}$ is diagonalizable and its eigenvectors and eigenvalues are respectively

$$\mathbf{U}_k(\mathbb{B}) = \mathbf{U}_k(\mathbb{T}) \qquad 1 \leq k \leq N$$

$$\lambda_k(\mathbb{B}) = 1 - \alpha\lambda_k(\mathbb{T}) = 1 - 4\alpha\left(\sin\frac{k\pi}{2(N+1)}\right)^2 \qquad 1 \leq k \leq N.$$

Then the eigenvalues of $\mathbb{B}$ are all simple; furthermore:

$$\lambda_k(\mathbb{B}) \leq 1 \quad \Leftrightarrow \quad 0 \leq \alpha \leq 1/2.$$

By Theorem 5.8 and the following considerations about the DDS $\mathbf{W}_{k+1} = \mathbb{B}\mathbf{W}_k$, we can say that the equilibrium $\mathbf{0}$ is stable if $0 \leq \alpha \leq 1/2$ and attractive if $0 < \alpha \leq 1/2$.

An initial rounding error $\mathbf{E}_0$ ($E_{j,0}$ is the error in the numerical evaluation of $f(js)$) generates an error $\mathbf{E}_k$ in step k. This error solves the DDS $\mathbf{E}_{k+1} = \mathbb{B}\mathbf{E}_k$ with initial datum $\mathbf{E}_0$. In conclusion, if the numerical instability conditions hold, then unavoidable (however small) errors will be amplified by successive iterations, while if the numerical stability condition holds they will remain "manageable". $\square$

Example 5.37 (Backward Euler method for the heat equation). Again with reference to Example 1.21 and our notations, an approximated solution of problem (5.22) can be obtained using the backward Euler method, that is, by the numerical scheme[8]

$$\begin{cases} (1 + 2\alpha)\, V_{j,k+1} - \alpha V_{j-1,k+1} - \alpha V_{j+1,k+1} = V_{j,k} \\ V_{0,k+1} = V_{N+1,k+1} = 0 \end{cases} \tag{5.24}$$

In terms of DDS, the numerical scheme (5.24) generates an orbit of the vector-valued DDS $\mathbb{A}\mathbf{V}_{k+1} = \mathbf{V}_k$ starting from $\mathbf{V}_0$, sampling of the initial datum f, with $\mathbb{A} = \mathbb{I}_N + \alpha\mathbb{T}$, where $\mathbb{T}$ is the tridiagonal matrix (5.20), $\alpha = h/s^2$, h is the time-step discretization, s is the spatial one.

$$\mathbb{A} = \begin{bmatrix} 1 + 2\alpha & -\alpha & & & & & \\ -\alpha & 1 + 2\alpha & -\alpha & & & & \\ & -\alpha & 1 + 2\alpha & \ddots & & & \\ & & \ddots & \ddots & \ddots & & \\ & & & \ddots & \ddots & -\alpha & \\ & & & & -\alpha & 1 + 2\alpha & -\alpha \\ & & & & & -\alpha & 1 + 2\alpha \end{bmatrix}$$

Now each decisive step is numerically non-trivial because it corresponds to inverting the tridiagonal matrix $\mathbb{A}$.

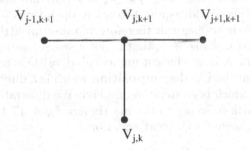

Fig. 5.11 Finite difference discretization for backward Euler method (5.18)

Thanks to the analysis of Example 5.34 we observe that $\mathbb{A}$ is diagonalizable and the eigenvectors and eigenvalues are respectively

$$\mathbf{U}_k\left(\mathbb{A}\right) = \mathbf{U}_k\left(\mathbb{T}\right)   1 \leq k \leq N$$

$$\lambda_k\left(\mathbb{A}\right) = 1 + \alpha\lambda_k\left(\mathbb{T}\right) = 1 + 4\alpha\left(\sin\frac{k\pi}{2\left(N+1\right)}\right)^2   1 \leq k \leq N .$$

Then

$$\lambda_k\left(\mathbb{A}\right) > 1   1 \leq k \leq N, \ \forall \alpha > 0 .$$

Therefore $\mathbb{A}$ is invertible for every $\alpha > 0$, and the eigenvalues of the inverse matrix satisfy $0 < \lambda_k\left(\mathbb{A}^{-1}\right) = \left(\lambda_k\left(\mathbb{A}\right)\right)^{-1} < 1$ for each k.
Theorem 5.8 ensures that for the DDS $\mathbf{V}_{k+1} = \mathbb{A}^{-1}\mathbf{V}_k$ the equilibrium $\mathbf{0}$ is stable and attractive for all $\alpha > 0$.
In this case the rounding error $\mathbf{E}_k$ in the evaluation of $\mathbf{V}_k$ is governed by the DDS $\mathbf{E}_{k+1} = \mathbb{A}^{-1}\mathbf{E}_k$ with initial datum $\mathbf{E}_0$ (rounding error of the initial datum), therefore the implicit Euler scheme is numerically stable for each $\alpha > 0$, ie for each $h > 0$ and $s > 0$.
In summary, we have brought back the solution of the heat equation to a cascade of linear algebraic problems.   □

We observe that $\mathbb{T}$, $\mathbb{A}$, $\mathbb{A}^{-1}$ are positive definite matrices for each α, while $\mathbb{B}$ is definite positive only if $0 < \alpha < 1/4$.
In practice, to solve the system $\mathbb{A}\mathbf{V}_{k+1} = \mathbf{V}_k$, one does not invert the $N \times N$ matrix $\mathbb{A}$, because this would require too many operations (of the order of N^2). Furthermore $\mathbb{A}$ is tridiagonal, therefore it takes up little space in the memory ($3N - 2$ data), unlike $\mathbb{A}^{-1}$, that is not a sparse matrix (N^2 data).
Rather than inverting $\mathbb{A}$ more efficient numerical algorithms can be used. We cite only two of them: the **LU decomposition** which is a direct method, and the **SOR method** which is an iterative method. We illustrate them, for the sake of simplicity, with reference to the matrix $\mathbb{A} = \mathbb{I}_N + \alpha\mathbb{T}$ that appears in the implicit Euler scheme for the heat equation.

Example 5.38 (LU method).   Given the matrix $\mathbb{A} = \mathbb{I}_N + \alpha\mathbb{T}$ with $\mathbb{T}$ defined by (5.20), we look for two matrices, a lower triangular one $\mathbb{L}$, and an upper triangular $\mathbb{U}$ respectively of the form

$$\mathbb{L} = \begin{bmatrix} 1 & 0 & 0 & \cdots & 0 \\ l_1 & 1 & \ddots & & \vdots \\ 0 & \ddots & \ddots & \ddots & 0 \\ \vdots & & \ddots & \ddots & 0 \\ 0 & \cdots & 0 & l_{N-1} & 1 \end{bmatrix} \qquad \mathbb{U} = \begin{bmatrix} y_1 & z_1 & 0 & \cdots & 0 \\ 0 & y_2 & \ddots & & \vdots \\ 0 & \ddots & \ddots & \ddots & 0 \\ \vdots & & \ddots & \ddots & z_{N-1} \\ 0 & \cdots & 0 & 0 & y_N \end{bmatrix}$$

such that the equality $\mathbb{A} = \mathbb{L}\mathbb{U}$ holds. To determine the elements l_i, y_i and z_i of the two matrices, we perform the product $\mathbb{L}\mathbb{U}$: setting the result equal to

$\mathbb{A}$, we obtain:

$$y_1 = (1 + 2\alpha) \tag{5.25}$$

$$y_i = (1 + 2\alpha) - \frac{\alpha^2}{y_{i-1}} \qquad i = 2, \ldots, N$$

$$z_l = -\alpha$$

$$l_i = -\frac{\alpha}{y_i} \qquad i = 1, \ldots, N-1$$

Therefore, the only quantities to calculate and store are the y_i for $i = 2, \ldots, N$, from which we deduce l_i and z_i.

The starting problem $\mathbb{A}\mathbf{V}_{k+1} = \mathbf{V}_k$ may be rewritten in the form $\mathbb{L}(\mathbb{U}\mathbf{V}_{k+1}) = \mathbf{V}_k$: this system is equivalent to the two subproblems

$$\mathbb{U}\mathbf{V}_{k+1} = \mathbf{W}_k \qquad\qquad \mathbb{L}\mathbf{W}_k = \mathbf{V}_k$$

where $\mathbf{W}_k$ is an intermediate vector.

The actual solution of the two algebraic systems is obtained as follows. From the first equation of the system $\mathbb{L}\mathbf{W}_k = \mathbf{V}_k$ the value of $W_{1,k}$ is deduced while the other values of the components of the vector $\mathbf{W}_k$ are determined by the recursive relations

$$W_{j,k} = V_{j,k} + \frac{\alpha W_{j-1,k}}{y_{j-1}} \qquad j = 2, \ldots, N .$$

Finally, by solving the system $\mathbb{U}\mathbf{V}_{k+1} = \mathbf{W}_k$ one immediately determines the value

$$V_{N,k+1} = \frac{W_{N,k+1}}{y_N}$$

and backwards the remaining components of the vector $\mathbf{V}_{k+1}$:

$$V_{j,k+1} = \frac{W_{j,k} + \alpha V_{j+1,k+1}}{y_j} \qquad j = 1, \ldots, N-1 . \qquad \square$$

Example 5.39 (SOR method). The **LU method** illustrated above is a direct method for solving a linear algebraic system $\mathbb{A}\mathbf{V}_{k+1} = \mathbf{V}_k$ in the sense that the unknowns are determined in a single step. Alternatively, one can consider iterative methods.

The SOR (Successive Over-Relaxation) method can be considered a particular version of Gauss-Seidel, which in turn is a development of the Jacobi method. The starting point of all three of these methods is the observation that the system

$$(1 + 2\alpha) V_{j,k+1} - \alpha V_{j-1,k+1} - \alpha V_{j+1,k+1} = V_{j,k}$$

can be rewritten in the equivalent form

$$V_{j,k+1} = \frac{1}{1 + 2\alpha} \left(V_{j,k} + \alpha \left(V_{j-1,k+1} + V_{j+1,k+1} \right) \right) \tag{5.26}$$

This simple rearrangement isolates the diagonal terms of the left side.

The **Jacobi method** starts from an approximated initial value for $V_{j,k+1}$ with $1 \leq j \leq N - 1$, replacing it in the right-hand side of (5.26) to obtain a new approximate value for $V_{j,k+1}$. This process is iterated until the approximation does not change significantly from one iteration to the next: when that happens, one stops the iterations because the solution has been reached. More precisely: if $V_{j,k+1}^m$ is the m-th iterate of $V_{j,k+1}$ and $V_{j,k+1}^0$ is the initial approximation, it is expected that

$$\lim_m V_{j,k+1}^m = V_{j,k+1} \ .$$

Once one has calculated $V_{j,k+1}^m$ one can determine $V_{j,k+1}^{m+1}$ from

$$V_{j,k+1}^{m+1} = \frac{1}{1 + 2\alpha} \left(V_{j,k} + \alpha \left(V_{j-1,k+1}^m + V_{j+1,k+1}^m \right) \right) \tag{5.27}$$

and the process is iterated until the error

$$\left\| \mathbf{V}_{k+1}^{m+1} - \mathbf{V}_{k+1}^m \right\|^2 = \sum_j \left(V_{j,k+1}^{m+1} - V_{j,k+1}^m \right)^2$$

becomes sufficiently small as to justify no further iterations. At this point $V_{j,k+1}^{m+1}$ can be considered the value of $V_{j,k+1}$. It is possible to verify that the Jacobi method converges to the correct solution for each $\alpha > 0$.

The **Gauss-Seidel method** represents a refinement of the Jacobi method. The starting point is the observation that when computing $V_{j,k+1}^{m+1}$ in (5.27) the value $V_{j-1,k+1}^{m+1}$ is known: then this value is used in place of $V_{j-1,k+1}^m$ and therefore (5.27) becomes

$$V_{j,k+1}^{m+1} = \frac{1}{1 + 2\alpha} \left(V_{j,k} + \alpha \left(V_{j-1,k+1}^{m+1} + V_{j+1,k+1}^m \right) \right) . \tag{5.28}$$

Summarizing, we can say that the Gauss-Seidel method uses an approximant as soon as this becomes available, thus increasing, compared to the Jacobi method, the speed of convergence. The greater efficiency of the method of Gauss-Seidel is also due to the fact that the update of iterates occurs immediately by rewriting the single iterate calculated in the previous step[9], while applying the Jacobi method, one must separately store the old iterate since all are involved in the calculation of iterates to the next step. Finally, we observe that even in the case of the Gauss-Seidel method one can prove the convergence for each value of $\alpha > 0$.

The **SOR method** represents a refinement of the Gauss-Seidel method. One starts from the trivial identity

$$V_{j,k+1}^{m+1} = V_{j,k+1}^m + \left(V_{j,k+1}^{m+1} - V_{j,k+1}^m \right) .$$

---

[9] In memory $V_{j-1,k+1}^{m+1}$ is overwritten on $V_{j-1,k+1}^m$ before computing $V_{j,k+1}^{m+1}$.

As $m \rightarrow +\infty$ implies $V_{j,k+1}^m \rightarrow V_{j,k+1}$, one may think that the term $\left(V_{j,k+1}^{m+1} - V_{j,k+1}^m \right)$ represents a corrective term that must be added to $V_{j,k+1}^m$ to make it closer to the real value $V_{j,k+1}$. One can try to over-correct: precisely, $V_{j,k+1}^{m+1}$ is calculated in two steps from relations

$$Y_{j,k+1}^{m+1} = \frac{1}{1+2\alpha} \left(V_{j,k} + \alpha \left(V_{j-1,k+1}^{m+1} + V_{j+1,k+1}^m \right) \right)$$
$$V_{j,k+1}^{m+1} = V_{j,k+1}^m + \omega \left(Y_{j,k+1}^{m+1} - V_{j,k+1}^m \right)$$

where $\omega > 1$ is called **over-relaxation parameter**. Notice that the value $Y_{j,k+1}^{m+1}$ is the one provided by the Gauss-Seidel method for $V_{j,k+1}^{m+1}$: however, with the SOR method the term $Y_{j,k+1}^{m+1} - V_{j,k+1}^m$ is regarded as a correction to be made to $V_{j,k+1}^m$ in order to obtain $V_{j,k+1}^{m+1}$. The SOR method converges to the solution for each value of $\alpha > 0$ provided $0 < \omega < 2$. One can also prove the existence of a single value of the parameter ω between 1 and 2 for which the speed of convergence is maximal. This value depends on the size of the matrix under study and the type of its elements.   □

Now consider a simple but instructive example: the system of linear algebraic equations

$$\mathbb{E}\mathbf{X} = \mathbf{B} \quad \text{where} \quad \mathbb{E} = \begin{bmatrix} 1 & 1 \\ 1 & 1.01 \end{bmatrix} \quad \mathbf{X} - \begin{bmatrix} X_1 \\ X_2 \end{bmatrix} \quad \mathbf{B} = \begin{bmatrix} B_1 \\ B_2 \end{bmatrix}.$$

Since $\det \mathbb{E} \neq 0$ we know that, whatever the choice of $\mathbf{B}$, there exists one and only one solution of the system. Nevertheless the system is **ill-conditioned** in the following sense: to small variations of $\mathbf{B}$ can correspond large variations of the solution.
For instance, if $\mathbf{B} = \begin{bmatrix} 2 & 2.01 \end{bmatrix}^T$ then $\mathbf{X} = \begin{bmatrix} 1 & 1 \end{bmatrix}^T$.
If instead $\mathbf{B} = \begin{bmatrix} 2 & 1.9 \end{bmatrix}^T$ then $\mathbf{X} = \begin{bmatrix} 12 & -10 \end{bmatrix}^T$.
This occurs because (although $\mathbb{E}$ is positive definite and strictly positive) the ratio between the maximum and minimum eigenvalue of $\mathbb{E}$ is very large; in fact, solving the characteristic equation $\lambda^2 - 2.01\lambda + 0.01 = 0$ one obtains the eigenvalues $\lambda_{1.2} = \left(201 \pm \sqrt{40001} \right)/200$, namely:

$$\lambda_1 = 4.9875 \times 10^{-3} \qquad \lambda_2 = 2.005$$

from which we obtain

$$\frac{\lambda_2}{\lambda_1} \sim 400$$

Notice that λ_2 can be estimated without actually calculating it (see Remark 5.20): $2 \leq \lambda_2 \leq 2.01$.
Ill-conditioned problems need to be addressed with appropriate manipulations preliminary to the numerical solution, in order to avoid gross errors. Let us formalize this argument by discussing the sensitivity of the solutions of the

linear system

$$MX = B \tag{5.29}$$

to perturbations of the given vector $\mathbf{B}$. Consider the perturbed problem

$$M(X + \delta X) = (B + \delta B) \tag{5.30}$$

where $\delta\mathbf{B}$ is a perturbation of $\mathbf{B}$ and $\mathbf{X} + \delta\mathbf{X}$ is the solution corresponding to the perturbed datum $\mathbf{B} + \delta\mathbf{B}$. We observe that, thanks to linearity, by (5.29) and (5.30) it follows $M\,\delta\mathbf{X} = \delta\mathbf{B}$.

Theorem 5.40 (Estimate of the relative error due to perturbations of the data in the solution of a linear algebraic system). *If* M *is an invertible matrix, (5.29) and (5.30) imply*

$$\frac{\|\delta\mathbf{X}\|}{\|\mathbf{X}\|} \leq \mathcal{X}(M)\frac{\|\delta\mathbf{B}\|}{\|\mathbf{B}\|}$$

where $\mathcal{X}(M) = \|M\|\,\|M^{-1}\|$ *is called* **condition number** *for the non-singular matrix* M *and* $\|M\| = \max\limits_{\mathbf{X}\neq 0}(\|M\mathbf{X}\| / \|\mathbf{X}\|)$ *is the norm of* M *as a linear transformation in* $\mathbb{R}^n$ *(see Definition 5.31).*

Proof. By (5.29) and (5.30) it follows

$$\delta\mathbf{X} = M^{-1}(B + \delta B) - X = M^{-1}(B + \delta B - B) = M^{-1}\delta B$$

$$\frac{\|\delta\mathbf{X}\|}{\|\mathbf{X}\|} \leq \|M^{-1}\|\frac{\|\delta\mathbf{B}\|}{\|\mathbf{X}\|} \leq \|M^{-1}\|\,\|M\|\frac{\|\delta\mathbf{B}\|}{\|M\|\,\|\mathbf{X}\|} \leq \mathcal{X}(M)\frac{\|\delta\mathbf{B}\|}{\|\mathbf{B}\|}. \qquad \square$$

Remark 5.41. If M is a real symmetric and positive definite matrix, then $\mathcal{X}(M) = \lambda_{\max}(M)/\lambda_{\min}(M)$.

Let us examine the consequences of Theorem 5.40 and Remark 5.41 in some concrete examples.

We observe that the real symmetric tridiagonal matrix of Example 5.34, despite being positive definite, is ill-conditioned:

$$\frac{\lambda_{\max}(\mathbb{T})}{\lambda_{\min}(\mathbb{T})} \sim \left(\frac{N+1}{\pi}\right)^2$$

Instead the matrix $\mathbb{B} = \mathbb{I} - \alpha\mathbb{T}$ is well-conditioned if $0 < \alpha < 1/8$ since in that case

$$0 < \frac{1}{2} < \lambda_{\min}(\mathbb{B}) < \lambda_{\max}(\mathbb{B}) < 1.$$

The matrix $\mathbb{A}^{-1} = (\mathbb{I} + \alpha\mathbb{T})^{-1}$ is well-conditioned if $\alpha > 0$ since in that case

$$0 < \frac{1}{5} < \lambda_{\min}(\mathbb{A}^{-1}) < \lambda_{\max}(\mathbb{A}^{-1}) < 1$$

therefore, the algebraic problem to be solved in the implicit Euler scheme is well-conditioned.

Remark 5.42. Consider the problem $\Delta u = f$ on the square $(0,a) \times (0,a)$ with homogeneous Dirichlet condition ($u = 0$ on the edge of the square) using the formula of the five-point Laplacian (see Exercise 1.16 of which we use the notations). By splitting each side of the square in 101 intervals, one gets a grid at the internal nodes of which $(100 \times 100 = 10\,000)$ a linear relationship must be satisfied. Ultimately, one must solve the linear algebraic system $\mathbb{M}\mathbf{U} = \mathbf{F}$ where $\mathbb{M}$ has $10\,000 \times 10\,000 = 100\,000\,000$ elements. Even if each matrix element only had one bit (in practice there are more) we would have a memory occupation of 100 megabytes only to store the discretization of the data, unless one takes advantage of algorithms suitable for sparse matrices, as is the case of $\mathbb{M}$ or parallel algorithms. It is therefore essential to take full advantage of the matrix structure $\mathbb{M}$ when its size is large as we did with the decomposition **LU** or the **SOR** method.

Remark 5.43. The problem of determining the option price described by the Black and Scholes model can be solved numerically using the methods described in this section for the heat equation after the transformation of Exercise 1.17.

Exercise 5.5. Given the tridiagonal matrix

$$\mathbb{M} = \begin{bmatrix} 5 & 2 & 0 & 0 & \cdots & 0 \\ 2 & 5 & 2 & 0 & \cdots & 0 \\ 0 & 2 & 5 & 2 & \cdots & 0 \\ 0 & 0 & 2 & 5 & \cdots & \vdots \\ \vdots & \vdots & \vdots & \vdots & \ddots & 2 \\ 0 & 0 & 0 & \cdots & 2 & 5 \end{bmatrix}$$

a) estimate (in a simple way!) its dominant eigenvalue;
b) prove that it is positive definite and compute the eigenvalues.

5.9 Summary exercises

Exercise 5.6. Given the matrices

$$\begin{bmatrix} 1 & 0 \\ 1 & 1 \end{bmatrix} \quad \begin{bmatrix} 1 & -1 \\ 1 & 1 \end{bmatrix} \quad \begin{bmatrix} 1 & 2 & 3 \\ 3 & 2 & 1 \\ 1 & 1 & 1 \end{bmatrix} \quad \begin{bmatrix} 1 & 2 & 3 \\ 3 & 2 & 1 \\ 1 & 2 & 2 \end{bmatrix} \quad \begin{bmatrix} 1 & 0 & 0 \\ 3 & 1 & 0 \\ 1 & 2 & 2 \end{bmatrix}$$

determine the Jordan canonical form and the corresponding basis.

Exercise 5.7. Prove that if $\mathbb{M}$ is a 2×2 matrix which admits the decomposition $\mathbb{M} = \mathbb{S} + \mathbb{T}$ with $\mathbb{S}$ semi-simple and $\mathbb{T}$ nilpotent, then $\mathbb{M}^k = \mathbb{S}^k + k\mathbb{T}\mathbb{S}^{k-1}$.

Exercise 5.8. Prove that if $\mathbb{A}$ is a square matrix of order n, and $\lambda = a + ib$ is an eigenvalue of $\mathbb{A}$ (with $a, b \in \mathbb{R}$, $b \neq 0$), then also $\overline{\lambda} = a - ib$ is an eigenvalue of $\mathbb{A}$ and the eigenvectors associated to λ and $\overline{\lambda}$ are complex and conjugate.

For $n = 2$, show that there exists a real basis with respect to which the matrix $\mathbb{A}$ is represented as $\begin{bmatrix} a & -b \\ b & a \end{bmatrix}$, ie a homothety composed with a rotation matrix.

Exercise 5.9. Let us consider a Leslie matrix $\mathbb{L}$ and assume (more realistically than we did in Section 5.5) that the fertility rate of the last age group is zero: $\varphi_n = 0$. In this case Theorem 5.13 does not apply: each power $\mathbb{L}^k$ has the last column of zero entries.

$$\mathbb{L} = \left[\begin{array}{ccccc|c} \blacksquare & \blacksquare & \blacksquare & \blacksquare & \blacksquare & 0 \\ \blacksquare & 0 & 0 & 0 & 0 & 0 \\ 0 & \blacksquare & 0 & 0 & 0 & 0 \\ 0 & 0 & \blacksquare & 0 & 0 & 0 \\ 0 & 0 & 0 & \blacksquare & 0 & 0 \\ \hline 0 & 0 & 0 & 0 & \blacksquare & 0 \end{array}\right].$$

Nevertheless Theorem 5.14 applies.

Try to say something more precise about the dynamics of $\mathbf{Y}_{k+1} = \mathbb{L}\mathbf{Y}_k$ by considering the Leslie matrix $\mathbb{B}$ of order $n - 1$ obtained by by removing the last row and the last column.

Exercise 5.10. If one considers several age classes in the Leslie model, then it is appropriate to set to zero the fertility rate of the youngest class, e.g. $\varphi_1 = 0$, or of some younger classes of the population.

Study the dynamics of the Leslie DDS when: $\varphi_j \geq 0$ $j = 1, \ldots, n$, $\varphi_{n-1} > 0$, $\sigma_j > 0$ $j = 1, \ldots, n - 1$. The worst case from the point of view of the strict positivity of the matrix is suggested (for $n = 6$) by the following matrix:

$$\mathbb{L} = \left[\begin{array}{ccccc|c} 0 & 0 & 0 & 0 & \blacksquare & 0 \\ \blacksquare & 0 & 0 & 0 & 0 & 0 \\ 0 & \blacksquare & 0 & 0 & 0 & 0 \\ 0 & 0 & \blacksquare & 0 & 0 & 0 \\ 0 & 0 & 0 & \blacksquare & 0 & 0 \\ \hline 0 & 0 & 0 & 0 & \blacksquare & 0 \end{array}\right].$$

Exercise 5.11. Compute the dominant eigenvalue (or at least estimate its numerical value) of the following matrices:

$$\mathbb{A} = \begin{bmatrix} 1 & 2 & 3 & 4 \\ 4 & 3 & 2 & 1 \\ 5 & 5 & 0 & 0 \\ 2 & 2 & 3 & 3 \end{bmatrix} \qquad \mathbb{B} = \begin{bmatrix} 1 & 1 & 4 & 2 \\ 1 & 1 & 2 & 2 \\ 2 & 1 & 0 & 2 \\ 2 & 2 & 0 & 0 \end{bmatrix}.$$

Exercise 5.12. Determine the non-recursive expression of the solutions of the DDS $\mathbf{X}_{k+1} = \mathbb{A}\mathbf{X}_k$ with $\mathbf{X}_0 = \begin{bmatrix} 1 & 0 & -1 \end{bmatrix}^T$ where $\mathbb{A} = \begin{bmatrix} 0 & 1 & 0 \\ 0 & 0 & -1 \\ 2 & -1 & 0 \end{bmatrix}$.

6

Markov chains

Markov[1] chains are abstract models which describe processes of extreme importance in the applications of probability theory.

We will consider a finite set of events assuming that they follow one another in dependence on a discrete parameter (which in many cases is time) and occur with a probability that depends only on the event occurred previously.

In a deterministic dynamical system the state at time $k+1$ is uniquely determined by the one at time k. In this chapter instead we study the case where the probability distribution of various events at time $k+1$ is determined from the one at time k. We will assume that this distribution depends uniquely on that at time k. We shall also assume that this dependency is the same for each k.

We will denote by $P(E)$ the probability of realization of an event E; so, $P(E)$ denotes always a real number that belongs to the interval $[0,1]$. For a summary of the elementary identities useful in probability we refer to Appendix C.

6.1 Examples, definitions and notations

Before introducing some formal definitions, we consider some examples.

Let us review the problem of the "gambler ruin", already discussed in Example 1.17 and solved in Exercise 2.36 as a two-steps scalar equation. Here we present a less elementary model in order to address an example already studied, through the techniques of this section.

Example 6.1. Two players $\mathcal{A}$ and $\mathcal{B}$ possess respectively a and b Euros, $a > 0, b > 0$.

$\mathcal{A}$ and $\mathcal{B}$ play heads or tails with a fair coin (the probability p of heads is equal to the probability q of tails: $p = q = 1/2$): the one which wins a round gets an Euro from the opponent. The one which "ruins" the opponent by obtaining all the $a+b$ Euros wins the game. Set $a+b = s$.

---

[1] Andrei A. Markov, 1856-1922. One of the founders of modern probability.

E. Salinelli, F. Tomarelli: *Discrete Dynamical Models.*
UNITEXT – La Matematica per il 3+2 76
DOI 10.1007/978-3-319-02291-8_6, © Springer International Publishing Switzerland 2014

To describe the game at a given round, we consider the number X_k of Euros that player $\mathcal{A}$ owns ($\mathcal{B}$ owns $s - X_k$ Euros) before playing the $(k+1)$-th round.

It is convenient to display the possible outcomes of the game in a diagram: see Fig. 6.1.

To calculate the probability that $\mathcal{A}$ owns $a + 1$ coins after three tosses, we have to assess the probability of each of the possible "paths" ending in three steps in $a + 1$:

$$a \xrightarrow{\mathcal{A} \text{ loses}} a - 1 \xrightarrow{\mathcal{A} \text{ wins}} a \xrightarrow{\mathcal{A} \text{ wins}} a + 1 \ : \text{ probability } \left(\frac{1}{2}\right)^3$$

$$a \xrightarrow{\mathcal{A} \text{ wins}} a + 1 \xrightarrow{\mathcal{A} \text{ loses}} a \xrightarrow{\mathcal{A} \text{ wins}} a + 1 \ : \text{ probability } \left(\frac{1}{2}\right)^3$$

$$a \xrightarrow{\mathcal{A} \text{ wins}} a + 1 \xrightarrow{\mathcal{A} \text{ wins}} a + 2 \xrightarrow{\mathcal{A} \text{ loses}} a + 1 \ : \text{probability } \left(\frac{1}{2}\right)^3 .$$

Since the three paths represent three different stories (disjoint events) the required probability is given by the sum of the probability of each path, that is, by 3/8.

The possible states are the values that X_k can assume during the game, namely $0, 1, \ldots, s$.

It may be helpful to visualize graphically the possible transitions between the various states: with reference to Fig. 6.2, we use oriented arcs to represent the probability of transition from a value to another of the sum owned by $\mathcal{A}$. To each arc corresponds a **transition probability**[2]. Each one of these probabilities depends on the coin but does not depend on the number of rounds played before.

Having assumed that the coin is not rigged and the game is fair, the probability of moving from j to $j + 1$ or from j to $j - 1$ Euros is 1/2, when $0 < j < s$,

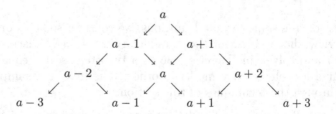

Fig. 6.1 Possible behavior for the capital of $\mathcal{A}$ during three consecutive tosses

---

[2] The reader familiar with probability can observe that the probability of transition from state E to state F is an example of a conditional probability, i.e. the probability of falling into F being sure to start from state E.

Fig. 6.2 Graph that shows the possible transitions and their probabilities

while the probability of moving from 0 to 0 or from s to s Euros is 1 (these last are states that one can reach, but can not get out of).

Compute the probability that $\mathcal{A}$ wins the game. This corresponds to the occurrence of an event among the infinitely many possible (but not all) game developments:

> $\mathcal{A}$ wins in the first round
> $\mathcal{A}$ wins in the second round
> $\mathcal{A}$ wins in the third round
>
> $\vdots$

Each one of these events is realized by a finite number of paths in the graph (this number grows rapidly as the winning-time increases). The probability of each path is given by the product of the probabilities of occurrence of all transitions in the path. The probability of winning is the sum of the probabilities of these events. If $P(j)$, $j = 0, 1, \ldots, s$, denotes the probability of winning starting with j Euros, we obtain

$$P(j) = \frac{1}{2}(P(j-1) + P(j+1)) \qquad j \geq 1 \qquad (6.1)$$

a formula which has a geometric interpretation: each set of three coordinates $(j-1, P(j-1))$, $(j, P(j))$, $(j+1, P(j+1))$ corresponds to three points aligned in the plane. Moreover, from $P(0) = 0$ and $P(s) = 1$ it follows that all the pairs $(j, P(j))$ belong to a straight line passing through the origin:

$$P(j) = \frac{j}{s}.$$

Therefore, the probability of winning for $\mathcal{A}$ (equal to the ruin probability for $\mathcal{B}$), replacing our initial data, is

$$P(a) = \frac{a}{a+b}.$$

Despite the fairness of the game, the probabilities favor the player which holds a more substantial initial capital, as it was already observed. Nevertheless we recall that $\mathcal{A}$, if victorious, wins b Euros, $\mathcal{B}$ instead wins a).

A complete description of the evolution of the game can be obtained by using linear algebra tools to express succinctly all the information contained in Fig. 6.2:

if $\mathbf{P}_k = \begin{bmatrix} P_k(0) & P_k(1) & \ldots & P_k(s) \end{bmatrix}^T$ is the vector of the probabilities $P_k(t)$ to have t Euros at time k, we have an example of Markov chain with $s+1$ states described by

$$\mathbf{P}_{k+1} = \mathbb{M}\mathbf{P}_k \qquad \text{where} \qquad \mathbb{M} = \begin{bmatrix} 1 & 1/2 & 0 & 0 & \cdots & 0 & 0 \\ 0 & 0 & 1/2 & 0 & \cdots & 0 & 0 \\ 0 & 1/2 & 0 & 1/2 & \cdots & 0 & 0 \\ 0 & 0 & 1/2 & \ddots & \ddots & \vdots & \vdots \\ \vdots & \vdots & \vdots & \ddots & 0 & 1/2 & 0 \\ 0 & 0 & 0 & \cdots & 1/2 & 0 & 0 \\ 0 & 0 & 0 & \cdots & 0 & 1/2 & 1 \end{bmatrix}.$$

With this notation, if one starts with j Euros, then the probability of winning at time k, namely $P_k(s)$, is the last component of the vector $\mathbb{M}^k \mathbf{P}_0$, where $\mathbf{P}_0 = \begin{bmatrix} 0 & \cdots & 1 & \cdots & 0 \end{bmatrix}^T$, whose unique nonzero component is the j-th. $\qquad\square$

Example 6.2. Two urns contain colored marbles. The first contains 2 white marbles and 3 black ones, the second urn contains 4 white marbles and 5 black ones. A ball is drawn from the first urn, its color is recorded and it is reinserted in its urn: if white the subsequent drawing is carried out from the first urn, if black, from the second one. Then the process is iterated.

One asks what the probability is to draw a white marble at the fifth draw.

This is an example of a two-state Markov chain: a state is the draw of a white marble, the other is the draw of a black marble. If the k-th drawn marble is white, then at the $(k+1)$-st draw the probability to draw white is $2/5$, the one to draw black is $3/5$. If the k-st marble is black, then at the $(k+1)$-th draw the probability of drawing white is $4/9$, black is $5/9$.

To solve the problem, we consider all the possible cases and define P_k as the probability that the k-th draw is white and $Q_k = 1 - P_k$ the probability that it is black.

Since we extract from the first urn, we know that $P_1 = 2/5$ and $Q_1 = 3/5$. If the k-th draw is white, then the probability that the $(k+1)$-th one is still white is $2/5$; if the k-th draw is black, then the probability that $(k+1)$-th is white is $4/9$. Since they are disjoint events whose union gives all the cases in which one can get white at the $(k+1)$-th draw, one has:

$$P_{k+1} = \frac{2}{5}P_k + \frac{4}{9}Q_k \tag{6.2}$$

and remembering that $P_k + Q_k = 1$

$$P_{k+1} = \frac{4}{9} - \frac{2}{45}P_k.$$

It is a linear equation with constant coefficients. By (2.4) and paying attention to the fact that we start from $k = 1$, we obtain the general solution:

$$P_k = \frac{20}{47} + \left(-\frac{2}{45}\right)^{k-1}(P_1 - 20/47), \qquad k \in \mathbb{N}\setminus\{0\}.$$

Then $\lim_k P_k = 20/47 \in \mathbb{R}$, that is P_k tends to the equilibrium.

Alternatively, one could proceed with the same rules but starting by drawing from the second urn. However, the previous argument proves that the choice of the first urn and the result of the first draw does not influence the probability P_k for "large" values of k. Hence, if we had started from the second urn not much would have changed: P_k tends to the same value L whatever the urn from which one starts. However $P_1 = 2/5$ if one starts from the first urn, and $P_1 = 4/9$ if one starts from the second urn.

If we set $\mathbf{P}_k = \begin{bmatrix} P_k \ Q_k \end{bmatrix}^T$ we can translate (6.2) as:

$$\mathbf{P}_{k+1} = \mathbb{M}\mathbf{P}_k \qquad k \geq 1 \qquad \text{where} \qquad \mathbb{M} = \begin{bmatrix} 2/5 & 4/9 \\ 3/5 & 5/9 \end{bmatrix}.$$

If there had been marbles with more than two colors, we would have had a Markov chain with more than two states and we would have used matrices of order equal to the number of colors. ◻

In general, a **Markov chain** is a model for describing a series of experiments, in each of which the possible results (for instance, draw of a white or black marble) are the same. However, the probability of the same event (for instance, white marble) depends only on the outcome of the previous experiment. When the parameter is time, it is customary to say that the Markov chains describe phenomena that depend on the past only through the present. At this point we give a formal definition. From now on the parameter k that describes the process is always considered as time.

Definition 6.3. *A **finite and homogeneous Markov chain**[3] is a process described by a finite set $\Omega = \{s_1, s_2, \ldots, s_n\}$ of distinct states (or events) s_i such that for every ordered pair of states s_j, s_i, the **transition probability** m_{ij} from state s_j to state s_i is given and it is independent of k.*
Necessarily the transition probability of a Markov chain fulfills the relations:

$$\boxed{0 \leq m_{ij} \leq 1 \quad i,j = 1,\ldots,n; \qquad \sum_{i=1}^{n} m_{ij} = 1 \quad \forall j = 1,\ldots,n.}$$ (6.3)

The probability of visiting state s_i at time $k+1$ starting from state s_j at time k, depends only on s_j, that is on the state visited at time k and not on the

---

[3] We will study only the homogeneous case where $\mathbb{M}$ is constant. In the non-homogeneous cases we have $\mathbb{M} = \mathbb{M}(k)$.

states that are visited before k:

$$m_{ij} = P\left(X_{k+1} = s_i \mid X_k = s_j, \ X_{k-1} = s_{j_{k-1}}, \ \ldots, \ X_1 = s_{j_1}, \ X_0 = s_{j_0}\right) =$$
$$= P\left(X_{k+1} = s_i \mid X_k = s_j\right).$$

As done for Example 6.1 in Fig. 6.2, a Markov chain can be described and visualized by an oriented graph, namely a set of vertices (the events) and arrows (connecting events). At each vertex A connected by an arrow to another vertex (including A itself) and to each arrow is associated a positive probability of transition, whereas paths corresponding to zero transition probability are not traced. For a more detailed discussion of graphs, see Chap. 7.

From the algebraic point of view, it is convenient to represent a Markov chain by a **transition matrix** (or **stochastic matrix**) $\mathbb{M} = [m_{ij}]$ whose elements are the transition probabilities m_{ij} from state s_j to state s_i.

Relation (6.3) ensures that $\mathbb{M}$ is a positive matrix and the elements of each column sum up to 1.

Lemma 6.4. *Let* $\mathbb{M} = [m_{ij}]$ *be a positive matrix of dimension* n. *The following conclusions hold:*

(1) $\mathbb{M}$ is stochastic if and only if:

1 *is an eigenvalue of* $\mathbb{M}^T$ *with associated eigenvector* $\mathbf{1} = \begin{bmatrix} 1 \ 1 \ \ldots \ 1 \end{bmatrix}^T$;

(2)   if $\mathbb{M}$ *is stochastic, then each eigenvalue* λ_s *fulfills* $|\lambda_s| \leq 1$.

Proof. (1) It is sufficient to observe that the stochasticity condition for a positive matrix $\mathbb{M}$ is equivalent to $\mathbb{M}^T \cdot \mathbf{1} = \mathbf{1}$ and to remember that $\mathbb{M}$ and $\mathbb{M}^T$ have the same eigenvalues.

(2) If $\mathbf{V}^s$ is an eigenvector associated to λ_s then:

$$|\lambda_s| \sum_{i=1}^{n} |V_i^s| = \sum_{i=1}^{n} |\lambda_s V_i^s| = \sum_{i=1}^{n} \left| \sum_{j=1}^{n} m_{ij} V_j^s \right| \leq \sum_{i=1}^{n} \sum_{j=1}^{n} m_{ij} |V_j^s| =$$

$$= \sum_{j=1}^{n} \left(\sum_{i=1}^{n} m_{ij} \right) |V_j^s| = \sum_{j=1}^{n} |V_j^s|.$$

Since $\sum_{r=1}^{n} |V_r^s| \neq 0$ we obtain the conclusion. $\square$

Notice that a stochastic matrix can admit more eigenvalues of unit modulus. In addition, there may also be double, zero or negative eigenvalues, as it happens for the following stochastic matrices:

$$\begin{bmatrix} 1 & 0 \\ 0 & 1 \end{bmatrix} \qquad \begin{bmatrix} 0 & 1 \\ 1 & 0 \end{bmatrix} \qquad \begin{bmatrix} 1/2 & 1/2 \\ 1/2 & 1/2 \end{bmatrix}$$

There may also be complex eigenvalues if $n \geq 3$ (see Exercise 6.13).[4]

---

[4] If $n = 2$ and $\mathbb{M} = \begin{bmatrix} a & b \\ c & d \end{bmatrix}$, with $a, b, c, d \geq 0$, the characteristic polynomial $\mathcal{P}(\lambda) = \lambda^2 - (a+d)\lambda + (ad - bc)$ has discriminant $(a - d)^2 + 4bc \geq 0$. There-

Let $\mathbb{M} = [m_{ij}]$ be a stochastic matrix and $\mathbf{P}_k = \begin{bmatrix} P_k^1 & P_k^2 & \cdots & P_k^n \end{bmatrix}^T$ the vector of probabilities of each state at time k (P_k^j is the probability to observe state s_j at time k). For every $k \in \mathbb{N}$:

$$0 \le P_k^j \le 1, \ j = 1, \ldots, n; \qquad \sum_{j=1}^{n} P_k^j = 1 \quad \forall k \in \mathbb{N}. \qquad (6.4)$$

From the assumptions, for each $k \in \mathbb{N}$

$$P_{k+1}^j = P_k^1 m_{j1} + P_k^2 m_{j2} + \cdots + P_k^n m_{jn}$$

therefore the n difference equations that describe the probability of the n admissible states at time $k + 1$ as functions of the corresponding probability of the previous time, constitute a one-step vector-valued linear homogeneous discrete dynamical system[5]:

$$\boxed{\mathbf{P}_{k+1} = \mathbb{M}\mathbf{P}_k \qquad \forall k \in \mathbb{N}}$$

$$(6.5)$$

Lemma 6.5. *Suppose* $\mathbb{M}$ *is stochastic and* $\mathbf{P}$ *is a sequence that satisfies (6.5). If (6.4) holds for a fixed value* $\tilde{k}$, *then (6.4) holds true for all* $k > \tilde{k}$.

Proof. Relation (6.5) says that $\mathbf{P}_{k+1}$ is a linear combination of the columns of $\mathbb{M}$ whose coefficients are the components of $\mathbf{P}_k$.
From $m_{ij} \ge 0$ and $P_k^j \ge 0 \quad \forall i, j$ it follows $P_{k+1}^j \ge 0 \quad \forall j$. Furthermore

$$\sum_{j=1}^{n} P_{k+1}^j = \sum_{j=1}^{n}\left(\sum_{h=1}^{n} m_{jh} P_k^h \right) = \sum_{h=1}^{n}\left(\sum_{j=1}^{n} m_{jh} P_k^h \right) = \sum_{h=1}^{n}\left(\sum_{j=1}^{n} m_{jh} \right) P_k^h =$$

$$= \sum_{h=1}^{n} P_k^h = 1.$$

It follows

$$P_{k+1}^j \le 1 \quad \forall j.$$

The validity of (6.4) has been proven for $k = \tilde{k} + 1$. The general case ($k > \tilde{k}$) follows by induction. □

In summary, the lemma just proved states that a stochastic matrix transforms a **probability distribution vector** into another probability distribution vector. In general, a Markov chain will be described by the vector-valued

---

fore, all the eigenvalues are real. Moreover, by Descartes's rule, they are positive if $ad > bc$, one positive and one zero if $ad = bc$, one positive and one negative if $ad < bc$.

[5] **Caution**: in some treatises it is assumed that m_{ij} represents the transition probability from i to j and then $P_{k+1}^j = \sum_{h=1}^{n} m_{hj} P_k^h$ that is $\mathbf{P}_{k+1} = \mathbb{M}^T \mathbf{P}_k$; in such cases the sum of the elements of each row of $\mathbb{M}$ is 1.

DDS (6.5) where $\mathbb{M}$ is a stochastic matrix and with initial datum which is a probability distribution (**initial probability distribution**), i.e.

$$0 \leq P_0^j \leq 1, \quad j = 1, \ldots, n; \qquad \sum_{j=1}^{n} P_0^j = 1 . \qquad (6.6)$$

The dynamics of this system takes place in the proper subset S of $\mathbb{R}^n$:

$$S = \left\{ \mathbf{V} \in \mathbb{R}^n : V^j \geq 0 \; \forall j, \; \sum_{j=1}^{n} V^j = 1 \right\},$$

that is, all orbits are contained in S.
S is called an n-**simplex**. S is convex, closed and bounded, and coincides with the set of all the probability distributions on a set of n states. Lemma 6.5 can be reformulated as follows:

the simplex S is *invariant* for the dynamics of the vector-valued DDS (6.5) in the sense that if $\mathbf{P} \in S$ then $\mathbb{M}\mathbf{P} \in S$.

We will study in detail the dynamics of system (6.5) restricted to S.
One-step vector-valued dynamical systems as (6.5) have been studied in Chap. V. Their explicit solution is:

$$\boxed{\mathbf{P}_k = \mathbb{M}^k \mathbf{P}_0 \qquad \forall k \in \mathbb{N}}$$

Since $\{s_1, \ldots, s_n\}$ describes a disjoint collection of all possible states, the generic element $m_{ij}^{(k)}$ of $\mathbb{M}^k$ represents the **transition probability from state j to state i in k steps**, namely:

$$m_{ij}^{(k)} = P\left(X_{n+k} = s_i \mid X_n = s_j\right) . \qquad (6.7)$$

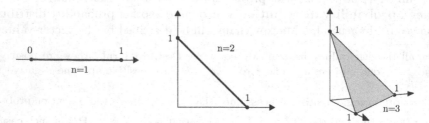

Fig. 6.3 n simplices, $n = 1, 2, 3$

In general, for all k and h in $\mathbb{N}$, the following relationships hold:

$$\text{(Chapman-Kolmogorov equations)} \qquad m_{ij}^{(k+h)} = \sum_{r=1}^{n} m_{ir}^{(k)} m_{rj}^{(h)}.$$

In the context of Markov chains we introduce a probabilistic description for the equilibrium of the corresponding vector discrete dynamical system: the invariant probability distribution.

Definition 6.6. *Given a finite Markov chain on a set of n states with transition matrix* $\mathbb{M}$, *a vector* $\mathbf{P} = \begin{bmatrix} P^1 \ P^2 \ \dots \ P^n \end{bmatrix}^T$ *is called **invariant probability distribution** if*

$$\boxed{\quad \mathbb{M}\mathbf{P} = \mathbf{P}, \qquad P^j \geq 0 \qquad and \qquad \sum_{j=1}^{n} P^j = 1 \quad}$$

As already mentioned, a Markov chain with n states is a linear vector-valued discrete dynamical system, whose orbits (and initial data) are constrained to belong to the n-simplex S in $\mathbb{R}^n$. From this point of view, an invariant probability distribution is an equilibrium: the associated stationary orbit is characterized by the fact that the probability of finding the system in a given state is not time dependent.

Example 6.7. If $\mathbb{M} = \begin{bmatrix} 0 & 1 \\ 1 & 0 \end{bmatrix}$ then the eigenvalues are ± 1, and $\begin{bmatrix} 1/2 \ 1/2 \end{bmatrix}^T$ is the (unique) invariant probability distribution. Furthermore if $\mathbf{P}_{k+1} = \mathbb{M}\mathbf{P}_k$, then $\begin{bmatrix} 1/2 \ 1/2 \end{bmatrix}^T$ is an equilibrium (unique in the 2-simplex of the probability distributions on 2 states), and starting from $\mathbf{P}_0 = \begin{bmatrix} 1 \ 0 \end{bmatrix}^T$, $\mathbf{P}_k$ evolves according to a 2 periodic orbit : $\mathbf{P}_{2k} = \mathbf{P}_0$, $\mathbf{P}_{2k+1} = \begin{bmatrix} 0 \ 1 \end{bmatrix}^T$. $\qquad\square$

Remark 6.8. If a stochastic matrix $\mathbb{M}$ admits an eigenvector $\mathbf{W} > \mathbf{0}$ corresponding to the eigenvalue 1, then $\mathbf{W}/\sum_j W_j$ is an invariant probability distribution for $\mathbb{M}$.

When 1 is a simple eigenvalue of the transition matrix $\mathbb{M}$ and there are no other eigenvalues of modulus one, the asymptotic analysis of the corresponding Markov chain is completely described by the following theorem.

Theorem 6.9. *Let* $\mathbb{M}$ *be a stochastic matrix,* $\mathbf{V}^1, \dots, \mathbf{V}^n$ *a basis of eigenvectors (or a Jordan basis of generalized eigenvectors, possibly) of* $\mathbb{M}$ *to which correspond the eigenvalues* $\lambda_1, \dots, \lambda_n$ *verifying* $\lambda_1 = 1$, $|\lambda_j| < 1$, $\forall j \neq 1$. *Then it is possible to choose* $\mathbf{V}^1 > \mathbb{O}$ *and, therefore, it is not restrictive to assume* $\mathbf{V}^1$ *is a probability distribution*

$$\sum_j V_j^1 = 1 \tag{6.8}$$

a choice that we will make in the following.

In addition, each initial probability distribution $\mathbf{P}_0$ can be represented as

$$\mathbf{P}_0 = \sum_{j=1}^{n} c_j \mathbf{V}^j \qquad with \ \ c_1 = 1. \tag{6.9}$$

The corresponding solution $\mathbf{P}_k$ of the dynamical system (6.5) verifies

$$\mathbf{P}_k = \mathbf{V}^1 + c_2 \lambda_2^k \mathbf{V}^2 + \cdots + c_n \lambda_n^k \mathbf{V}^n \tag{6.10}$$

if $\mathbf{V}^1, \ldots, \mathbf{V}^n$ are all eigenvectors (or in general, if in the basis there are also generalized eigenvectors,

$$\mathbf{P}_k = \mathbf{V}^1 + \mathbf{W}_k \tag{6.11}$$

where $\mathbf{W}_k$ is a finite sum of terms with growth at most $k^{m_j-1}\lambda_j^k$, where $|\lambda_j| < 1$, m_j is the geometric multiplicity of λ_j) and the coefficients are independent of k. Furthermore:

$$\lim_k \mathbf{P}_k = \mathbf{V}^1. \tag{6.12}$$

Proof. Given $\mathbf{P}_0$, for suitable (unique) c_j's we have $\mathbf{P}_0 = \sum_{j=1}^{n} c_j \mathbf{V}^j$. Taking into account Remark 5.9, we obtain representations (6.10) and (6.11), that imply

$$\lim_k \mathbf{P}_k = c_1 \mathbf{V}^1, \qquad hence \ \ c_1 V_j^1 \geq 0 \ \ \forall j. \tag{6.13}$$

Since $\sum_{h=1}^{n} P_k^h = 1$, $\forall k$, we obtain

$$1 = \lim_k \sum_{h=1}^{n} P_k^h = c_1 \sum_{j=1}^{n} V_j^1$$

i.e. $c_1 \neq 0$, $\sum_{h=1}^{n} V_h^1 \neq 0$ and

$$c_1 = \frac{1}{V_1^1 + V_2^1 + \cdots + V_n^1}. \tag{6.14}$$

By (6.4) and (6.13), it follows that all the components of $c_1 \mathbf{V}^1$ are non-negative with at least one strictly positive, therefore $\mathbf{V}^1$ can be chosen as a probability distribution; it follows $c_1 = 1$ and (6.12). $\qquad\square$

We have thus proved that every initial probability distribution $\mathbf{P}_0$ has a non-trivial component along $\mathbf{V}^1$. More precisely, the component of $\mathbf{P}_0$ along $\mathbf{V}^1$ coincides with $\mathbf{V}^1$ if this eigenvector is selected (this is always possible) as a probability distribution (compare this conclusion with the one of Theorem 5.19 valid in the case of vector-valued DDS whose dynamics is not confined to the simplex S of the probability distributions).

Remark 6.10. The information $c_1 = 1$ in Theorem 6.9 may at first sight seem surprising. It actually becomes quite natural when one considers that, for each stochastic matrix M, the uniform probability distribution $[1/n \, 1/n \cdots 1/n]^T$ is an eigenvector corresponding to the eigenvalue 1 for M^T, and in the present case this eigenvalue is simple for M^T because it is simple for M. Therefore, since every initial probability distribution verifies $\mathbf{P}_0 > \mathbf{0}$, it is possible to repeat the argument in the proof of Theorem 5.19 (without using the Perron-Frobenius Theorem which was used only for the simplicity of the eigenvalue of maximum modulus, which here is equal to 1 and is simple by assumption) getting that all the generalized eigenvectors $\mathbf{V}^j$ different from $\mathbf{V}^1$ are orthogonal to $[1/n \, 1/n \cdots 1/n]^T$ which in turn is orthogonal to the simplex of all probability distributions.

The situation described in Theorem 6.9 (simple dominant eigenvalue equal to 1 associated to a positive eigenvector) is particularly simple and well described. It is useful to know how to identify it in general.

Definition 6.11. *A **regular Markov chain** is a Markov chain with transition matrix M for which there exists a k such that M^k is strictly positive (see Definition 5.11).*

Therefore, a Markov chain is regular if there exists a positive integer k such that the transition probability in k steps between two states is positive.

Example 6.12. The stochastic matrix

$$M = \begin{bmatrix} 0 & 1/2 \\ 1 & 1/2 \end{bmatrix}$$

is regular because

$$M^2 = \begin{bmatrix} 0 & 1/2 \\ 1 & 1/2 \end{bmatrix} \begin{bmatrix} 0 & 1/2 \\ 1 & 1/2 \end{bmatrix} = \begin{bmatrix} 1/2 & 1/4 \\ 1/2 & 3/4 \end{bmatrix}.$$

The stochastic matrix

$$M = \begin{bmatrix} 1 & 1/2 \\ 0 & 1/2 \end{bmatrix}$$

is not regular; in fact

$$M^2 = \begin{bmatrix} 1 & 1/2 \\ 0 & 1/2 \end{bmatrix} \begin{bmatrix} 1 & 1/2 \\ 0 & 1/2 \end{bmatrix} = \begin{bmatrix} 1 & 3/4 \\ 0 & 1/4 \end{bmatrix}$$

and assuming

$$M^{k-1} = \begin{bmatrix} 1 & p_{k-1} \\ 0 & 1 - p_{k-1} \end{bmatrix}$$

one obtains

$$M^k = M \cdot M^{k-1} = \begin{bmatrix} 1 & 1/2 \\ 0 & 1/2 \end{bmatrix} \begin{bmatrix} 1 & p_{k-1} \\ 0 & 1 - p_{k-1} \end{bmatrix} = \begin{bmatrix} 1 & (p_{k-1} + 1)/2 \\ 0 & (1 - p_{k-1})/2 \end{bmatrix}. \qquad \square$$

The following theorem helps to recognize whether a stochastic matrix is regular.

Theorem 6.13. *If a Markov chain is regular, then its transition matrix* $\mathbb{M}$ *has exactly one eigenvalue equal to 1 while all the others have modulus strictly less than 1. The eigenvector associated to this dominant eigenvalue is strictly positive.*

Proof. We know (see Lemma 6.4) that 1 is an eigenvalue of $\mathbb{M}$ and that there are no eigenvalues of modulus greater than 1. By the Perron-Frobenius Theorem (see Theorem 5.12 and 5.13) if the chain is regular, then $\mathbb{M}$ has a positive and simple dominant eigenvalue which necessarily is equal to 1. The corresponding eigenvector, by the same theorem, is strictly positive hence there exists a strictly positive multiple the sum of whose components is 1 (i.e. it is a probability distribution). $\square$

Theorems 6.9 and 6.13 show that the DDS associated to regular Markov chains have remarkable qualitative and quantitative properties: the general framework will be summarized in Theorem 6.39.

Example 6.14. The stochastic matrix

$$\mathbb{M} = \begin{bmatrix} 1 & 0 & 1/3 \\ 0 & 1 & 1/3 \\ 0 & 0 & 1/3 \end{bmatrix}$$

is not regular, having eigenvalues $\lambda_1 = \lambda_2 = 1$ and $\lambda_3 = 1/3$. $\square$

Definition 6.15. *An **absorbing state** of a Markov chain with n states is a state s_j such that $m_{jj} = 1$.*

Explicitly, a state s_j is absorbing if its occurrence at time k guarantees that one can get s_j at time $k+1$.

Remark 6.16. If s_j is an absorbing state, the j-th column of the transition matrix $\mathbb{M}$ is of the type

$$\begin{bmatrix} 0 \dots 0 \ 1 \ 0 \dots 0 \end{bmatrix}^T \tag{6.15}$$

↑
j-th position

Moreover, to s_j corresponds an eigenvalue of $\mathbb{M}$ equal to 1 that has (6.15) as associated eigenvector. This eigenvector is an invariant probability distribution.

Definition 6.17. *An **absorbing Markov chain** is a Markov chain that verifies the two following conditions:*

- *there exists at least an absorbing state;*
- *from each state one can reach an absorbing state, i.e. for each j there exists i and an integer $k > 0$ such that s_i is absorbing and $m_{ij}^{(k)} > 0$.*

Example 6.18. The following matrix

$$M = \begin{bmatrix} 0 & 0 & 0 \\ 1 & 0 & 0 \\ 0 & 1 & 1 \end{bmatrix}$$

is associated to an absorbing Markov chain.   □

Example 6.19. For the Markov chain with transition matrix

$$M = \begin{bmatrix} 1 & 0 & 0 \\ 0 & 1/2 & 1/2 \\ 0 & 1/2 & 1/2 \end{bmatrix}$$

the first state is absorbing, but there are no transitions from the first state to the second or third ones, or vice versa: the corresponding Problem 6.24 is *decoupled*. Hence the Markov chain is not absorbing.   □

Example 6.20. The Markov chain with transition matrix

$$M = \begin{bmatrix} 1 & 0 & 1/2 & 1/3 \\ 0 & 1 & 1/2 & 2/3 \\ 0 & 0 & 0 & 0 \\ 0 & 0 & 0 & 0 \end{bmatrix}$$

is absorbing.   □

Example 6.21. Let us consider the absorbing Markov chain with 5 states (the first two are absorbing) and transition matrix:

$$M = \begin{bmatrix} 1 & 0 & 0.2 & 0 & 0.1 \\ 0 & 1 & 0.2 & 0.3 & 0.2 \\ 0 & 0 & 0.2 & 0.3 & 0.3 \\ 0 & 0 & 0.2 & 0.4 & 0.2 \\ 0 & 0 & 0.2 & 0 & 0.2 \end{bmatrix}.$$

Notice that one can can stay out of states 1 and 2 for an arbitrarily long time, but in each evolution, for each k, at every state the probability to run into state 1 or state 2 is positive.   □

Example 6.22. Let us consider the non-absorbing Markov chain with four states, the first one of which is absorbing, and transition matrix:

$$M = \begin{bmatrix} 1 & 0.2 & 0 & 0 \\ 0 & 0.2 & 0 & 0 \\ 0 & 0.3 & 0.5 & 0.5 \\ 0 & 0.3 & 0.5 & 0.5 \end{bmatrix}.$$

In this case, if one passes in the third or fourth state, the first one becomes inaccessible.   □

Remark 6.23. Notice that the presence of even a single absorbing state in a Markov chain with at least 2 states ensures that the chain can not be regular: in fact, the corresponding transition matrix $\mathbb{M}$ is not strictly positive because there are null elements in the column that corresponds to the absorbing state. Then, every integer power of $\mathbb{M}$ will present at least a null element since the above mentioned column is an eigenvector of $\mathbb{M}$.

Exercise 6.1. Determine which of the Markov chains associated with the following transition matrices are regular:

$$
\mathbb{M}_1 = \begin{bmatrix} 1/2 & 0 & 1/5 \\ 0 & 1 & 2/5 \\ 1/2 & 0 & 2/5 \end{bmatrix} ; \quad
\mathbb{M}_2 = \begin{bmatrix} 1/2 & 0 & 1/5 \\ 0 & 0 & 2/5 \\ 1/2 & 1 & 2/5 \end{bmatrix} ; \quad
\mathbb{M}_3 = \begin{bmatrix} 1/2 & 0 & 0 \\ 0 & 1/3 & 3/5 \\ 1/2 & 2/3 & 2/5 \end{bmatrix} .
$$

Exercise 6.2. Determine to which transition matrices of the previous exercise correspond absorbing Markov chains.

6.2 Asymptotic analysis of models described by absorbing Markov chains

In this paragraph we will illustrate some tools for the qualitative study of Markov chains.

The recursive law $\mathcal{P}_{k+1} = \mathbb{M}\mathcal{P}_k$ that defines a Markov chain (that is a vector-valued linear homogeneous DDS) allows us to calculate the probabilities of the various states at time $k+1$ when their probability at time k is known (we recall that, $\forall k \in \mathbb{N}$, $\mathcal{P}_k$ is the vector with n components that determines the probability distribution corresponding to the n states of the system). In these situations it is interesting for the applications to solve the following problem:

Problem 6.24. Given an initial state s_i or, more in general, a probability distribution $\mathcal{P}_0$ for the initial state[6], compute or at least estimate the probability that the evolution is concluded in a predetermined final state s_j, that is once the state is reached it is maintained forever (whatever the history is which leads to that state).

We study Problem 6.24 only for absorbing chains: as will be evident from Theorem 6.26, for an absorbing chain Problem 6.24 is meaningful only if the final state s_j is absorbing. In fact, in this case the probability of remaining indefinitely in a non-absorbing state s_j is zero.

We will denote with $\mathcal{P}$ the sequence of probability vectors: $\mathcal{P}_{k+1} = \mathbb{M}\mathcal{P}_k$ where $\mathcal{P}_0$ is assigned and $\mathbb{M}$ has order n.

---

[6] Prescribing the initial state s_i, $i \in \{1, 2, \ldots, n\}$, coincides with the assignment of the initial probability distribution $\mathbf{P}_0 = \begin{bmatrix} 0 & \ldots & 0 & 1 & 0 & \ldots & 0 \end{bmatrix}^T$ where the unique non-trivial component is the i-th one.

To solve the problem one could determine all the possible evolutions that, starting from $\mathcal{P}_0$, lead to s_j, calculate the corresponding probabilities and then add them. This technique is practically prohibitive.

However, if $\mathcal{P}_0$ is assigned, it is possible to prove the existence of

$$\mathcal{P}_\infty = \lim_k \mathcal{P}_k$$

and to calculate it, and the problem is solved by the value of the j-th component of $\mathcal{P}_\infty$.

Example 6.25. If

$$\mathbb{M} = \begin{bmatrix} 1/2 & 1/2 \\ 1/2 & 1/2 \end{bmatrix}$$

then there are no absorbing states. However, there is an invariant probability distribution (see Definition 6.6) which is strictly positive:

$$\mathcal{P} = \begin{bmatrix} 1/2 & 1/2 \end{bmatrix}^T.$$

Since $\mathbb{M}^k = \mathbb{M}$ for each $k \geq 1$, the associated dynamics is trivial: $\mathcal{P}_k = \mathbb{M}^k \mathcal{P}_0 = \mathcal{P}$ for each $k \geq 1$. Observe that, whatever the initial distribution is, the event "ultimately staying in a given state" has null probability.   □

We start by saying that if $\mathbb{M}$ is the transition matrix of an absorbing Markov chain, then it is not restrictive to assume that the absorbing states are the first m $(1 \leq m < n)$, possibly rearranging the states (by a coordinate change in $\mathbb{R}^m$ arising from a permutation of the basis, see Exercises 6.6 and 6.7). The probability vector $\mathcal{P}_k$ can be decomposed into two vectors $\mathbf{P}_k$ and $\mathbf{Q}_k$, where $\mathbf{P}_k$ corresponds to the first m components of $\mathcal{P}_k$ (probabilities of the absorbing states), $\mathbf{Q}_k$ corresponds to the last h components of $\mathcal{P}_k$ (probabilities of the non-absorbing states). So, if we set $h = n - m \geq 1$, we have

$$\mathbb{M} = \begin{bmatrix} \mathbb{I}_m & \mathbb{B} \\ \mathbb{O}_{h,m} & \mathbb{A} \end{bmatrix} \qquad \mathcal{P}_k = \begin{bmatrix} \mathbf{P}_k \\ \mathbf{Q}_k \end{bmatrix} \tag{6.16}$$

where $\mathbb{I}_m$ is the identity matrix of order m, $\mathbb{O}_{h,m}$ is the null matrix of order $h \times m$, $\mathbb{A}$ is square of order h and can coincide with the null matrix $\mathbb{O}_h$, while $\mathbb{B}$ (of order $m \times h$) can not be the null matrix because the chain is absorbing.

Theorem 6.26. *Let $\mathbb{M}$ be the $n \times n$ stochastic matrix of an absorbing Markov chain with m absorbing states and decomposed as in (6.16).*
If one starts from a non-absorbing state or, more in general, from a probability distribution $\mathcal{P}_0 = \begin{bmatrix} \mathbf{P}_0 & \mathbf{Q}_0 \end{bmatrix}^T$, then there exists the probability distribution $\mathcal{P}_\infty = \begin{bmatrix} \mathbf{P}_\infty & \mathbf{Q}_\infty \end{bmatrix}^T$ where

$$\mathbf{P}_\infty = \lim_k \mathbf{P}_k \qquad \mathbf{Q}_\infty = \lim_k \mathbf{Q}_k = \mathbf{0}.$$

More precisely, if $h = n - m$ is the number of non-absorbing states, one obtains

$$\mathbf{P}_k = \mathbf{P}_0 + \mathbb{B}\left(\mathbb{I}_h - \mathbb{A}^k\right)\left(\mathbb{I}_h - \mathbb{A}\right)^{-1}\mathbf{Q}_0 \qquad \mathcal{P}_k = \begin{bmatrix} \mathbf{P}_k & \mathbb{A}^k\mathbf{Q}_0 \end{bmatrix}^T$$

$$\mathbf{P}_\infty = \mathbf{P}_0 + \mathbb{B}\left(\mathbb{I}_h - \mathbb{A}\right)^{-1}\mathbf{Q}_0 \qquad\qquad \mathcal{P}_\infty = \begin{bmatrix} \mathbf{P}_\infty & \mathbf{0} \end{bmatrix}^T$$

In particular, starting from $\begin{bmatrix} \mathbf{P}_0 & \mathbf{Q}_0 \end{bmatrix}^T$

- *the probability to remain eventually[7] in a non-absorbing state is zero;*
- *the probability to remain eventually[7] in an absorbing state s_j is equal to the j-th component of $\mathbf{P}_\infty$;*
- *if $\mathbb{A} = \mathbb{O}$, the dynamic is trivial: $\mathbf{P}_k = \mathbf{P}_0 + \mathbb{B}\mathbf{Q}_0 \quad k \geq 1, \quad \mathbf{Q}_k = \mathbf{0} \quad k \geq 1$.*

Before to prove the theorem, it is appropriate to introduce a definition and a lemma.

Definition 6.27. *A square matrix $\mathbb{A}$ is called sub-stochastic if:*

- *$\mathbb{A} \geq \mathbb{O}$;*
- *the sum of each column of $\mathbb{A}$ is ≤ 1;*
- *at least one of previous sums is < 1.*

Lemma 6.28. If $\mathbb{A}$ is the sub-matrix of an absorbing stochastic matrix $\mathbb{M}$ which acts on all the non-absorbing states, then $\mathbb{A}$ is sub-stochastic and all the eigenvalues of $\mathbb{A}$ have modulus strictly less than 1.

Proof. Thanks to the estimates (5.16) and (5.17) on the dominant eigenvalue (whose existence is ensured by the Perron-Frobenius Theorem, in its third formulation) by the maximum of columns sum, the dominant eigenvalue $\lambda_\mathbb{A}$ fulfils $0 \leq \lambda_\mathbb{A} \leq 1$. By contradiction, if $\lambda_\mathbb{A} = 1$ then $\lambda_\mathbb{A}$ is associated to the invariant probability distribution $\mathbf{P}$ of $\mathbb{A}$ (normalized eigenvector of $\mathbb{A}$) and to the invariant probability distribution $\widetilde{\mathbf{P}} = \begin{bmatrix} \mathbf{0} & \mathbf{P} \end{bmatrix}^T$ of $\mathbb{M}$:

$$\mathbb{M}\widetilde{\mathbf{P}} = \begin{bmatrix} \mathbb{I}_m\mathbf{0} & \mathbb{A}\mathbf{P} \end{bmatrix}^T = \begin{bmatrix} \mathbf{0} & \mathbb{A}\mathbf{P} \end{bmatrix}^T = \widetilde{\mathbf{P}}$$

$$\mathbb{M}^k\widetilde{\mathbf{P}} = \widetilde{\mathbf{P}} \qquad \forall k.$$

The existence of $\widetilde{\mathbf{P}}$ contradicts the property that all the h non-absorbing states must communicate with at least one among the m absorbing states, that is the existence of k_0 such that at least one among the the first m components of $\mathbb{M}^{k_0}\widetilde{\mathbf{P}}$ is strictly positive. $\qquad\square$

Notice that a sub-stochastic matrix can have eigenvalues with modulus equal to 1:

$$\begin{bmatrix} 1 & 1/2 \\ 0 & 1/3 \end{bmatrix}$$

---

[7] Here eventually has this precise meaning: there exists $\widetilde{k}$ s.t. the property holds for any $k \geq \widetilde{k}$.

Proof of Theorem 6.26. The Markov chain is described by the relation

$$\begin{bmatrix} \mathbf{P}_{k+1} \\ \mathbf{Q}_{k+1} \end{bmatrix} = \begin{bmatrix} \mathbb{I}_m & \mathbb{B} \\ \mathbb{O} & \mathbb{A} \end{bmatrix} \begin{bmatrix} \mathbf{P}_k \\ \mathbf{Q}_k \end{bmatrix}.$$

Notice that it is correct to algebrically operate as if $\mathbf{P}_{k+1}$, $\mathbf{Q}_{k+1}$, $\mathbb{I}_m$, $\mathbb{B}$, $\mathbb{A}$ were numbers, since it boils down to vectors and matrices whose dimensions satisfy the relations required for sums and products to be well defined.

In particular, the dynamical system associated to the Markov chain can be decoupled into two discrete dynamical systems

$$\begin{cases} \mathbf{P}_{k+1} = \mathbf{P}_k + \mathbb{B}\mathbf{Q}_k \\ \mathbf{Q}_{k+1} = \mathbb{A}\mathbf{Q}_k. \end{cases}$$

By the second one obtains

$$\mathbf{Q}_k = \mathbb{A}^k \mathbf{Q}_0, \qquad \forall k \in \mathbb{N}. \tag{6.17}$$

Since the first m states are all the absorbing states of the Markov chain, all the columns of $\mathbb{B}$ cannot all be zero simultaneously, hence the matrix $\mathbb{A}$ is sub-stochastic. By Lemma 6.28, all the eigenvalues of $\mathbb{A}$ have modulus strictly less than 1. Then $\lim_k \mathbb{A}^k = \mathbb{O}$ and by (6.17) it follows[8]

$$\lim_k \mathbf{Q}_k = \mathbf{0}.$$

By substituting (6.17) in the first discrete dynamical system, we obtain

$$\mathbf{P}_{k+1} = \mathbf{P}_k + \mathbb{B}\mathbb{A}^k \mathbf{Q}_0$$

whose solution is (as can be verified by induction)

$$\mathbf{P}_k = \mathbf{P}_0 + \mathbb{B} \left(\sum_{j=0}^{k-1} \mathbb{A}^j \right) \mathbf{Q}_0 = \mathbf{P}_0 + \mathbb{B} \left(\mathbb{I}_h - \mathbb{A}^k \right) \left(\mathbb{I}_h - \mathbb{A} \right)^{-1} \mathbf{Q}_0.$$

The matrix $\mathbb{I}_h - \mathbb{A}$ is always invertible because all the eigenvalues of $\mathbb{A}$ have modulus strictly less than 1. Taking the limit for $k \to +\infty$, one obtains

$$\mathbf{P}_\infty = \mathbf{P}_0 + \mathbb{B} \left(\mathbb{I}_h - \mathbb{A} \right)^{-1} \mathbf{Q}_0. \qquad \square$$

Previous theorem gives a complete description of the dynamics associated to an absorbing Markov chain: if we starts from an absorbing state s_j i.e. $P_0^j = 1$, then we remain there indefinitely. If, instead, we do not start from an absorbing state, namely $\mathbf{P}_0 = \mathbf{0}$, then the dynamics is much richer and more interesting for the applications and, for each choice of $\mathcal{P}_0 = [\, \mathbf{0} \ \mathbf{Q}_0 \,]^T$, the set of invariant probability distributions attracts the corresponding trajectories.

---

[8] Notice that the calculation of the limit in the first equation, taking into account this result ($\lim_k \mathbf{Q}_k = \mathbf{0}$) and if one supposes $\exists \lim_k \mathbf{P}_k$, leads to the identity $\lim_k \mathbf{P}_{k+1} = \lim_k \mathbf{P}_k$ that gives no information.

6.3 Random walks, duels and tennis matches

We apply the results of the previous section to some examples.

Example 6.29. Consider a drunk coming out of a pub: to his left there is a lake, to the right there is his own house. We assume the drunk never stands still and the probability of making a step toward home is 0.5 while the probability of taking a step in the opposite direction, towards the lake, is 0.5. He continues to walk randomly until either he gets home or ends up in the lake! Just to reduce the size of the problem without changing the content, we have reduced the distances involved: home is a step away from the pub, the lake two.

Then there are four states, a, b, c, and d. The vector that describes the probability of each state at time k is $\mathbf{W}_k = \begin{bmatrix} a_k & b_k & c_k & d_k \end{bmatrix}^T$, while the transition matrix and the Markov chain are respectively

$$\mathbb{T} = \begin{bmatrix} 1 & 0.5 & 0 & 0 \\ 0 & 0 & 0.5 & 0 \\ 0 & 0.5 & 0 & 0 \\ 0 & 0 & 0.5 & 1 \end{bmatrix} \qquad \mathbf{W}_{k+1} = \mathbb{T}\mathbf{W}_k.$$

Note the analogy with the "heads or tails" game with fair coin if one of the players starts with a sum equal to the stake and the other with a double sum. We use an argument (different from the one used for the coin tosses) which is suitable for generalizations. It is useful to separate the absorbing states from the others by placing the fourth state before the second and third one:

$$A_1 = a , \quad A_2 = d , \quad B_1 = b , \quad B_2 = c.$$

In the new coordinates we have

$$\mathbb{M} = \begin{bmatrix} 1 & 0 & 0.5 & 0 \\ 0 & 1 & 0 & 0.5 \\ 0 & 0 & 0 & 0.5 \\ 0 & 0 & 0.5 & 0 \end{bmatrix} \qquad \mathcal{P}_{k+1} = \mathbb{M}\mathcal{P}_k \qquad (6.18)$$

and, at time k,

$$P_k^1 = \text{prob. of } A_1, \quad P_k^2 = \text{prob. of } A_2, \quad P_k^3 = \text{prob. of } B_1, \quad P_k^4 = \text{prob. of } B_2.$$

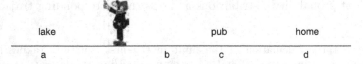

Fig. 6.4 Random walk of a drunk

The second equation
$$P_{k+1}^2 = P_k^2 + 0.5P_k^4$$
says that the probability to get home after $k+1$ steps is the sum of the probability of getting home after k steps and 0.5 times the probability of being in $b = B_1$ after k steps. Setting

$$M = \left[\begin{array}{c|c} \mathbb{I}_2 & \mathbb{B} \\ \hline \mathbb{O} & A \end{array}\right] = \left[\begin{array}{cc|cc} 1 & 0 & 0.5 & 0 \\ 0 & 1 & 0 & 0.5 \\ \hline 0 & 0 & 0 & 0.5 \\ 0 & 0 & 0.5 & 0 \end{array}\right] \tag{6.19}$$

and

$$\mathbf{P} = \begin{bmatrix} P^1 \\ P^2 \end{bmatrix} \qquad \mathbf{Q} = \begin{bmatrix} P^3 \\ P^4 \end{bmatrix} \qquad \mathcal{P} = \begin{bmatrix} \mathbf{P} & \mathbf{Q} \end{bmatrix}^T = \begin{bmatrix} P^1 & P^2 & P^3 & P^4 \end{bmatrix}^T$$

we rewrite in a block form the starting DDS

$$\begin{bmatrix} \mathbf{P}_{k+1} \\ \mathbf{Q}_{k+1} \end{bmatrix} = \left[\begin{array}{c|c} \mathbb{I}_2 & \mathbb{B} \\ \hline \mathbb{O} & A \end{array}\right] \begin{bmatrix} \mathbf{P}_k \\ \mathbf{Q}_k \end{bmatrix}.$$

In such cases, that is if $\mathbf{P}_0 = \mathbf{0}$, the probability of ending up in an absorbing state are, with reference to the notation of (6.19):

$$\lim_k \mathbf{P}_k = \mathbb{B} \left(\mathbb{I}_2 - A\right)^{-1} \mathbf{Q}_0$$

$$\mathbb{B} = \begin{bmatrix} 0.5 & 0 \\ 0 & 0.5 \end{bmatrix} \qquad \mathbb{I}_2 - A = \begin{bmatrix} 1 & -0.5 \\ -0.5 & 1 \end{bmatrix}.$$

Furthermore, we obtain:

$$\left(\mathbb{I}_2 - A\right)^{-1} = \begin{bmatrix} 4/3 & 2/3 \\ 2/3 & 4/3 \end{bmatrix}.$$

So, if we are sure ... that the walk starts at the pub, we have

$$\mathbf{P}_0 = \begin{bmatrix} 0 & 0 \end{bmatrix}^T \qquad \mathbf{Q}_0 = \begin{bmatrix} 0 & 1 \end{bmatrix}^T$$

$$\mathbb{B} \left(\mathbb{I}_2 - A\right)^{-1} \mathbf{Q}_0 = \begin{bmatrix} \dfrac{1}{3} & \dfrac{2}{3} \end{bmatrix}^T.$$

In conclusion, starting from the pub, the probability of arriving home is 2/3 while that of being "absorbed" by the lake is 1/3.   □

Example 6.30. Imagine a sequence from western movies where a duel takes place simultaneously among three gunmen A, B, C (so we call it *truel* in the sequel). Suppose A is a good shooter, having a percentage of hit targets of 70%, B is less precise and has a percentage of 50%, while C has a very poor percentage of 30%. The truel takes place in subsequent rounds: in each one, everyone shoots one of the other two chosen in such a way as to maximize his chances of winning.

As the best strategy, everyone tries to hit the better of the two remaining opponents. Therefore in the first round A tries to hit B, whereas B and C try to hit A. If more than one survives, one switches to the next round. We wonder which of the three has the best chances to survive, and, more generally, what are the probabilities of the various possible conclusions (with one or no winner).

The problem presents interesting analogies with the study of elections rounds with more than two candidates (or political parties) and analyzing it mathematically leads to some surprising conclusions.

It is convenient to study the general case, indicating with a, b and c the probabilities of hitting the target chosen respectively by A, B and C, with the constraints

$$0 < c < b < a < 1 .$$

Each state is a set of possible survivors after a round. The *absorbing states* are:

$$E_1 \text{ no one survives}$$
$$E_2 \text{ only } A \text{ survives}$$
$$E_3 \text{ only } B \text{ survives}$$
$$E_4 \text{ only } C \text{ survives.}$$

The *non-absorbing states* are: s_1 only A and C survive, s_2 only B and C survive, s_3 everyone survive. It can not occur that only A and B survive (but as initial state), because if they are all alive, no one tries to hit C.

Let P_k^1, P_k^2, P_k^3 and P_k^4 be the probabilities of E_1, E_2, E_3 and E_4 after k turns, and Q_k^1, Q_k^2 and Q_k^3 the probabilities of s_1, s_2 and s_3.

We calculate P_{k+1}^1. There are three cases to consider that correspond to killing occurrence after round $k+1$: no survivors after round k, with probability P_k^1; only A and C survive after round k and are both killed in round $k+1$, with probability $Q_k^1 \cdot a \cdot c$; only B and C survive after round k and both are killed in round $k+1$, with probability $Q_k^2 \cdot b \cdot c$. Summing up the probabilities for the three cases:

$$P_{k+1}^1 = P_k^1 + Q_k^1 \cdot a \cdot c + Q_k^2 \cdot b \cdot c.$$

To obtain P_{k+1}^2 we consider two cases: only A survives after k rounds; only A and C survive after k rounds, and in the $(k+1)$-th round A hits C and C overshoots A:

$$P_{k+1}^2 = P_k^2 + a\,(1-c)\,Q_k^1.$$

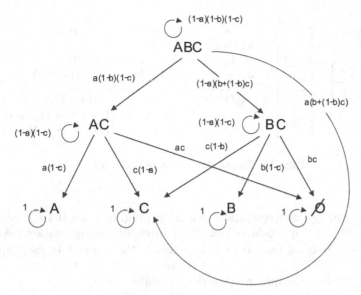

Fig. 6.5 Transition probability in a truel

Similarly we get:

$$P_{k+1}^3 = P_k^3 + b\,(1-c)\,Q_k^2$$
$$P_{k+1}^4 = P_k^4 + c\,(1-a)\,Q_k^1 + c\,(1-b)\,Q_k^2 + a\,(b+(1-b)\,c)\,Q_k^3$$
$$Q_{k+1}^1 = (1-a)\,(1-c)\,Q_k^1 + a\,(1-b)\,(1-c)\,Q_k^3$$
$$Q_{k+1}^2 = (1-b)\,(1-c)\,Q_k^2 + (1-a)\,(b+(1-b)\,c)\,Q_k^3$$
$$Q_{k+1}^3 = (1-a)\,(1-b)\,(1-c)\,Q_k^3.$$

To perform the calculations attention must be paid to assessing the probability that B or C or both hit A. This occurs in two ways: in the first one B hits A (and therefore does not care about what C does), in the second one B fails and C hits. Therefore the probability is $b+(1-b)\,c$.

Note that it is very useful (in this example, but especially in more complicated situations with many states) to use a chart to establish what are all the possible paths that lead to a particular final state (see Fig. 6.5). In practice, first one draws the diagram of the possible paths, then one calculates the probabilities of each transition (matrix $\mathbb{M}$ of Fig. 6.6).

From Theorem 6.26 we know that $\det\,(\mathbb{I}_3 - \mathbb{A}) \neq 0$. Moreover, the direct verification of this property is immediate because $\mathbb{A}$ is upper triangular and therefore, so are $\mathbb{I}_3 - \mathbb{A}$ and $(\mathbb{I}_3 - \mathbb{A})^{-1}$:

$$\det\,(\mathbb{I}_3 - \mathbb{A}) =$$
$$[1 - (1-a)\,(1-c)]\,[1 - (1-b)\,(1-c)]\,[1 - (1-a)\,(1-b)\,(1-c)]$$

$$
\mathbf{M} = \left[\begin{array}{c|c} \mathbb{I}_4 & \mathbb{B} \\ \hline \mathbb{O}_{4,3} & \mathbb{A} \end{array}\right] = \left[\begin{array}{c|ccc} \mathbb{I}_4 & \begin{matrix} ac & bc & 0 \\ a\,(1-c) & 0 & 0 \\ 0 & b\,(1-c) & 0 \\ c\,(1-a) & c\,(1-b) & a\,(b+(1-b)\,c) \end{matrix} \\ \hline \mathbb{O}_{4,3} & \begin{matrix} (1-a)\,(1-c) & 0 & a\,(1-b)\,(1-c) \\ 0 & (1-b)\,(1-c) & (1-a)\,(b+(1-b)\,c) \\ 0 & 0 & (1-a)\,(1-b)\,(1-c) \end{matrix} \end{array}\right]
$$

Fig. 6.6 Transition matrix of the truel

and each round bracket contains a quantity ranging from 0 to 1 because $a, b, c \in (0, 1)$. So the products contained in every square bracket provide again numbers ranging from 0 to 1, and this is the amount in each square bracket.

If, returning to the example, we consider the data

$$
a = 0.7 \qquad b = 0.5 \qquad c = 0.3
$$

we get (it is advisable to deal with numerical calculation programs)

$$
\mathbb{I}_3 - \mathbb{A} = \begin{bmatrix} 0.79 & 0 & -0.245 \\ 0 & 0.65 & -0.195 \\ 0 & 0 & 0.895 \end{bmatrix}
$$

$$
(\mathbb{I}_3 - \mathbb{A})^{-1} = \begin{bmatrix} 1.2658 & 0 & 0.34651 \\ 0 & 1.5385 & 0.3352 \\ 0 & 0 & 0.1173 \end{bmatrix}.
$$

Using the formula of Theorem 6.26, starting from the initial situation s_3, that is, all alive and $\mathbf{Q}_0 = \begin{bmatrix} 0 & 0 & 1 \end{bmatrix}^T$, one obtains

$$
\mathbf{P}_\infty = \mathbb{B}\,(\mathbb{I}_3 - \mathbb{A})^{-1}\,\mathbf{Q}_0 = \begin{bmatrix} 0.26582 & 0.23078 & 0.12305 \\ 0.62024 & 0 & 0.16979 \\ 0 & 0.53848 & 0.11732 \\ 0.11392 & 0.23078 & 0.58984 \end{bmatrix} \begin{bmatrix} 0 \\ 0 \\ 1 \end{bmatrix} = \begin{bmatrix} 0.12305 \\ 0.16979 \\ 0.11732 \\ 0.58984 \end{bmatrix}.
$$

Let us summarize and interpret the results approximating to three decimal digits.

The probability that none survives is 0.123, the probability that A wins is 0.170, the probability that B wins is 0.117 and the probability C that wins is 0.589.

The result is less surprising than it may appear at first sight: one needs only to reflect on the fact that, after the first round, A is alive only in 35% of cases and B in 50% of cases, whereas C is certainly alive, indeed C wins the first round in 45.5 cases out of 100.

If B does not show up to the duel, things are different: one starts from s_1 and obtains ($\mathbb{M}$ is the same, just change $\mathbf{Q}_0$ choosing $\mathbf{Q}_0 = [1\,0\,0]^T$)

$$\mathbb{B}\,(\mathbb{I}_3 - A)^{-1}\,\mathbf{Q}_0 = \begin{bmatrix} 0.26582 & 0.23078 & 0.12305 \\ 0.62024 & 0 & 0.16979 \\ 0 & 0.53848 & 0.11732 \\ 0.11392 & 0.23078 & 0.58984 \end{bmatrix} \begin{bmatrix} 1 \\ 0 \\ 0 \end{bmatrix} = \begin{bmatrix} 0.26582 \\ 0.62024 \\ 0 \\ 0.11392 \end{bmatrix}.$$

In this case the event that none survives has 0.266 probability, A wins with 0.620 probability, B does not win because he is not participating[9] and C wins with a 0.114 probability. Clearly C hopes that B participates in the duel.

Even in elections with subsequent rounds of balloting there may be paradoxical situations in which weak candidates take advantage of the strongest candidates by the presence of candidates of disturbance.[10]   □

Example 6.31. Let us consider a *tennis match* between two players A and B: the player who first wins 2 *sets* out of three wins the match. Each set is divided into several games, and to win a set one of the two players must win 6 games with a difference of at least two (if this does not happen, at 5 – 5 even, usually one passes to the so-called tie-break).

In every *game* the sequence of scores before the victory is 15, 30, 40. Thus, the first possible partial results of a game can be the following (in lexicographic order):

$$0 - 15 \quad 0 - 30 \quad 0 - 40 \quad 15 - 0 \quad 15 - 15 \quad 15 - 30 \quad 15 - 40$$

$$30 - 0 \quad 30 - 15 \quad 30 - 30 \quad 30 - 40 \quad 40 - 0 \quad 40 - 15 \quad 40 - 30$$

In addition, if starting from a score of $40 - 30$ the losing player wins the point, we reach a "tie": this situation will change to that of "advantage" for one of the two players, and then it's back to parity or the game ends with the victory of the player who was ahead.

One way to take into account the different levels of the two players is to assign two values p_A and $p_B = 1 - p_A$ to the probability of winning a point in a game by A and by B respectively. A game is therefore a system whose states are the possible partial results beyond the situation of equality and victory of one of the two players.

The transition from one state to another depends (besides the amount p_A) only on the state of departure, but not on the previous states. The possible developments of a game are therefore described by a Markov chain.

We simply represented by a graph the possible transitions in a game (see Fig. 6.7).

---

[9] ... nor he loses.

[10] These situations can be encouraged or discouraged depending on the regulations and protocols of voting system.

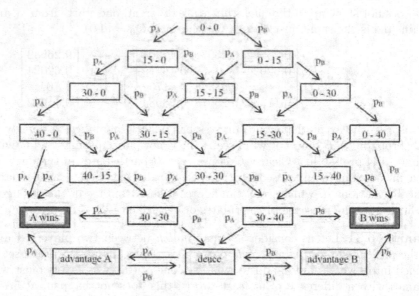

Fig. 6.7 Graph of possible developments of a game with the transition probabilities between the different states. Deuce denotes either the score 40-all or any subsequent parity situations

As the notion of absorbing state, the definitions of **transient state** and **recurrent state** that we will introduce in the next chapter (Definition 7.47) are very useful in the analysis of the possible outcomes of a match.   □

6.4 More on asymptotic analysis

With reference to Definition 6.6, we address the problem of existence, uniqueness and stability of the invariant probability distribution.

Remark 6.32. The existence of an invariant probability distribution for a generic finite and homogeneous Markov chain is an immediate consequence of Theorem 5.14 and Remark 6.8, since an eigenvector ≥ 0 must be also > 0.

We present also a result on this subject based on a constructive proof that does not refer to the theory of positive matrices.

Theorem 6.33. (*of Markov–Kakutani*) *A transition matrix on a finite set of states Ω has always at least an invariant probability distribution.*

Proof. The existence of an invariant probability distribution for $\mathbb{M}$ coincides with the existence of a fixed point for the continuous function $\mathbb{M} : S \to S$.

Given $\mathbf{V} \in S$, for every $k \in \mathbb{N} \setminus \{0\}$ we define

$$\mathbf{V}_k = \frac{1}{k} \sum_{h=0}^{k-1} \mathbb{M}^h \mathbf{V}.$$

Then $\mathbf{V}_k \in S$, $\forall k$, thanks to Lemma 6.5.
By the Bolzano-Weierstrass Theorem, there exist a subsequence $\mathbf{V}_{k_l}$ and $\mathbf{W} \in S$ such that

$$\lim_l \mathbf{V}_{k_l} = \mathbf{W}.$$

Still $\mathbf{W}$ is an invariant probability distribution because, after cancelation of equal terms in the two summations,

$$\mathbf{V}_{k_l} - \mathbb{M}\mathbf{V}_{k_l} = \frac{1}{k_l} \left(\sum_{h=0}^{k_l-1} \mathbb{M}^h \mathbf{V} - \sum_{h=0}^{k_l-1} \mathbb{M}^{h+1} \mathbf{V} \right) =$$

$$= \frac{1}{k_l} \left(\mathbf{V} - \mathbb{M}^{k_l} \mathbf{V} \right)$$

and then, passing to the limit $l \to +\infty$ in both sides and taking into account that $k_l \to +\infty$, $\|\mathbf{V}\|_{\mathbb{R}^n} \le 1$ and $\|\mathbb{M}^{k_l} \mathbf{V}\|_{\mathbb{R}^n} \le 1$, it follows $(\mathbf{V} - \mathbb{M}^{k_l} \mathbf{V})/k_l \to 0$, therefore $\mathbf{W} - \mathbb{M}\mathbf{W} = 0$. $\qquad\square$

Remark 6.34. Another more direct but less elementary proof of the Markov–Kakutani Theorem is the following. Since S is closed, bounded and convex in $\mathbb{R}^n$ and $\mathbb{M} : S \to S$ is continuous, the existence of the fixed point follows by Brouwer's Theorem.

Remark 6.35. Previous considerations ensure in any case, even if neither $\mathbb{M}$ nor any of its powers are strictly positive, the existence (but not the uniqueness) of an invariant probability distribution (positive but not necessarily strictly positive).

As for the uniqueness, in general it is easy to construct examples with more than one invariant probability distribution: consider the case of chains with more than one absorbing state.
Furthermore, if $\overline{\mathbf{P}}$, $\overline{\mathbf{Q}}$ are invariant probabilities, then $t\overline{\mathbf{P}} + (1-t)\overline{\mathbf{Q}}$ is an invariant probability $\forall t \in [0, 1]$, hence the invariant probabilities of $\mathbb{M}$ form a closed and convex set included in the n-simplex (if there are n states). Thus, if uniqueness fails, there are infinitely many invariant probability distributions, and the DDS associated to the Markov chain has infinitely many equilibria.
Thanks to Theorem 6.26 the closed and convex set of all invariant probability distributions of an absorbing Markov chain is an attractor in the sense of Definition 3.42 (Notice that this definition befits, without any modification, the case of vector-valued DDS).

For a non-absorbing Markov chain, the set of invariant probabilities is not attractive; consider for example the stochastic matrix

$$\begin{bmatrix} 1 & 0 & 0 \\ 0 & 0 & 1 \\ 0 & 1 & 0 \end{bmatrix} \tag{6.20}$$

with 2 invariant probability distributions ($\widetilde{\mathbf{P}} = [1\,0\,0]^T$, $\overline{\mathbf{P}} = [0\ 1/2\ 1/2]^T$): starting from the initial probability $\mathbf{P}_0 = [1/2\ 1/2\ 0]^T$, (6.20) generates the periodic dynamic $\mathbf{P}_k$ verifying $\mathbf{P}_{2h} = [1/2\ 1/2\ 0]^T$, $\mathbf{P}_{2h+1} = [1/2\ 0\ 1/2]^T$; in this case the limits of the second and third column of the k-th powers of (6.20) as $k \to +\infty$ do not exist.

We conclude with some summary results that link the asymptotic behavior and the invariant probability distributions.

Theorem 6.36. *Given a Markov chain with n states and transition matrix* $\mathbb{M} = [m_{ij}]$, *if there exists the limit of a column of* $\mathbb{M}^k$ *as* $k \to +\infty$, *namely*

$$\exists j \in \{1, 2, \ldots, n\}: \qquad \forall i \in \{1, 2, \ldots, n\} \qquad \exists \lim_k m_{ij}^{(k)} = \mathfrak{P}_i \tag{6.21}$$

then $\mathfrak{P} = \begin{bmatrix} \mathfrak{P}_1 & \mathfrak{P}_2 & \cdots & \mathfrak{P}_n \end{bmatrix}^T$ *is an invariant probability distribution.*

Proof. Formula (6.21) implies: $\mathfrak{P}_j \geq 0 \ j = 1, \ldots, n$,

$$\sum_{i=1}^{n} \mathfrak{P}_i = \sum_{i=1}^{n} \lim_k m_{ij}^{(k)} = \lim_k \sum_{i=1}^{n} m_{ij}^{(k)} = 1 \tag{6.22}$$

and

$$\mathfrak{P}_i = \lim_k m_{ij}^{(k)} = \lim_k m_{ij}^{(k+1)} = \lim_k \sum_{s=1}^{n} m_{is} m_{sj}^{(k)} =$$

$$= \sum_{s=1}^{n} \lim_k m_{is} m_{sj}^{(k)} = \sum_{s=1}^{n} m_{is} \mathfrak{P}_s \, .$$

Therefore $\mathbb{M}\mathfrak{P} = \mathfrak{P}$. $\qquad\qquad\qquad\qquad\qquad\qquad\qquad\qquad\qquad\qquad\qquad$ □

Example 6.37. The stochastic matrix

$$\mathbb{M} = \begin{bmatrix} 1 & 0 & 0 \\ 0 & 1/2 & 1/2 \\ 0 & 1/2 & 1/2 \end{bmatrix}$$

verifies $\lim_k \mathbb{M}^k = \mathbb{M} = \mathbb{M}^h$ for every h. Moreover $\mathbb{M}$ has two invariant probability distributions $[1\ 0\ 0]^T$ and $[0\ 1/2\ 1/2]^T$ that coincide respectively with the first and last columns of $\lim_k \mathbb{M}^k$. $\qquad\qquad$ □

Example 6.38. The stochastic matrix

$$\mathbb{A} = \begin{bmatrix} 1 & 0 & 0 \\ 0 & 1/3 & 2/3 \\ 0 & 2/3 & 1/3 \end{bmatrix}$$

verifies

$$\mathbb{A}^k = \mathbb{M} + \frac{1}{6}\left(-\frac{1}{3}\right)^{k-1}\begin{bmatrix} 0 & 0 & 0 \\ 0 & -1 & 1 \\ 0 & 1 & -1 \end{bmatrix} \qquad k \geq 1$$

and $\lim_k \mathbb{A}^k = \mathbb{M}$ where $\mathbb{M}$ is the matrix of Example 6.37. Furthermore $\mathbb{M}$ has two invariant probability distributions $[1\ 0\ 0]^T$ and $[0\ 1/2\ 1/2]^T$ that coincide respectively with the first and the last columns of $\lim_k \mathbb{A}^k$. □

Under certain conditions we can prove the uniqueness of the invariant probability distribution and a sort of converse of Theorem 6.36, as shown by the following result.

Theorem 6.39. *Each regular Markov chain (Definition 6.11) with n states has a unique invariant probability distribution $\widetilde{\mathbf{P}}$: the positive eigenvector (probability distribution) associated to the dominant and simple eigenvalue 1. If $\mathbb{M}$ is the stochastic matrix of a regular Markov chain and $\mathbf{P}$ is the corresponding invariant probability distribution, then all the eigenvectors and generalized eigenvectors of $\mathbb{M}$ different from (multiplies of) $\mathbf{P}$ are orthogonal to the uniform probability distribution $[1/n\ 1/n \cdots 1/n]^T$ which is the (strictly positive) dominant eigenvector $\mathbb{M}^T$. Moreover, for every initial probability distribution $\mathbf{P}_0$*

$$\lim_k \mathbb{M}^k \mathbf{P}_0 = \widetilde{\mathbf{P}}. \tag{6.23}$$

In particular, all the columns of $\mathbb{M}^k$ converge to $\widetilde{\mathbf{P}}$ as k tends to $+\infty$:

$$\lim_k \mathbb{M}^k = [\widetilde{\mathbf{P}}\,|\,\widetilde{\mathbf{P}}\,|\cdots|\,\widetilde{\mathbf{P}}\,]. \tag{6.24}$$

Proof. All claims except (6.24) are a reformulation of Theorems 6.9, 6.13 and Remark 6.10. Alternatively, they are an immediate consequence of Theorem 5.19 on vector-valued DDS when considering the restriction of the dynamics generated by $\mathbb{M}$ to the simplex S of probability distributions.

To prove (6.24) it is sufficient to replace $\mathbf{P}_0$ with $\mathbf{e}_h$ (certainty to start in state h) into (6.23):

$$h\text{-th column of } \mathbb{M}^k = \mathbb{M}^k \mathbf{e}_h \to \widetilde{\mathbf{P}}. \qquad \square$$

The Markov–Kakutani Theorem guarantees the existence of at least one invariant probability distribution in each case. Theorem 6.39 provides a sufficient condition (regular chain) for the uniqueness of the invariant probability distribution, and ensures that such distribution is a globally asymptotically stable equilibrium for the DDS $(\mathbb{M}, S)$ described by the regular Markov chain in the simplex S. Example 6.7 shows that the regularity of the chain is not a necessary condition for the uniqueness. There are also examples of Markov chains that have cycles or periodic orbits: see (6.20).

6.5 Summary exercises

Exercise 6.3. Verify through a direct calculation that the stochastic matrix

$$\mathbb{M} = \begin{bmatrix} a & b \\ 1-a & 1-b \end{bmatrix} \qquad 0 < a < 1, \quad 0 < b < 1$$

admits a unique strictly positive invariant probability vector.

Exercise 6.4. Given the three-state Markov chain with transition matrix

$$\mathbb{M} = \begin{bmatrix} 1 & 1/2 & 1/4 \\ 0 & 1/2 & 1/2 \\ 0 & 0 & 1/4 \end{bmatrix}$$

determine $m_{23}^{(k)}$ for every $k \in \mathbb{N}$.

Exercise 6.5. (*Wheatstone bridge*) Consider the electrical circuit of Fig. 6.8 consisting of 5 switches A, B, C, D and E, which can be closed with respective probabilities a, b, c, d and e (therefore the probability that each of the switches is open is respectively $1-a$, $1-b$ ...).
Assuming that all switches are *independent*, calculate the probability that a current passes in the circuit. This probability is called *circuit reliability*.

Exercise 6.6. A stochastic matrix $\mathbb{T}$ whose elements are 0 or 1 and such that $\det \mathbb{T} = \pm 1$, is called a **permutation matrix**. Explain the reason for the name.

Exercise 6.7. Prove that the change of coordinates (in the states' space $\mathbb{R}^n$) given by a permutation transforms stochastic matrices in stochastic matrices.

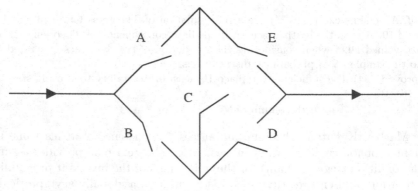

Fig. 6.8 Wheatstone bridge

It should be noted that this property has been implicitly used in the study of random walks and the duel, to separate absorbing states from the others.

Exercise 6.8. Three machines produce respectively r_1, r_2 and r_3 pieces, corresponding to a total of $r = r_1 + r_2 + r_3$ pieces, of which d_1, d_2 and d_3 are faulty. A piece is randomly extracted among those products and it is faulty. Calculate the probability that the extracted piece was produced respectively by machine 1, 2 or 3.

Exercise 6.9. Random walks in the mountains. We introduce a variant into Example 6.29 of a drunkard's random walk: we suppose that because of the slope of the road the probability of making a step toward home is 0.3, and to take a step towards the lake is 0.7. What is the probability of concluding the walk dry?

Exercise 6.10. With reference to Example 6.30, starting from s_3 and retaining the restrictions $0 < c < b < a < 1$, are there choices for the probabilities a, b, c that support A?

Exercise 6.11. Write the transition matrix related to the possible developments of a game in a tennis match by ordering states in lexicographical order.

Exercise 6.12. Determine the probability that starting from a score of $0 - 40$ player A will win the game.

Exercise 6.13. Show through examples that a stochastic matrix may have complex eigenvalues (if $n \geq 3$).

Exercise 6.14. Consider a set of volleyball, starting from the score $14 - 15$ between two teams A and B, and service is up to A. If $p \in [0, 1]$ is the probability that A wins a single dribble, illustrate with a graph the possible next states of the set and their transition probabilities; write the corresponding transition matrix; compute the probability that A wins the set.

Exercise 6.15. A variant of the truel: the participants **shoot one at a time**, starting gallantry with the least accurate shooter still alive, and continuing where necessary. Analyze the evolution of the duel assuming

$$
\begin{array}{lll}
A & \text{infallible shooter} & a = 1 \\
B & \text{average shooter} & b = 2/3 \\
C & \text{clumsy shooter} & c = 1/3
\end{array}
$$

and that each duelist adopts the best strategy.
To whom will C (that starts the game) fire the first shot to?

7

Positive matrices and graphs

In this chapter we further investigate the properties of positive matrices already studied in Sect. 5.4, introducing the notion of permutation matrices and of irreducible matrices. We examine the links with adjacency matrices of graph theory and illustrate some applications to networks, by showing how Web surfing can be described as the discrete dynamical system associated to a suitable Markov chain.

The main result of this chapter is the Frobenius Theorem which specifies the properties of the spectrum of the positive irreducible matrices. According to such properties, irreducible matrices are divided into two classes: the primitive matrices and the cyclic ones.

7.1 Irreducible matrices

For the notations concerning matrices and vectors we refer to Appendix D and to the notions of weakly positive, positive, and strictly positive matrix in Definition 5.11.

It will be useful to consider special changes of basis: the ones that rearrange the elements of a basis of $\mathbb{R}^n$ and are thus associated in a natural way to permutation matrices.

Definition 7.1. *A **permutation matrix** is a square matrix* $\mathbb{T} = [t_{ij}]$ *having only 0 and 1 as elements and a single 1 in each row and in each column.*

Remark 7.2. If $\mathbb{T}$ is a permutation matrix, then its transpose $\mathbb{T}^T$ is a permutation matrix too, and they are both orthogonal (namely $\mathbb{T}^{-1} = \mathbb{T}^T$, by which $|\det \mathbb{T}| = 1$).

A permutation matrix $\mathbb{T}$ corresponds in a natural way to the permutation map τ acting on the first n natural numbers that verifies $\tau(j) = i$ if and only if $t_{ij} = 1$. The change of basis (see Appendix D) induced by $\mathbb{T}$, e.g.

$$\mathbf{y} = \mathbb{T}\mathbf{x} \qquad \mathbb{B} = \mathbb{T}\mathbb{A}\mathbb{T}^{-1} = \mathbb{T}\mathbb{A}\mathbb{T}^T,$$

E. Salinelli, F. Tomarelli: *Discrete Dynamical Models.*
UNITEXT – La Matematica per il 3+2 76
DOI 10.1007/978-3-319-02291-8_7, © Springer International Publishing Switzerland 2014

operates as follows on the canonical basis $\{\mathbf{e}_1, \ldots, \mathbf{e}_n\}$, on the generic vector $\mathbf{x}$ and on the generic square matrix $\mathbb{A}$:

- it permutes the components of the basis of $\mathbb{R}^n$ rearranging them in the sequence $\mathbf{u}_1, \ldots, \mathbf{u}_n$, where $\mathbf{u}_i = \mathbf{e}_j$ if $t_{ij} = 1$;
- it permutes the components of $\mathbf{x}$ (if $t_{ij} = 1$ then $y_i = x_j$);
- it swaps the rows of the matrix $\mathbb{A}$ through the permutation τ associated to $\mathbb{T}$ and swaps the columns too through the same permutation τ ($\mathbb{B}_{\tau(i), \tau(j)} = \mathbb{A}_{i,j}$).
  In particular, off-diagonal elements of $\mathbb{A}$ are relocated off-diagonal; the elements of the principal diagonal of $\mathbb{A}$ are relocated on the principal diagonal: explicitly (with no summation convention on repeated indices)

$$\text{if } t_{ij} \neq 0 \text{ then } b_{ii} = a_{jj} \, .$$

Notice that zero entries are neither created nor destroyed by the coordinate change associated to a permutation. This property does not hold for more general changes of coordinates even if they are orthogonal, as illustrated by the following example.

Example 7.3. The orthogonal change of coordinates (namely the counterclockwise rigid rotation of $\pi/4$ radians) associated to the matrix:

$$\mathbb{U} = \begin{bmatrix} \sqrt{2}/2 & -\sqrt{2}/2 \\ \sqrt{2}/2 & \sqrt{2}/2 \end{bmatrix}$$

transforms the vector $\mathbf{x} = \begin{bmatrix} 1 & 1 \end{bmatrix}^T$ into the vector $\mathbf{y} = \mathbb{U}\mathbf{x} = \begin{bmatrix} 0 & \sqrt{2} \end{bmatrix}^T$, and transforms the matrix

$$\mathbb{A} = \begin{bmatrix} 1 & 1 \\ 1 & 1 \end{bmatrix}$$

into the matrix

$$\mathbb{B} = \mathbb{U}\mathbb{A}\mathbb{U}^{-1} = \begin{bmatrix} 0 & 0 \\ 0 & 2 \end{bmatrix}. \qquad \square$$

Exercise 7.1. Verify that a permutation matrix is **doubly stochastic** (that is both $\mathbb{T}$ and $\mathbb{T}^T$ are stochastic).

Exercise 7.2.

- Compute the number of permutations of n elements.
- Compute the number of cyclic permutations of n elements.
- Compute the number of circular permutations of n elements.

We recall that: a permutation τ is called cyclic if there exists $t \in \{1, \ldots, n\}$ such that $\tau(j) = j + t \mod n$, $\forall j$; a circular permutation of n objects corresponds to their arrangement on a circle (the places where you can have n guests for dinner served at a round table: relevant is not the place a guest occupies, rather the relative position).

Definition 7.4. *A square matrix* A *of order* $n > 1$ *is called* **reducible** *if there exist an integer* r *with* $0 < r < n$ *and a permutation matrix* T *such that the matrix* $\widetilde{A} = TAT^T$ *is of the kind*

$$\widetilde{A} = \begin{bmatrix} B & O \\ C & D \end{bmatrix}$$

where B *and* D *are square matrices of order respectively* r *and* $n - r$, *and* O *denotes the null matrix (with* r *rows and* $n - r$ *columns).*
A square matrix A *is called* **irreducible** *if it is not reducible.*

Remark 7.5. It is easy to show (see Exercise 7.7) that a square matrix A of order n is reducible if and only if there exists a permutation matrix T such that the matrix $\widetilde{A} = T^T AT$ is of the kind

$$\widetilde{A} = \begin{bmatrix} B & C \\ O & D \end{bmatrix}$$

with B and D square matrices.

Notice that the null matrix O may not be a square matrix.

Example 7.6. The following matrices are reducible:

$$A_1 = \begin{bmatrix} 1 & 0 \\ 1 & 1 \end{bmatrix}; \quad A_2 = \begin{bmatrix} 1 & 0 \\ 0 & 1 \end{bmatrix}; \quad A_3 = \begin{bmatrix} 1 & 1 \\ 0 & 1 \end{bmatrix}; \quad A_4 = \begin{bmatrix} -1 & 0 \\ 0 & 1 \end{bmatrix}.$$

A_1, A_2, A_4 are already in the form required by the definition: observe that zeroes are allowed to be present in the matrix C of Definition 7.4. A_3 can be transformed into A_1 by the permutation matrix

$$T = \begin{bmatrix} 0 & 1 \\ 1 & 0 \end{bmatrix}. \qquad \qquad \square$$

Example 7.7. The following matrices are irreducible:

$$A_5 = \begin{bmatrix} 1 & 1 \\ 1 & 0 \end{bmatrix}; \quad A_6 = \begin{bmatrix} 0 & 1 \\ 1 & 1 \end{bmatrix}; \quad A_7 = \begin{bmatrix} 0 & 1 \\ 1 & 0 \end{bmatrix}; \quad A_8 = \begin{bmatrix} 0 & 1 \\ -1 & 0 \end{bmatrix}. \quad \square$$

Example 7.8. The following matrices are irreducible:

$$A_9 = \begin{bmatrix} 1 & 1 & 0 \\ 1 & 1 & 1 \\ 1 & 1 & 1 \end{bmatrix} \qquad A_{10} = \begin{bmatrix} 1 & 1 & 1 \\ 1 & 1 & 1 \\ 0 & 1 & 1 \end{bmatrix}. \qquad \qquad \square$$

Exercise 7.3. (a) Prove that not all permutation matrices (of order $n > 1$) are irreducible, by showing examples of reducible permutation matrices.
(b) In which cases is a permutation matrix irreducible?

Examples of irreducible permutation matrices together with their eigenvalues:

$$\begin{bmatrix} 0 & 1 \\ 1 & 0 \end{bmatrix} \qquad \begin{bmatrix} 0 & 1 & 0 \\ 0 & 0 & 1 \\ 1 & 0 & 0 \end{bmatrix} \qquad \begin{bmatrix} 0 & 0 & 1 \\ 1 & 0 & 0 \\ 0 & 1 & 0 \end{bmatrix}$$

$\lambda = \pm 1 \qquad\qquad \lambda = e^{2h\pi i/3} \quad h = 0,1,2 \qquad \lambda = e^{2h\pi i/3} \quad h = 0,1,2$

$$\begin{bmatrix} 0 & 1 & 0 & 0 \\ 0 & 0 & 1 & 0 \\ 0 & 0 & 0 & 1 \\ 1 & 0 & 0 & 0 \end{bmatrix} \qquad \begin{bmatrix} 0 & 0 & 0 & 1 \\ 0 & 0 & 1 & 0 \\ 1 & 0 & 0 & 0 \\ 0 & 1 & 0 & 0 \end{bmatrix} \qquad \begin{bmatrix} 0 & 0 & 0 & 1 \\ 1 & 0 & 0 & 0 \\ 0 & 1 & 0 & 0 \\ 0 & 0 & 1 & 0 \end{bmatrix}.$$

$\lambda = \pm 1, \ \pm i \qquad\qquad \lambda = \pm 1, \ \pm i \qquad\qquad \lambda = \pm 1, \ \pm i$

Examples of reducible permutation matrices together with their eigenvalues:

$$\begin{bmatrix} 1 & 0 \\ 0 & 1 \end{bmatrix} \qquad \begin{bmatrix} 1 & 0 & 0 \\ 0 & 1 & 0 \\ 0 & 0 & 1 \end{bmatrix} \qquad \begin{bmatrix} 0 & 0 & 1 \\ 0 & 1 & 0 \\ 1 & 0 & 0 \end{bmatrix}$$

$\lambda = 1$ double $\qquad \lambda = 1$ triple $\qquad \lambda = 1, \ 1, \ -1$

$$\begin{bmatrix} 1 & 0 & 0 \\ 0 & 0 & 1 \\ 0 & 1 & 0 \end{bmatrix} \qquad \begin{bmatrix} 0 & 1 & 0 \\ 1 & 0 & 0 \\ 0 & 0 & 1 \end{bmatrix} \qquad \left[\begin{array}{cc|cc} 0 & 1 & 0 & 0 \\ 1 & 0 & 0 & 0 \\ \hline 0 & 0 & 0 & 1 \\ 0 & 0 & 1 & 0 \end{array}\right].$$

$\lambda = 1, \ 1, \ -1 \qquad \lambda = 1, \ 1, \ -1 \qquad \lambda = \pm 1$ double

Notice the invariance property with respect to suitable rotations of the spectrum in the irreducible case (consistently with the subsequent Theorem 7.13); this property may fail in the reducible case. It is immediate to verify that the four reducible permutation matrices of order 3 have the eigenvalue 1 of multiplicity at least 2. This is a general property, valid for reducible permutation matrices of any order since, in Definition 7.4, $\mathbb{C} = \mathbb{O}$ and the blocks on the diagonal are permutation matrices, hence stochastic too.

Remark 7.9. The matrices $\mathbb{A}$ and $\mathbb{B}$ in Example 7.3 show that a change of basis (even if it is orthogonal, but it is not associated with a permutation of the basis) can transform an irreducible matrix in a reducible one.

Some useful criteria follow by Definition 7.4:

- an irreducible square matrix of order $n > 1$ can have neither a row nor a column of zeroes;
- a reducible square matrix of order $n > 1$ has at least $n - 1$ zero elements (see Exercise 7.6).

Despite the last statement, the presence of n or more null elements does not imply reducibility: for instance the following matrix of order 3 is irreducible:

$$\begin{bmatrix} 0 & 1 & 0 \\ 1 & 0 & 1 \\ 0 & 1 & 0 \end{bmatrix} . \tag{7.1}$$

Exercise 7.4. Prove that if $\mathbb{A}$ and $\mathbb{B}$ are square matrices of the same order, where $\mathbb{A}$ is irreducible, $\mathbb{A} > \mathbb{O}$ and $\mathbb{B} \gg \mathbb{O}$, then $\mathbb{A}\mathbb{B} \gg \mathbb{O}$.

We emphasize that in the definition of irreducibility (Definition 7.4) no condition is imposed on the sign of the elements. Nevertheless in the following we will deal almost exclusively with non-negative matrices for the important role they play in the applications.
Notice that a strictly positive matrix is irreducible, whereas the converse is not true, as showed by the following exercise.

Exercise 7.5. Verify that the matrix

$$A = \begin{bmatrix} 0 & 1/2 & 1/3 \\ 1/4 & 1/2 & 1/3 \\ 3/4 & 0 & 1/3 \end{bmatrix}$$

is not strictly positive but is irreducible (use Definition 7.4).

Taking into account the action of permutation matrices on diagonal elements of a matrix, it is clear that:

- a matrix which has zeroes at most on the main diagonal is irreducible;
- a triangular (in particular diagonal) matrix of order $n > 1$ is reducible.

Therefore it is useful to have criteria for deciding whether a matrix with some zeroes in the principal diagonal and elsewhere, is irreducible. A first answer for nonnegative matrices is given by the following Theorem 7.10, while a complete answer will be given by Theorem 7.29 without restrictions on the sign of the elements.

Theorem 7.10. *If $\mathbb{A} \geq \mathbb{O}$ is irreducible and of order n, then:*

$$(\mathbb{I} + \mathbb{A})^{n-1} \gg \mathbb{O} .$$

Proof. It is sufficient to show that

$$(\mathbb{I} + \mathbb{A})^{n-1} \mathbf{x} \gg \mathbf{0} \qquad \forall \mathbf{x} > \mathbf{0}. \tag{7.2}$$

Since any zero component of $\mathbf{z} = (\mathbb{I} + \mathbb{A})\mathbf{x}$ must also be a zero of $\mathbf{x}$, condition (7.2) is verified if for each positive vector $\mathbf{x} > \mathbf{0}$ the vector $\mathbf{z} = (\mathbb{I} + \mathbb{A})\mathbf{x}$ has less null entries than $\mathbf{x}$.

By contradiction assume that $\mathbf{x}$ and $\mathbf{z}$ have the same number of zeroes, and then, up to appropriate permutations, they are of the form

$$\mathbf{x} = \begin{bmatrix} \mathbf{u} \\ \mathbf{0} \end{bmatrix}, \qquad \mathbf{z} = \begin{bmatrix} \mathbf{w} \\ \mathbf{0} \end{bmatrix}, \qquad \mathbf{u} \gg 0, \ \mathbf{w} \gg 0$$

where $\mathbf{u}$ and $\mathbf{w}$ are column vectors of the same dimension. By setting

$$\mathbb{A} = \begin{bmatrix} \mathbb{A}_{11} & \mathbb{A}_{12} \\ \mathbb{A}_{21} & \mathbb{A}_{22} \end{bmatrix}$$

where the order of $\mathbb{A}_{11}$ is equal to the dimension of $\mathbf{u}$, the relation $\mathbf{z} = (\mathbb{I} + \mathbb{A})\,\mathbf{x}$ can be rewritten as

$$\begin{cases} \mathbf{u} + \mathbb{A}_{11}\mathbf{u} = \mathbf{w} \\ \mathbb{A}_{21}\mathbf{u} = \mathbf{0}, \end{cases}$$

hence $\mathbf{u} \gg 0$ implies $\mathbb{A}_{21} = \mathbb{O}$ against the condition that $\mathbb{A}$ is irreducible. $\square$

By the previous theorem we immediately get the following result that will be made precise in Theorem 7.29:

Corollary 7.11. *If $\mathbb{A} > \mathbb{O}$ is an irreducible matrix of order n, then for any pair of indices $i, j \in \{1, \dots, n\}$ there exists a positive integer $q = q\,(i, j)$, $q \leq n$, such that*

$$a_{ij}^{(q)} > 0. \tag{7.3}$$

More precisely, q satisfies:

$$q \leq m - 1 \quad \textit{if } i \neq j; \qquad\qquad q \leq m \quad \textit{if } i = j,$$

*where m is the degree of the **minimal polynomial** of $\mathbb{A}$, namely the polynomial ψ with leading coefficient equal to 1 and of least degree such that $\psi(\mathbb{A}) = \mathbb{O}$.*

Proof. By Theorem 7.10

$$\mathbb{I} + \sum_{k=1}^{n-1} \binom{n}{k} \mathbb{A}^k = (\mathbb{I} + \mathbb{A})^{n-1} \gg \mathbb{O}. \tag{7.4}$$

By the last relation, if $i \neq j$ then the nonnegative numbers $a_{ij}, a_{ij}^{(2)}, \dots, a_{ij}^{(n-1)}$ cannot be all zeroes.

Moreover, $(\mathbb{I} + \mathbb{A})^{n-1} \gg \mathbb{O}$ and $\mathbb{A} > \mathbb{O}$ imply $\mathbb{A}\,(\mathbb{I} + \mathbb{A})^{n-1} \gg \mathbb{O}$ (see Exercise 7.4), that is the nonnegative numbers $a_{ii}, a_{ii}^{(2)}, \dots, a_{ii}^{(n)}$ cannot be all zeroes.

Finally, if ψ (the minimal polynomial of $\mathbb{A}$) has degree m strictly lower than the characteristic polynomial of $\mathbb{A}$, then there are unique polynomials q and r such that

$$(1 + \lambda)^{n-1} = q\,(\lambda)\,\psi\,(\lambda) + r\,(\lambda) \qquad\qquad \text{degree of } r \ < m.$$

The equality $\psi(\mathbb{A}) = \mathbb{O}$ entails

$$(\mathbb{I} + \mathbb{A})^{n-1} = q(\mathbb{A})\, \psi(\mathbb{A}) + r(\mathbb{A}) = r(\mathbb{A}) =$$

$$= \sum_{k=0}^{m-1} c_k \mathbb{A}^k = c_0 \mathbb{I} + \sum_{k=1}^{m-1} c_k \mathbb{A}^k .$$

This relation together with $(\mathbb{I} + \mathbb{A})^{n-1} \gg \mathbb{O}$ imply that if $i \neq j$ the nonnegative numbers a_{ij}, $a_{ij}^{(2)}$, ..., $a_{ij}^{(m-1)}$ cannot be all zeroes.

Furthermore, $r(\mathbb{A}) \gg \mathbb{O}$ implies $\mathbb{A}\, r(\mathbb{A}) \gg \mathbb{O}$ (see Exercise 7.4), i.e. the nonnegative numbers a_{ii}, $a_{ii}^{(2)}$, ..., $a_{ii}^{(m)}$ cannot be all zero. $\qquad\qquad\square$

Remark 7.12. *If* $\mathbb{A} \geq \mathbb{O}$ *and there exists* $k \in \mathbb{N} \setminus \{0, 1\}$ *such that* $\mathbb{A}^k \gg \mathbb{O}$ *then* $\mathbb{A}$ *is irreducible.* In fact, by definition of reducible matrix, the null NE (or SW) minor persists in each power, therefore reducible matrices cannot have strictly positive powers.

Note also that the opposite implication is not true, despite Corollary 7.11: for instance,

$$\mathbb{A} = \begin{bmatrix} 0 & 1 \\ 1 & 0 \end{bmatrix} \tag{7.5}$$

is irreducible, therefore it verifies (7.3), but it has no strictly positive power. In fact:

$$\mathbb{A}^2 = \begin{bmatrix} 1 & 0 \\ 0 & 1 \end{bmatrix} \quad \text{and} \quad \begin{cases} q = 1 & \text{if } i \neq j \\ q = 2 & \text{if } i = j \end{cases} ; \quad \mathbb{A}^{k+2} = \mathbb{A}^k.$$

In the years 1909–1912, exploiting the concept of irreducibility Frobenius extended Perron's result of 1907 (the one that for the sake of simplicity is called Perron-Frobenius Theorem in Chap. 5, see Theorems 5.12 and 5.13) as follows (about its proof see Gantmacher [8]).

Theorem 7.13 (Frobenius). *A positive and irreducible square matrix* $\mathbb{A}$ *has a real positive eigenvalue* $\lambda_{\mathbb{A}}$ *of algebraic multiplicity 1, and the modulus of any other eigenvalue* λ *does not exceed this:* $\lambda_{\mathbb{A}} \geq |\lambda|$. *A unique (up to multiplicative constants) strictly positive eigenvector is associated to the eigenvalue* $\lambda_{\mathbb{A}}$.

Furthermore, if $\mathbb{A}$ *has* h *eigenvalues whose modulus is* $\lambda_{\mathbb{A}}$ *then they are all simple and they are the roots of the equation*

$$\lambda^h - (\lambda_{\mathbb{A}})^h = 0.$$

Precisely, the set of eigenvalues (spectrum) of $\mathbb{A}$ *is invariant under the rotation in the complex plane by an angle* $2\pi/h$.

Now we illustrate some important consequences of Theorem 7.13.

Theorem 7.14. *A positive and irreducible matrix* $\mathbb{A}$ *cannot have two positive linearly independent eigenvectors.*

Moreover any eigenvector of $\mathbb{A}$ *different from* $\mathbf{V}^{\mathbb{A}}$ *is orthogonal to* $\mathbf{V}^{\mathbb{A}^T} \gg \mathbb{O}$, *the dominant eigenvector of* $\mathbb{A}^T$. *Generalized eigenvectors have the same property.*

Proof. Since each eigenvalue of $\mathbb{A}$ corresponding to a positive eigenvector has to be real and positive, the proof is identical to the one of Theorem 5.19 except that it uses Theorem 7.13 in place of Theorem 5.12 for both $\mathbb{A}$ and $\mathbb{A}^T$. □

Theorem 7.13 justifies the introduction of the following

Definition 7.15. *Let* $\mathbb{A} > \mathbb{O}$ *be an irreducible matrix with h eigenvalues* λ_1, λ_2, ..., λ_h *of maximum modulus:*

$$\lambda_1 = |\lambda_2| = \cdots = |\lambda_h| \ .$$

Then $\mathbb{A}$ *is called:*

- **primitive** *if* $h = 1$;
- **cyclic** *if* $h > 1$.

The matrix (7.5) is irreducible but not primitive.
The link between the notion of primitiveness and the analysis done in Chap. 5 is clarified by the following result (for the proof see for instance [8]).

Theorem 7.16. *A matrix* $\mathbb{A} > \mathbb{O}$ *is primitive if and only if some power of* $\mathbb{A}$ *is strictly positive:*

$$\exists k \in \mathbb{N} \backslash \{0\} : \quad \mathbb{A}^k \gg \mathbb{O} \ .$$

Theorem 7.17. *If* $\mathbb{A}$ *is a primitive matrix of order n,* $\mathbf{V}^{\mathbb{A}}$ *is the positive eigenvector of norm 1 associated to the dominant eigenvalue $\lambda_{\mathbb{A}}$ and we order eigenvalues as* $\lambda_{\mathbb{A}} = \lambda_1 > |\lambda_2| \geq \cdots \geq |\lambda_n|$, *then:*

- *if* $\lambda_2 \neq 0$ *then as* $k \to +\infty$

$$\mathbb{A}^k = (\lambda_{\mathbb{A}})^k \, \mathbf{V}^{\mathbb{A}} \left(\mathbf{V}^{\mathbb{A}}\right)^T + O\left(k^{m_2-1}|\lambda_2|^k\right)$$

  where m_2 is the algebraic multiplicity of λ_2;
- *if* $\lambda_2 = 0$ *then for* $k \geq n - 1$

$$\mathbb{A}^k = (\lambda_{\mathbb{A}})^k \, \mathbf{V}^{\mathbb{A}} \left(\mathbf{V}^{\mathbb{A}}\right)^T .$$

Moreover, each solution of the DDS $\mathbf{X}_{k+1} = \mathbb{A}\mathbf{X}_k$ *verifies*

$$(\lambda_{\mathbb{A}})^{-k} \, \mathbf{X}_k = c_{\mathbb{A}} \mathbf{V}^{\mathbb{A}} + \sigma(k)$$

where $\lambda_{\mathbb{A}}$ is the dominant eigenvalue of $\mathbb{A}$, $c_{\mathbb{A}}$ is the coefficient of the initial datum $\mathbf{X}_0$ along the component $\mathbf{V}^{\mathbb{A}}$ with respect to the Jordan basis of $\mathbb{A}$ and $\sigma(k)$ is an infinitesimal for $k \to +\infty$.

Moreover for each initial datum $\mathbf{X}_0 > 0$ *the solution* $\mathbf{X}_k$ *verifies*

$$\lim_k \frac{\mathbf{X}_k}{\|\mathbf{X}_k\|} = \frac{\mathbf{V}^{\mathbb{A}}}{\|\mathbf{V}^{\mathbb{A}}\|}.$$

Proof. The proof is an immediate consequence of Theorems 5.18 and 5.19 and the specifications concerning the infinitesimal $\sigma(k)$ are consequences of Remark 5.9.

Example 7.18. The strictly positive matrix

$$\mathbb{A} = \begin{bmatrix} 1/2 & 1/2 \\ 1/2 & 1/2 \end{bmatrix}$$

satisfies $\mathbb{A}^k = \mathbb{A}$. Its eigenvalues are $\lambda_{\mathbb{A}} = 1$ and $\lambda_2 = 0$, with

$$\mathbf{V}^{\mathbb{A}} = \left[\sqrt{2}/2 \ \sqrt{2}/2 \right]^T.$$

It is immediate to verify that:

$$\mathbf{V}^{\mathbb{A}} \left(\mathbf{V}^{\mathbb{A}} \right)^T = \left[\sqrt{2}/2 \ \sqrt{2}/2 \right]^T \left[\sqrt{2}/2 \ \sqrt{2}/2 \right] = \mathbb{A}. \qquad \square$$

Exercise 7.6. Prove that if a square matrix $\mathbb{A}$ of order n is reducible then it has at least $n - 1$ null entries.

Exercise 7.7. Prove that a square matrix $\mathbb{A}$ of order n is reducible if and only if there exists a permutation matrix $\mathbb{T}$ such that the matrix $\widetilde{\mathbb{A}} = \mathbb{T}^T \mathbb{A} \mathbb{T}$ is of the kind

$$\widetilde{\mathbb{A}} = \begin{bmatrix} \mathbb{B} & \mathbb{C} \\ \mathbb{O} & \mathbb{D} \end{bmatrix}$$

where $\mathbb{B}$ and $\mathbb{D}$ are square matrices.
Notice the different position of the null matrix $\mathbb{O}$ with respect to Definition 7.4.

Exercise 7.8. Prove that a square matrix $\mathbb{A}$ of order $n > 1$ is reducible if the set of indices $\{1, 2, \ldots, n\}$ can be split in two disjoint sets $\{i_1, i_2, \ldots, i_r\}$ and $\{j_1, j_2, \ldots, j_{n-r}\}$ with $0 < r < n$, such that

$$a_{i_\alpha j_\beta} = 0 \qquad \alpha = 1, \ldots, r, \quad \beta = 1, \ldots, n - r .$$

Exercise 7.9. Prove that a square matrix $\mathbb{A}$ of order n is reducible if and only if the corresponding linear operator $\mathbb{A} : \mathbb{R}^n \to \mathbb{R}^n$ has an invariant subspace of dimension r, with $r < n$ made of canonical coordinates.

Exercise 7.10. Verify that the matrix

$$\mathbb{A} = \begin{bmatrix} 1/4 & 1/2 & 1/3 \\ 1/4 & 1/2 & 1/3 \\ 1/2 & 0 & 1/3 \end{bmatrix}$$

is irreducible.

Exercise 7.11. Exhibit examples proving that the sum of irreducible matrices can be a reducible matrix.
Exhibit examples proving that the sum of reducible matrices can be an irreducible matrix.

7.2 Graphs and matrices

Now we introduce the formal definition of graph and analyze some related properties exploiting the analogies with irreducible matrices. Then we examine the application to graphs of the previous results about matrices. Unlike the simple examples already met in Chap. 6, the graphs that appear when modeling real problems (such as train schedules, airlines, telecommunications networks, supply chains) have enormous size: the large amount of nodes and edges renders their graphical representation very difficult or practically impossible. For this reason it is important to define and study the main qualitative and structural properties of graphs, to deal with them not only as geometrical objects but as structures of Linear Algebra.

Definition 7.19. *A **directed graph** G is a pair (V, L) where $V = \{v_1, \ldots, v_n\}$ is a finite set of elements called **nodes** (or **vertices**) and $L = \{l_1, \ldots, l_m\} \subseteq V \times V$ is a set of ordered pairs of these nodes, called (directed) **edges** (or **arcs**, or **arrows**).*
*If (v_s, v_k) is an edge of a graph G, then the node v_s is called **tail** of the edge, the node v_k is called **head** of the edge and the edge (v_s, v_k) is called **outgoing** at v_s and **ingoing** at v_k.*
By definition, the edge (v_s, v_k) is different from the edge (v_k, v_s).
*We define **subgraph** of a graph $G = (V, L)$ a graph $\mathcal{G} = (\mathcal{V}, \mathcal{L})$ such that $\mathcal{V} \subseteq V$, $\mathcal{L} \subseteq L$ (in particular $\mathcal{L}$ contains only pairs of elements in $\mathcal{V}$).*

It is customary to display a graph by representing the nodes as points and the edges as arrows that connect these points (see Fig. 7.1) and directed from the tail to the head of the edge.

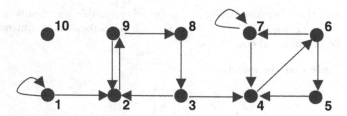

Fig. 7.1 An example of directed graph: the orientation of the edges is given by the order of the pairs in the relation L

The graph in Fig. 7.1 has nodes $\{1, 2, \ldots, 9, 10\}$. The ordered pairs $(1, 2)$, $(1, 1)$, $(6, 5)$ and $(5, 4)$ are edges, whereas the pairs $(2, 1)$, $(4, 3)$, $(4, 5)$ or $(6, 9)$ are not edges. The graph has 14 edges.

Definition 7.20. *Given a directed graph $G = (V, L)$ with $V = \{v_1, \ldots, v_n\}$ and $L = \{l_1, \ldots, l_m\}$, we call* **directed path** *C (from v_{i_0} to v_{i_p}) a sequence of nodes $C = \{v_{i_0}, v_{i_1}, \ldots, v_{i_p}\}$ such that $v_{i_k} \in V$ for $k = 0, \ldots, p$, and $(v_{i_{k-1}}, v_{i_k}) \in L$ for $k = 1, \ldots, p$.*
The nodes v_{i_0} and v_{i_p} are called **endpoints** *of the path C.*
The integer p is called **length** *of the path C.*
A **directed path** *$C = \{v_{i_0}, v_{i_1}, \ldots, v_{i_p}\}$ is called:*

- **simple** *if all its nodes are distinct from each other, i.e.*

$$v_{i_s} \neq v_{i_k} \qquad \forall s, k = 0, \ldots, p \ : \ s \neq k;$$

- **closed** *if its two endpoints coincide:*

$$v_{i_0} = v_{i_p};$$

- **(simple) cycle** *if the path $\{v_{i_0}, v_{i_1}, \ldots, v_{i_p}\}$ is closed and the path $\{v_{i_0}, v_{i_1}, \ldots, v_{i_{p-1}}\}$ is simple;*
- **(oriented) loop** *if it is a cycle with only one node.*

Referring to the graph in Fig. 7.1, the path $\{1, 2, 9, 8, 3\}$ is simple, the path $\{5, 4, 6, 5, 4, 6, 5\}$ is closed but it is not a cycle, the path $\{2, 9, 8, 3, 2\}$ is a cycle, the paths $\{1, 1\}$ and $\{7, 7\}$ are loops.
We specify the possible relationships between pairs of nodes.

Definition 7.21. *Given a directed graph $G = (V, L)$ we say that:*

- *a node $v_i \in V$ is* **connected** *with a node $v_j \in V$, and we write $v_i \to v_j$, if there is a directed path from v_i to v_j in G (we will write $v_i \overset{k}{\to} v_j$ when it is necessary to specify that starting from v_i one arrives in v_j in k steps);*
- *a node $v_i \in V$ is* **strongly connected** *with the node $v_j \in V$, and we write $v_i \leftrightarrow v_j$, if $v_i \to v_j$ and $v_j \to v_i$.*

A directed graph G is called **strongly connected** *if it has only one node or all its nodes are strongly connected to each other.*

In the graph of Fig. 7.1, node 2 is strongly connected with node 8, node 9 is connected with node 5 (but 5 is not connected with 9): so the graph is not strongly connected.

Given a directed graph $G = (V, L)$, the strong connectivity relationship is an equivalence relation on the set of nodes V. Then the set of nodes can be partitioned, in the sense that there are equivalence classes of nodes $V_1, V_2, \ldots,$

V_q such that $V = \bigcup\limits_{j=1}^{q} V_j$ and $V_i \cap V_j = \emptyset$ for $i \neq j$. The following definition characterizes every subgraph of G whose nodes belong to a single equivalence class.

Definition 7.22. *Given a directed graph $G = (V, L)$ we call **strongly connected component of** G each subgraph of G that is strongly connected and maximal (i.e. it is not strictly contained in another strongly connected subgraph of G).*

Example 7.23. The four strongly connected components of the graph in Fig. 7.1 are: $V_1 = \{1\}$, $V_2 = \{2, 9, 8, 3\}$, $V_3 = \{4, 5, 6, 7\}$ and $V_4 = \{10\}$.   □

Exercise 7.12. Given a directed graph $G = (V, L)$, prove that strong connectivity is an equivalence relationship on the set of its nodes V.

A further useful way to represent a directed graph is given by the following notion of adjacency matrix.

Definition 7.24. *The **adjacency matrix** of a directed graph $G = (V, L)$ with n nodes, is the square matrix $\mathbb{A} = [a_{ij}]$ of order n such that*

$$a_{ij} = \begin{cases} 1 & if \ (v_j, v_i) \in L \quad i.e. \quad v_j \xrightarrow{\ 1\ } v_i \\ 0 & if \ (v_j, v_i) \notin L \end{cases}.$$

On the other hand, to any given nonnegative ($a_{ij} \geq 0 \ \forall i, j$) matrix $\mathbb{A}$ of order n we can associate the directed graph $G_{\mathbb{A}} = (V_{\mathbb{A}}, L_{\mathbb{A}})$ where

$$V_{\mathbb{A}} = \{1, 2, \ldots, n\} \quad and \quad L_{\mathbb{A}} = \{(j, i) \in V_{\mathbb{A}} \times V_{\mathbb{A}} : a_{ij} > 0\} \ .$$

Example 7.25. The adjacency matrix of the graph in Fig. 7.1 is:

$$\begin{bmatrix} 1 & 0 & 0 & 0 & 0 & 0 & 0 & 0 & 0 & 0 \\ 1 & 0 & 1 & 0 & 0 & 0 & 0 & 0 & 1 & 0 \\ 0 & 0 & 0 & 0 & 0 & 0 & 0 & 1 & 0 & 0 \\ 0 & 0 & 1 & 0 & 1 & 0 & 1 & 0 & 0 & 0 \\ 0 & 0 & 0 & 0 & 0 & 1 & 0 & 0 & 0 & 0 \\ 0 & 0 & 0 & 1 & 0 & 0 & 0 & 0 & 0 & 0 \\ 0 & 0 & 0 & 0 & 0 & 1 & 1 & 0 & 0 & 0 \\ 0 & 0 & 0 & 0 & 0 & 0 & 0 & 0 & 1 & 0 \\ 0 & 1 & 0 & 0 & 0 & 0 & 0 & 0 & 0 & 0 \\ 0 & 0 & 0 & 0 & 0 & 0 & 0 & 0 & 0 & 0 \end{bmatrix}.$$

     □

Exercise 7.13. Given the matrix

$$\mathbb{A} = \begin{bmatrix} 2 & 1 & 0 & 1 & 0 & 3 \\ 0 & 0 & 4 & 0 & 0 & 0 \\ 0 & 5 & 0 & 0 & 0 & 0 \\ 0 & 0 & 0 & 3 & 0 & 0 \\ 0 & 3 & 0 & 0 & 2 & 0 \\ 1 & 0 & 2 & 0 & 0 & 1 \end{bmatrix}$$

draw the associated graph, identifying all its strongly connected components, if any.

Given a nonnegative matrix $\mathbb{A}$, every non-zero element $a_{ij} > 0$ can be considered as a weight associated to the edge of the graph $G_\mathbb{A}$ that connects j with i. More in general, given a directed path $C = \{i_0, i_1, \ldots, i_p\}$ in $G_\mathbb{A}$, with $p \geq 1$, we call **weight** of the path C the product

$$a_{i_1 i_0} \cdot a_{i_2 i_1} \cdot \ldots \cdot a_{i_p i_{p-1}} > 0.$$

If $\mathbb{A}^k = \left[a_{ij}^{(k)} \right]$ then it is possible to prove that for any $k \geq 2$

$$a_{ij}^{(k)} = \sum_{s=1}^{n} a_{is} a_{sj}^{(k-1)} = \sum_{\substack{i_0 = j,\ i_k = i \\ i_s \in V_\mathbb{A},\ \forall s}} a_{i_1 i_0} \cdot a_{i_2 i_1} \cdot \ldots \cdot a_{i_k i_{k-1}}$$

that is $a_{ij}^{(k)}$ is the sum of the weights of distinct paths of length k from j to i. This result immediately implies the following conclusion:

$a_{ij}^{(k)} > 0$ if and only if $G_\mathbb{A}$ has a directed path of length k from j to i.

In particular, if $\mathbb{A}$ is the adjacency matrix of a directed graph, then $a_{ij}^{(k)}$ represents exactly the (integer) number of distinct paths of length k between the nodes v_j and v_i.

Theorem 7.26. *A directed graph $G_\mathbb{A}$ is strongly connected if and only if for each i and j there exists $k = k(i, j)$ such that $a_{ij}^{(k)} > 0$.*

We specify the notion of irreducibility of a nonnegative square matrix, using the language of graphs.

Theorem 7.27. *A nonnegative square matrix $\mathbb{A} = [a_{ij}]$ of order $n > 1$ is irreducible if and only if the graph $G_\mathbb{A}$ associated to $\mathbb{A}$ is strongly connected.*

Proof. If $\mathbb{A}$ is irreducible, then for each pair of indices $i, j \in \{1, \ldots, n\}$ there exists a positive integer k (depending on i and j) such that $a_{ij}^{(k)} > 0$ (see Corollary 7.11) and this is equivalent to saying that each pair of nodes j, i in $V_\mathbb{A}$ is connected, namely the graph $G_\mathbb{A}$ is strongly connected.

To prove the converse, we show that if $\mathbb{A}$ is reducible, then $G_{\mathbb{A}}$ is not strongly connected. In fact, if $\mathbb{A}$ is reducible, then by a suitable permutation $\mathbb{T}$, it can be transformed into the matrix $\mathbb{E}$

$$\mathbb{E} = \mathbb{T}\mathbb{A}\mathbb{T}^{-1} = \begin{bmatrix} \mathbb{B} & \mathbb{O} \\ \mathbb{C} & \mathbb{D} \end{bmatrix}$$

with $\dim \mathbb{B} = r$, $0 < r < n$. For any choice of $i \leq r < j$ there are no paths of length 1 from v_j to v_i in the associated graph $\Gamma_{\mathbb{E}}$. Consequently, for any $i \leq r < j$, there are no paths from v_j to v_i, that is $G_{\mathbb{E}}$, that coincides with $G_{\mathbb{A}}$ up to the permutation associated to $\mathbb{T}$ of the indices which label the nodes, is not strongly connected. $\square$

From what we have just proved it follows that a matrix of order $n > 1$ is reducible if and only if the associated graph has at least two strongly connected components.

Remark 7.28. Theorems 7.26 and 7.27 together tell us that a nonnegative square matrix $\mathbb{A}$ of order $n > 1$ is irreducible if and only if

$$\forall i, j \;\; \exists q = q\,(i, j) \;:\; a_{ij}^{(q)} > 0, \tag{7.6}$$

that is the converse of Corollary 7.11 holds.

Thanks to the correspondence between graphs and matrices we can now prove a characterization of irreducible matrices that are not necessarily positive (see also Theorem 7.10).

Theorem 7.29. *A square matrix $\mathbb{A}$ of order $n > 1$ is irreducible if and only if*

$$(\mathbb{I} + |\mathbb{A}|)^{n-1} \gg \mathbb{O}$$

where $|\mathbb{A}| = [|a_{ij}|]$.

Proof. Observe that $\mathbb{A}$ is irreducible if and only if $|\mathbb{A}|$ is irreducible too.
If $\mathbb{A}$ is irreducible then the same holds for $|\mathbb{A}|$ and by Theorem 7.10 $(\mathbb{I} + |\mathbb{A}|)^{n-1} \gg \mathbb{O}$.
On the other hand, let $(\mathbb{I} + |\mathbb{A}|)^{n-1} \gg \mathbb{O}$ and assume by contradiction that $\mathbb{A}$ is reducible. Then also $|\mathbb{A}|$ is reducible and by Theorem 7.27 the directed graph $G_{|\mathbb{A}|}$ associated to $|\mathbb{A}|$ is not strongly connected. So, there exist $j \neq i$ such that j is not connected with i in $G_{|\mathbb{A}|}$. Therefore also the directed graph $G_{\mathbb{I}+|\mathbb{A}|}$ associated to $\mathbb{I} + |\mathbb{A}|$ is not strongly connected, hence the element at place (i, j) of $(\mathbb{I} + |\mathbb{A}|)^{n-1}$ is 0, a contradiction with the assumption. $\square$

Definition 7.30. *Given a directed graph $G = (V, L)$, we define the **period of the node** $v_i \in V$, and denote it by $d\,(v_i)$, as the greatest common divisor of the lengths of all closed paths passing through v_i:*

$$d\,(v_i) = GCD\left\{ k \in \mathbb{N} \backslash \{0\} : v_i \xrightarrow{k} v_i \right\}.$$

The period of a node where no path passes through is set equal to 0 by convention.

A node is called **aperiodic** *if it has period* 1.

Notice that a loop is associated to an aperiodic node.

Example 7.31. With reference to the graph of Fig. 7.1, we observe that:

- the node 10 has period 0;
- the nodes 1 and 7 have period 1;
  the nodes 4, 5, and 6 have period 1 too, because 7 has a loop and the others belong to closed paths containing the loop;
- the nodes 2 and 9 have period 2;
  the nodes 3 and 8 have period 2 too, because they are part of closed paths of lengths 4 and 6: $\{8,3,2,9,8\}$, $\{8,3,2,9,2,9,8\}$. □

Lemma 7.32. *Given a directed graph* $G = (V, L)$, *if two nodes belong to the same strongly connected component, then they have the same period.*

Proof. Let $v_i \in V$ and $v_j \in V$ be any two nodes in the same strongly connected component of G. It suffices to prove that $d(v_j)$ divides $d(v_i)$. To this end it is sufficient to consider only the paths γ of length k from v_i to v_i which do not contain v_j. By hypothesis, there exists a path η of length h from v_j to v_j passing through v_i. If we consider the sum of paths $\gamma + \eta$, then $d(v_j)$ divides both h and $h + k$, thus it divides k. □

The previous result allows us to introduce the next definition.

Definition 7.33. *The* **period of a strongly connected component** *of a directed graph is the period of any one among the nodes that belong to it.*

Recalling that irreducible nonnegative matrices are associated with strongly connected graphs, the following statement holds.

Definition 7.34. *Given a square, nonnegative and irreducible matrix* $\mathbb{A}$, *we define the* **period of the matrix** $\mathbb{A}$, *and denote it by* $d_{\mathbb{A}}$, *as the period* d *of any of the nodes of* $G_{\mathbb{A}}$. *The matrix* $\mathbb{A}$ *is called:*

- **aperiodic** *if* $d_{\mathbb{A}} = 1$;
- **periodic** *if* $d_{\mathbb{A}} > 1$.

Lemma 7.35. *Let* $\mathbb{A} > \mathbb{O}$ *be an irreducible square matrix of order* $n > 1$. *Then, for each* $i = 1, \ldots, n$ *there exists an integer* $k_0(i) > 0$ *such that* $a_{ii}^{(kd)} > 0$ *for every* $k \geq k_0(i)$, *where* $d = d_{\mathbb{A}}$ *is the period of* $\mathbb{A}$.

Proof. We denote by $H(i)$ the set $\left\{ k \in \mathbb{N} \setminus \{0\} : a_{ii}^{(k)} > 0 \right\}$. It represents the set of lengths of the paths that start from i and end at i in the graph $G_{\mathbb{A}}$ associated to $\mathbb{A}$. By definition, $d_{\mathbb{A}}$ is the the greatest common divisor of $H(i)$; moreover $H(i)$ is closed with respect to the sum of paths, in the sense that, for every $k_1 \in H(i)$ and $k_2 \in H(i)$, we have

$$a_{ii}^{(k_1+k_2)} \geq a_{ii}^{(k_1)} a_{ii}^{(k_2)} > 0 \quad \Rightarrow \quad (k_1 + k_2) \in H(i) .$$

So $H(i)$ contains all the positive multiples of its GCD except at most a finite number of these.   □

We know (see Definition 7.15) that a primitive matrix is a particular positive and irreducible matrix characterized by having a unique eigenvalue of maximum modulus that is simple and positive. The following result clarifies the link between primitive and irreducible matrices in terms of periodicity.

Theorem 7.36. *Let $A > 0$ be an irreducible square matrix of order $n > 1$. Then*

- A *is primitive if and only if it is aperiodic;*
- A *is cyclic if and only if it is periodic.*

Proof. Let A be primitive. Since there exists k_0 such that $A^k \gg 0$ for every $k \geq k_0$, it follows that each node of G_A is aperiodic because it belongs to a closed path of length k for any $k \geq k_0$.

On the other hand, assume that A is an aperiodic irreducible matrix. Then (Theorem 7.26) for every i and j there exist $k_0 = k_0(i,j)$ such that $a_{ij}^{(k_0)} > 0$ and (Lemma 7.35) $k_1(j) > 0$ such that $a_{jj}^{(k)} > 0$ for each $k \geq k_1(j)$.

Having chosen $k_2(j) = k_0 + k_1(j)$, for each integer $k \geq k_2(j)$ we have

$$a_{ij}^{(k)} \geq a_{ij}^{(k_0)} a_{jj}^{(k-k_0)} > 0 .$$

After setting $k_3 = \max_j \{k_2(j)\}$ we obtain $A^k \gg 0$ for each $k \geq k_3$, hence A is primitive.   □

Example 7.37. In the directed graph associated to the adjacency matrix

$$A = \begin{bmatrix} 0 & 0 & 1 & 0 & 0 & 1 \\ 1 & 0 & 0 & 0 & 0 & 0 \\ 0 & 1 & 0 & 0 & 0 & 0 \\ 1 & 0 & 0 & 0 & 0 & 0 \\ 0 & 0 & 0 & 1 & 0 & 0 \\ 0 & 0 & 0 & 0 & 1 & 0 \end{bmatrix} \tag{7.7}$$

the closed paths $\{v_1, v_2, v_3, v_1\}$ and $\{v_1, v_4, v_5, v_6, v_1\}$ pass through the node v_1, then the node is aperiodic. Since the graph is strongly connected, one concludes that it is aperiodic and matrix (7.7) is irreducible and primitive. One can verify that $A^k \gg 0$ if $k \geq 12$.   □

Exercise 7.14. Compute, if it exists, the period of the following matrices:

$$A_1 = \begin{bmatrix} 0 & 1 \\ 1 & 0 \end{bmatrix} \qquad A_2 = \begin{bmatrix} 1 & 0 \\ 0 & 1 \end{bmatrix} \qquad A_3 = \begin{bmatrix} 0 & 1 \\ 1 & 1 \end{bmatrix} .$$

$$A_4 = \begin{bmatrix} 0 & 1 & 0 \\ 1 & 0 & 1 \\ 0 & 1 & 0 \end{bmatrix} \qquad A_5 = \begin{bmatrix} 1 & 1 & 1 \\ 1 & 1 & 0 \\ 1 & 0 & 0 \end{bmatrix} .$$

Table 7.1. Properties of matrices: $|A|$ denotes the matrix with elements $|a_{ij}|$

If A is a square matrix of order $n > 1$ then
A **irreducible** $\iff$ $(\mathbb{I}_n + |A|)^{n-1} \gg \mathbb{O}$ $\iff$ $(\mathbb{I}_n + |A|)$ **primitive**

Table 7.2. Properties of positive matrices

If $A \geq 0$ square matrix of order $n > 1$ then
A **irreducible** $\iff$ $(\mathbb{I}_n + A)^{n-1} \gg \mathbb{O}$ $\quad \updownarrow \qquad\qquad\qquad\qquad\qquad \updownarrow$ G_A **strongly connected** $\iff$ $\forall i,j \ \exists k\,(i,j) \in \mathbb{N}\backslash\{0\}: a_{ij}^{(k)} > 0$
$\qquad\qquad\qquad\qquad\qquad \Uparrow$
$\exists k \in \mathbb{N}\backslash\{0\}: A^k \gg \mathbb{O} \iff A$ **primitive** $\iff A$ **irreducible and aperiodic**

Legend: $a_{ij}^{(k)}$ denotes the element on row i and column j in the matrix A^k

Exercise 7.15. Determine the periods of the strongly connected components of the graph associated to the matrix in Exercise 7.13.

Exercise 7.16. Prove that all the eigenvalues of a permutation matrix $\mathbb{T}$ have modulus equal to 1. More precisely, they are the roots of a polynomial of the type $\prod_{j=1}^{m} \left(\lambda^{k_j} - 1\right)$ where $k_1 + k_2 + \cdots + k_m = n$ and n is the order of $\mathbb{T}$. This polynomial is, up to sign, the characteristic polynomial of $\mathbb{T}$. Moreover, if the matrix $\mathbb{T}$ is irreducible, then all its eigenvalues are simple: they are the n-th complex roots of unity.

We summarize in two tables the relationships among the many definitions and properties introduced in this chapter, showing that the different terminologies that occur naturally in different contexts often describe equivalent properties with reference to the theorems of the text.

7.3 More on Markov Chains

The concepts developed in this chapter provide useful tools for the analysis of Markov chains introduced in Chap. 6. Referring to the definition of *strongly connected nodes of a graph* (Definition 7.21), when studying a Markov chain with transition matrix $\mathbb{M}$ it is natural to call **communicating states** those states that correspond to strongly connected nodes in the graph $G_\mathbb{M}$ associated to this chain, as specified by the following definition.

Definition 7.38. *Let $\mathbb{M} = [m_{ij}]$ be the transition matrix of a homogeneous Markov chain with n states $s_1, \ldots, s_n$. We say that:*

- *the state s_i **is accessible from the state** s_j if there exists $k \in \mathbb{N}\setminus\{0\}$ such that $m_{ij}^{(k)} > 0$ (in this case we will write $s_j \xrightarrow{k} s_i$ or briefly $s_j \rightarrow s_i$);*
- *the states s_i and s_j are **communicating** if $s_j \xrightarrow{k} s_i$ and $s_i \xrightarrow{h} s_j$ for suitable k, h (in this case we will write $s_i \longleftrightarrow s_j$).*

Being communicating states is an equivalence relation on the set of the n states $\{s_1, s_2, \ldots, s_n\}$ of a Markov chain. Therefore this set can be partitioned in disjoint equivalence classes (each one is called **communicating class**), such that each of these states is communicating with each other. We will refer indifferently to **communicating classes** of both the set of states or of the corresponding transition matrix.

A **communicating class** C is **closed** if $s_i \in C$ and $s_i \rightarrow s_j$ imply $s_j \in C$. A state s_i is absorbing if $\{s_i\}$ is a closed communicating class.

Example 7.39. In order to determine the communicating classes corresponding to the stochastic matrix

$$\mathbb{M} = \begin{bmatrix} 1/3 & 0 & 1/4 & 0 & 0 & 0 \\ 2/3 & 0 & 0 & 0 & 0 & 0 \\ 0 & 1 & 0 & 0 & 0 & 0 \\ 0 & 0 & 1/4 & 1/2 & 0 & 0 \\ 0 & 0 & 1/2 & 1/2 & 0 & 1 \\ 0 & 0 & 0 & 0 & 1 & 0 \end{bmatrix}$$

we can use the corresponding graph, thus obtaining the classes $\{1, 2, 3\}$, $\{4\}$ and $\{5, 6\}$: only the last one is closed. $\qquad\square$

Example 7.40. The following stochastic matrix

$$\mathbb{A} = \begin{bmatrix} 0 & 1 & 0 & 0 & 0 & 0 & 0 & 0 \\ 0 & 0 & 1 & 0 & 0 & 0 & 0 & 0 \\ 0 & 0 & 0 & 1 & 0 & 0 & 0 & 0 \\ 0 & 0 & 0 & 0 & 1/2 & 0 & 0 & 0 \\ 1 & 0 & 0 & 0 & 0 & 1 & 0 & 0 \\ 0 & 0 & 0 & 0 & 0 & 0 & 1 & 0 \\ 0 & 0 & 0 & 0 & 0 & 0 & 0 & 1 \\ 0 & 0 & 0 & 0 & 1/2 & 0 & 0 & 0 \end{bmatrix}$$

is irreducible and primitive (it has period 1 because the associated graph exhibits periods 4 and 5), and has a unique communicating class. Correspondingly, the associated graph $G_\mathbb{A}$ has a unique strongly connected component. The characteristic polynomial of $\mathbb{A}$ is:

$$\lambda^8 - \frac{1}{2}\lambda^4 - \frac{1}{2}\lambda^3 = \lambda^3(\lambda - 1)(\lambda^4 + \lambda^3 + \lambda^2 + \lambda + 1/2).$$

The eigenvalues are 1, 0, $-0,668809 \pm 0,338851\,i$, $0,168809 \pm 0,927891\,i$; they are all simple except 0 which is triple; in agreement with the Frobenius Theorem, they all have modulus less than the dominant eigenvalue 1.   □

Example 7.41. The stochastic matrix

$$\mathbb{B} = \begin{bmatrix} 0 & 1 & 0 & 0 & 0 & 0 & 0 & 0 & 0 \\ 0 & 0 & 1 & 0 & 0 & 0 & 0 & 0 & 0 \\ 0 & 0 & 0 & 1 & 0 & 0 & 0 & 0 & 0 \\ 0 & 0 & 0 & 0 & 1 & 0 & 0 & 0 & 0 \\ 1 & 0 & 0 & 0 & 0 & 0 & 0 & 0 & 0 \\ 0 & 0 & 0 & 0 & 0 & 0 & 1 & 0 & 0 \\ 0 & 0 & 0 & 0 & 0 & 0 & 0 & 1 & 0 \\ 0 & 0 & 0 & 0 & 0 & 0 & 0 & 0 & 1 \\ 0 & 0 & 0 & 0 & 0 & 1 & 0 & 0 & 0 \end{bmatrix} = \begin{bmatrix} \mathbb{F}_1 & \mathbb{O} \\ \mathbb{O} & \mathbb{F}_2 \end{bmatrix}$$

is reducible with two closed communicating classes, one (containing the first five states) is periodic with period 5, the other (consisting of the last four states) is periodic with period 4. The corresponding graph $G_\mathbb{B}$ has two strongly connected components (the one containing the first 5 nodes and the one containing the last 4).

The characteristic polynomial of $\mathbb{B}$ is $\left(\lambda^5 - 1\right)\left(\lambda^4 - 1\right)$; the eigenvalues are all the fourth and fifth complex roots of 1: they are all simple except $\lambda = 1$ that is double.

Notice the structure of the square blocks $\mathbb{F}_j$: they are in the elementary forms of Frobenius.   □

It is natural to introduce the next definition.

Definition 7.42. *An **irreducible Markov chain** is a Markov chain whose transition matrix is irreducible.*
*A **reducible Markov chain** is a Markov chain whose transition matrix is reducible.*

A relevant example of irreducible Markov chains is provided by absorbing Markov chains that, with appropriate ordering of the states, have the form (6.16) and whose asymptotic analysis is described in Theorem 6.26. Being reducible does not imply being absorbing (see Example 7.43).

By Remark 7.28 and Theorem 7.27 we know that a *Markov chain* is *irreducible* if the set of states is a single communicating class (hence a closed one) or, equivalently, if for each pair of states s_i and s_j there exists $k = k\,(i, j)$ such that $m_{ij}^{(k)} > 0$.

Notice that the matrix $\mathbb{M}$ in Example 7.39 is reducible.

Example 7.43. The Markov chain with transition matrix

$$M = \begin{bmatrix} 1/3 & 1/2 & 0 & 0 \\ 2/3 & 1/2 & 0 & 0 \\ 0 & 0 & 3/4 & 1/4 \\ 0 & 0 & 1/4 & 3/4 \end{bmatrix}$$

is reducible. Observe that if one assumes as initial datum the probability distribution vector with 1 in one of the first (respectively, last) two positions, the dynamics of the system will take place exclusively between the first (respectively, last) two states.   □

Exercise 7.17. Prove that every finite-dimensional transition matrix has at least one closed communicating class.

Definition 7.44. *Given an homogeneous Markov chain with n states $s_1, \ldots, s_n$, a **state** s_i is called **aperiodic** if there exists $k_0 \in \mathbb{N} \setminus \{0\}$ such that $m_{ii}^{(k)} > 0$ for each $k > k_0$.*
*A Markov chain is called **aperiodic chain** if all states are aperiodic.*

Therefore a state s_i is aperiodic if and only if the set $\left\{ k \in \mathbb{N} \setminus \{0\} : m_{ii}^{(k)} > 0 \right\}$ has no common divisors greater than 1.
If a Markov chain is aperiodic, then there exists $k \in \mathbb{N} \setminus \{0\}$ such that $m_{ii}^{(k)} > 0$ for every i.
With the terminology introduced, we may reformulate Lemma 7.32 and Theorem 7.36 as follow.

Lemma 7.45. *If M is irreducible and has an aperiodic state, then every state is aperiodic.*

Theorem 7.46. *Let M be the transition matrix of an irreducible and aperiodic Markov chain. Then there exists $k_0 \in \mathbb{N} \setminus \{0\}$ such that $m_{ij}^{(k)} > 0$ for each $k > k_0$ and i, j.*

If a Markov chain has a transition matrix that satisfies the assumptions of Theorem 7.13 (Frobenius) then the existence, the uniqueness and the global stability of the invariant probability distribution is ensured only for $h = 1$.

Definition 7.47. *Given a Markov chain with n states, a state s_i is called:*

- **transient** *if there exists a state s_j with $s_j \neq s_i$ such that $s_i \to s_j$ but $s_j \not\to s_i$;*
- **recurrent** *if it is not transient, i.e., if: $s_i \to s_j \Rightarrow s_j \to s_i$.*

In simple words, a state is recurrent if sooner or later the process will certainly come back to it, it will come back infinitely any times; a state is transient if the process after transiting can never return there.

Note that for a Markov chain (with a finite number of states):

1. each state is either recurrent or transient in an exclusive way;
2. an absorbing state is recurrent;
3. if a state is recurrent then it communicates with itself;
4. states cannot be all transient (because the dynamic has to make infinite steps);
5. a recurrent but not absorbing state can be visited infinitely many times.

For what concerns the partition of a Markov chain in communicating classes, the following conclusions hold:

1. in a communicating class the states are all transients or all recurrent; in the first case the class is said to be transient, in the second one recurrent;
2. each communicating class is closed if and only if is recurrent.

An important consequence of these remarks is the following statement.

Theorem 7.48. *Every state of a finite and homogeneous irreducible Markov chain is recurrent.*

Example 7.49. The recurrent states in Example 6.31 (see Fig. 6.6) are "A wins" and "B wins". Anything else is transient. Notice that the transient states "advantage A", "advantage B" and "deuce" can be visited several times by a trajectory, while all other transient states can be visited at most once.
We also observe that the chain is absorbing and its absorbing states are the recurrent states "A wins" and "B wins". This is true in general: the absorbing states of an absorbing chain are recurrent.
The transition matrix describing the possible run of a game already at "deuce" (by ordering the five absorbing states left to right as in Fig. 6.6) is

$$\begin{bmatrix} 1 & p_A & 0 & 0 & 0 \\ 0 & 0 & p_A & 0 & 0 \\ 0 & p_B & 0 & p_A & 0 \\ 0 & 0 & p_B & 0 & 0 \\ 0 & 0 & 0 & p_B & 1 \end{bmatrix}. \qquad \square$$

Summing up the above considerations the subsequent conclusions hold for the probability of an invariant Markov chain:

1. if s_i is a transient state and $\mathbf{P} = \begin{bmatrix} P_1 & P_2 & \cdots & P_n \end{bmatrix}^T$ is an invariant probability distribution for the transition matrix $\mathbb{M}$, then $P_i = 0$;
2. there exist q linearly independent invariant probability distribution vectors if and only if the chain has q classes of recurrent states;
3. a chain has a unique invariant probability distribution if and only if it has a unique class of recurrent states and

$$P_i = 0 \ \text{ if } s_i \text{ is transient}, \qquad P_i > 0 \ \text{ if } s_i \text{ is recurrent};$$

4. if a chain is irreducible then there exists a unique invariant probability distribution $\mathbf{P}$ with $\mathbf{P} \gg 0$;
5. if a chain is (irreducible and) primitive then the unique invariant probability distribution is globally attractive for the associated DDS.

Remark 7.50. The definition of reducible and irreducible matrix were introduced for square matrices of order $n > 1$ (Definition 7.4): the exclusion of the case $n = 1$ is natural if we refer to the properties of the decomposition into blocks of a matrix, as it is the case in Sect. 7.1. Nothing prevents to remove the restriction on the order, using a conventional definition. However, the opportunity of the convention varies depending on the context. In a purely algebraic approach (Sect. 7.1) it is convenient to consider irreducible all square matrices of order $n = 1$ so as to maintain the validity of Theorem 7.29 also for $n = 1$: but in the case of graphs such a convention would correspond to considering only graphs with loops at each vertex. For this reason, in the context of graphs and of Markov chains it is more natural to establish by convention that a matrix of order 1 is reducible (respectively irreducible) if its only element is null (respectively different from zero), so as to maintain the equivalence between strong connection of a graph and irreducibility of its adjacency matrix for the order 1.

7.4 Algorithm PageRank: why a good search engine seems to read the minds of those who questioned it

Let's look at a very successful application of the matrix properties discussed in this chapter: the algorithm PageRank used by the search engine *Google* designed in 1998 by Sergei Brin and Lawrence Page. The trademark Google reminds of a stratospheric number (google=10^{100}): the reason is that the scale of the problem that the search engines have to face and solve is huge.

The problem solved by Google for answering to the millions of daily queries is to provide answers each of which must order billions of web pages [1]. The essential requirement for the success of this procedure is an effective and fast sorting order of all pages that meet the criteria of the **query** in object. Once you have obtained an efficient method to sort the web pages in relation to their significance you can use the same sorting order in response to any individual query.

The exact mechanism of how Google works is not publicly known (of course many algorithmic details are secrets protected by copyright) and some aspects of its remarkable efficiency are explained only in part [5], [4], [11]. However, it is based on a procedure of sorting pages which is independent of their actual content and uses the topological structure of the Web, i.e. the kind of

---

[1] A useful reference about Web "vocabulary" is *www.webopedia.com*

connections that exist between the internet pages. The idea is to associate an **authority index** with each page that depends on the number of citations (hyperlinks or briefly links) obtained from other pages and the authority index of the latter: this procedure is called *PageRank algorithm*. We describe below a heuristic way to introduce and calculate the authority index, introducing a model of Internet surfing as a process described by a Markov chain.

A set of web pages P_1, ..., P_n can be represented as a directed graph $G = (V, L)$ where the vertices v_i coincide with pages P_i and $(v_j, v_i) \in L$ if and only if there exists a **link** from page P_j to page P_i.

We denote by $\mathbb{A} = [a_{ij}]$ the adjacency matrix of $G = (V, L)$. Then

$$a_{ij} = \begin{cases} 1 & \text{if there exists a link from } P_j \text{ to } P_i \\ 0 & \text{if there is no link from } P_j \text{ to } P_i \end{cases} .$$

Notice that the sum of the elements in the i-th row of $\mathbb{A}$ represents the number of pages having a link which leads to P_i, while the sum of the elements of the j-th column of $\mathbb{A}$ represents the number of pages which can be reached starting from P_j.

The dimension of the matrix $\mathbb{A}$ is gigantic, consider that already in the year 1997 it was estimated that there were about 100 millions of web pages, an amount that has increased dramatically in the time of writing this book and will be further increased by the time you read this page!

The authority index $x_j \geq 0$ of the j-th page P_j must meet two requirements: on the one hand it must be high if it refers to a page cited by many other pages, on the other it must be high when referring to a page cited by (possibly a few) very significant pages. So the mere counting of links pointing to a page is not adequate to represent the authority index (it does not satisfy the second requirement). It appears more appropriate to define the authority index x_j of page P_j so that it is proportional, with proportionality constant c equal for all pages, to the sum of the authority index of pages that refer to P_j through a link. We illustrate all these facts by introducing an example.

Example 7.51. Given four web pages P_1, P_2, P_3 and P_4, we suppose that page P_1 is cited by page P_2; page P_2 is cited by P_1 and P_4; page P_3 is cited only by P_2 and page P_4 is cited by all other pages. The authority indices x_1, x_2, x_3 and x_4 are positive solutions of the following linear algebraic system:

$$\begin{cases} x_1 = c\,x_2 \\ x_2 = c\,(x_1 + x_4) \\ x_3 = cx_2 \\ x_4 = c\,(x_1 + x_2 + x_3) \end{cases} . \tag{7.8}$$

If $\mathbb{A}$ is the adjacency matrix corresponding to the graph of the connections between web pages and $\mathbf{x} = \begin{bmatrix} x_1 & x_2 & x_3 & x_4 \end{bmatrix}^T$ is the vector whose components are the authority indices of the four pages, then the above problem can be rewritten in vector terms as $\mathbf{x} = c\mathbb{A}\mathbf{x}$ or, since $c \neq 0$, $\mathbb{A}\mathbf{x} = \lambda\mathbf{x}$ having set

$\lambda = 1/c$. Explicitly:

$$\mathbb{A} = \begin{bmatrix} 0 & 1 & 0 & 0 \\ 1 & 0 & 0 & 1 \\ 0 & 1 & 0 & 0 \\ 1 & 1 & 1 & 0 \end{bmatrix}. \tag{7.9}$$

Solving (7.8) we obtain:
$c_{\mathbb{A}} = 0.5651978$, dominant eigenvalue $\lambda_{\mathbb{A}} = 1.769292$, dominant eigenvector
$\mathbf{R}_{\mathbb{A}} = [\, 0.321595 \; 0.568996 \; 0.321595 \; 0.685125 \,]^T$.
The ordering of pages with respect to the authority index obtained in this
way is: $\tilde{R}_4 > \tilde{R}_2 > \tilde{R}_1 = \tilde{R}_3$.   □

The above example shows how *the search of the authority index for a given
set of web pages is brought back to the search of a positive eigenvalue λ with
associated positive eigenvector $\tilde{\mathbf{R}}$ of the adjacency matrix of the graph of the
Web connections between the pages under analysis*. Given the nature of the
problem, it is obviously desirable that the solution is unique or at least, if not
unique, that all the solutions we found lead to the same ordering by using the
rank $\tilde{R}_j$ as authority index of each page P_j.
We observe that the consideration of the adjacency matrix (which assigns
weight 1 to every page output from a given page) runs the risk to overesti-
mate the authority index of the pages that have many incoming links from
irrelevant pages and/or that are poorly selective in inserting links. It is therefore appropriate to consider the matrix $\mathbb{M} = [m_{ij}]$ whose elements satisfy

$$m_{ij} = \begin{cases} 1/N_j & \text{if there exists a link from } P_j \text{ to } P_i \\ 0 & \text{if there is no link from } P_j \text{ to } P_i \end{cases} \tag{7.10}$$

where N_j, for $j = 1, \ldots, n$, denotes the number of pages that one can reach
in one step starting from page P_j (namely N_j is the sum of the elements
belonging to the j-th column of matrix $\mathbb{A}$).

With this procedure, the matrix (7.9) in Example 7.51 is replaced by

$$\mathbb{M} = \begin{bmatrix} 0 & 1/3 & 0 & 0 \\ 1/2 & 0 & 0 & 1 \\ 0 & 1/3 & 0 & 0 \\ 1/2 & 1/3 & 1 & 0 \end{bmatrix}. \tag{7.11}$$

Solving $\mathbb{M}\mathbf{x} = \lambda\mathbf{x}$ one obtains: $c_{\mathbb{M}} = 1$, dominant eigenvalue $\lambda_{\mathbb{M}} = 1$, dominant eigenvector $\mathbf{R}_{\mathbb{M}} = [\, 0.133333 \; 0.399999 \; 0.133333 \; 0.333333 \,]^T$.
The page ordering with respect to the authority index obtained in this way
is: $R_2 > R_4 > R_1 = R_3$.
Notice that the use of a less refined matrix $\mathbb{A}$ has given greater authority to
the page P_4 which was linked by all the others but it linked only P_2. The
more efficient use of the matrix $\mathbb{M}$ attaches greater authority to page P_2 who
is linked only by P_1 and P_4 but has more links towards the others.

If from any web page one can reach at least another page, then the corresponding matrix $\mathbb{M}$ built according to (7.10) is stochastic.

In fact the matrix $\mathbb{M}$ is the transition matrix of a Markov chain describing a random walk between web pages assuming that surfing the Web we move randomly from one page to the other, that all the transitions are equally likely and are carried out only by clicking on the various links available. In this case the dominant eigenvalue of $\mathbb{M}$ is $\lambda_{\mathbb{M}} = 1$ and the computation of the authority index for all the web pages is equivalent to looking for a unique equilibrium $\widetilde{\mathbf{R}}$ (invariant probability distribution) for the associated Markov chain: $\widetilde{\mathbf{R}} = \mathbb{M}\widetilde{\mathbf{R}}$, $\widetilde{R}_j \geq 0$, $\sum_j \widetilde{R}_j$.

If $\mathbb{M}$ was (positive irreducible and) primitive, then (by Theorem 7.17) such a solution would be provided by the only positive eigenvector $\mathbf{V}^{\mathrm{M}}$ normalized (with respect to the sum of its components $\|\mathbf{V}^{\mathrm{M}}\|_1 = \sum_j V_j^{\mathrm{M}}$) and computable with the power method: this ideal situation can only describe a strongly connected Web graph (see Theorem 7.27).

Nevertheless it is natural to assume that (as in Example 7.51) every page is not linked with itself, that is, the matrix $\mathbb{M}$ has all zero entries in the main diagonal. Furthermore, it is not even true that the adjacency matrix of the whole web has a strictly positive power.

In the real case many significant clusters of web pages and also the entire network of the World Wide Web as a whole are not strongly connected. In particular, there are the the following problems:

- (p1) given a page P_j, *in general one cannot go from P_j to any other page*. In this case P_j is called **dangling page** and the j-th column of matrix $\mathbb{M}$ has only zero elements. Therefore $\mathbb{M}$ is not even stochastic and the corresponding graph $G_{\mathbb{M}}$ is not connected;
- (p2) even if from each page one can go to at least another page, *not necessarily one can go from any page to any other page*: in this case the matrix $\mathbb{M}$ is not irreducible and the corresponding graph $G_{\mathbb{M}}$ is not strongly connected.

There are several ways to remedy problems (p1), (p2): all these possibilities are based on the fact that the transit from one page to another does not happen just through links already present but also by entering the address of the pages on the keyboard. We just mention one option for each of those problems.

To solve problem (p1), one can assume that all dangling pages point to each page of the Web: this amounts to replacing the corresponding columns of zeroes in the matrix $\mathbb{M}$ with the vector $\frac{1}{n}\mathbf{1}_n = \begin{bmatrix} 1/n & 1/n & \cdots & 1/n \end{bmatrix}^T$, thus obtaining a stochastic matrix $\widetilde{\mathbb{M}}$:

$$\widetilde{\mathbb{M}} = \mathbb{M} + \frac{1}{n}\mathbf{1}_n\,\mathbf{B}^T$$

where the elements of the column vector $\mathbf{B}$ are $B_j = 1$ if P_j is dangling, $B_j = 0$ otherwise. Notice that this corresponds to assuming a uniform probability distribution for the transition to other pages without the use of links, but more general (and customized) hypotheses can be made.

The adjustment of the adjacency matrix just described, though it ensures to deal with a stochastic matrix $\widetilde{\mathbb{M}}$, it does not guarantee its irreducibility. To obtain this property (that is, solving problem (p2)) for the general case of the whole Web, denoting by $\mathbb{1}_n$ the square matrix of order n with every element set equal to 1, one can "perturb" the stochastic matrix $\widetilde{\mathbb{M}}$ as follows:

$$\mathbb{G} = d\widetilde{\mathbb{M}} + (1-d)\,\mathbb{E}$$
$$d \in (0,1) \qquad \mathbb{E} = \frac{1}{n}\mathbf{1}_n\,\mathbf{1}_n^T = \frac{1}{n}\mathbb{1}_n \tag{7.12}$$

thus obtaining a strictly positive transition matrix: $\mathbb{G} \gg \mathbb{O}$. So $\mathbb{G}$ is irreducible and primitive; furthermore $\mathbb{G}$ is stochastic because both $\widetilde{\mathbb{M}}$ and $\mathbb{E}$ are. By the Markov-Kakutani Theorem $\mathbb{G}$ admits an invariant probability distribution that is unique by the Perron-Frobenius Theorem. We denote by $\mathbf{R}$ this invariant probability distribution:

$$\mathbb{G}\,\mathbf{R} = \mathbf{R}, \qquad \mathbf{R} \geq 0, \qquad \sum_j \mathbf{R}_j = 1. \tag{7.13}$$

$\mathbf{R}$ is the dominant eigenvector of $\mathbb{G}$ and gives the ordering of its components R_j that can be validly used as alternative to the authority index $\widetilde{R}_j$ of page P_j: we emphasize that we are not much interested in the numerical value of the R_j's but in their decreasing order (see [4]).

$\mathbf{R}$ is the vector of the authority index of the web pages: the component R_j is the authority index of page P_j.

Theorems 6.36, 6.39 and 7.17 imply

$$\lim_k \mathbb{G}^k\,\mathbf{P}_0 = \mathbf{R} \qquad \text{for each } \mathbf{P}_0 \text{ probability distribution,} \tag{7.14}$$

all the columns of $\mathbb{M}^k$ converge to $\widetilde{\mathbf{R}}$ as k tends to $+\infty$:

$$\lim_k \mathbb{G}^k = [\,\widetilde{\mathbf{R}}\,|\,\widetilde{\mathbf{R}}\,|\,\cdots\,|\,\widetilde{\mathbf{R}}\,]. \tag{7.15}$$

These theoretical results allow computer programmers to tackle the real problem and its difficulties related to the size of the involved matrices (it is now estimated that there are more than $9 \cdot 10^9$ pages). In fact the computation algorithm of the dominant eigenvector by the power method is very fast: Theorem 7.17 assures that the speed is linked to the second eigenvalue of $\mathbb{G}$ (about the estimation of the second eigenvalue of matrix $\mathbb{G}$ by d see [11]).

Relations (7.14) and (7.15) describe the asymptotic analysis of the DDS associated to the regular Markov chain with stochastic matrix $\mathbb{G}$:

$$\mathbf{P}_{k+1} = \mathbb{G}\,\mathbf{P}_k \qquad \mathbf{P}_0 \text{ probability distribution.} \tag{7.16}$$

By the structure properties of the matrix $\mathbb{E}$ the homogeneous linear vector DDS (7.16) is equivalent to the non homogeneous linear vector-valued DDS

$$\mathbf{P}_{k+1} = d\widetilde{\mathbb{M}}\mathbf{P}_k + \frac{(1-d)}{n}\mathbf{1}_n. \tag{7.17}$$

Notice that the matrix $\widetilde{\mathbb{M}}$ is sparse, contrary to $\mathbb{G}$ whose elements are all strictly positive: therefore it is very useful to exploit the equivalence highlighted and implement the power method on (7.17) instead of using (7.16).

7.5 Summary exercises

Exercise 7.18. Determine if the stochastic matrix

$$\mathbb{M} = \begin{bmatrix} 0 & 0 & 0 & 1/2 \\ 1 & 0 & 1/4 & 0 \\ 0 & 1 & 0 & 1/2 \\ 0 & 0 & 3/4 & 0 \end{bmatrix}$$

is reducible or irreducible.

Exercise 7.19. Prove that the square matrix A of order n is reducible if and only if it corresponds to a linear map of $\mathbb{R}^n$ in $\mathbb{R}^n$ that transforms a nontrivial (i.e. different from $\{0\}$) proper subspace V generated by the canonical basis of $\mathbb{R}^n$ into itself.

Exercise 7.20. Assume that all the survival rates are strictly positive ($\sigma_j > 0$, $j = 1,\ldots,n-1$), then prove that the matrix in the Leslie model is irreducible if and only if $\varphi_n > 0$.

Exercise 7.21. Prove that if the matrix of the Leslie model is irreducible (e.g. $\varphi_n > 0$) then it is primitive if at least two consecutive age classes are fertile. Show that the condition is not necessary.

Exercise 7.22. In the model for the assignment of the authority index and the computation of the eigenvector $\mathbf{R}$ (Sect. 7.4 - Algorithm PageRank) we have implicitly assumed that the result does not depend on the (arbitrary) ordering of the pages, that is on the index j according to which the pages P_j are ordered. Prove the validity of this assumption.
Hint: to show the independence of $\mathbf{R}$ with respect to a permutation τ of the pages it is sufficient to exploit the invariance of the eigenvalues and eigenvectors with respect to transformation of $\mathbb{G}$ in $\mathbb{T}\mathbb{G}\mathbb{T}^T$ where $\mathbb{T}$ is the permutation matrix associated to τ.

Exercise 7.23. Referring to Example 7.51, the managers of page P_1 are dissatisfied with the authority (or ranking) attributed to this page: it is less than the one ascribed to page P_4. In an attempt to increase it, they create a further page P_5 with a link from P_5 to P_1 and a link from P_1 to P_5.
Can this action increase the authority of P_1 over that of P_4?

8

Solutions of the exercises

8.1 Solutions of the exercises in Chapter 1

Solution 1.1 If we trace the various lines, giving each one a conventional orientation, then on the $(k+1)$-th line there are k distinct and ordered points $p_1, p_2, \ldots, p_k$, that are the intersections of that line with the pre-existing lines. These points identify on the line $k-1$ segments and two half-lines. Each of these $k+1$ geometric objects disconnects one and only one of the regions that "it crosses", whereby

$$\begin{cases} N_{k+1} = N_k + k + 1 \\ N_0 = 1 \end{cases}$$

whose solution is $N_k = \dfrac{1}{2}k^2 + \dfrac{1}{2}k + 1$. This formula can be proved by induction. The technique for obtaining the formula is illustrated in Chap. 2.

To answer the question, just observe that the cuts should be carried out in such a way that each pair of cuts intersects inside the circle, but no more than two intersect at each of those points. Therefore, the maximum number is precisely the one given by the formula: $N_k = \dfrac{1}{2}k^2 + \dfrac{1}{2}k + 1$.

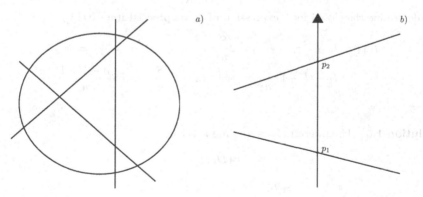

Fig. 8.1 If $k=2$ then: a) $N_{k+1} = 7$; b) $(k+1)$-th line

E. Salinelli, F. Tomarelli: *Discrete Dynamical Models.*
UNITEXT – La Matematica per il 3+2 76
DOI 10.1007/978-3-319-02291-8_8, © Springer International Publishing Switzerland 2014

Solution 1.2   We proceed by induction. We have:

$$S_1 = (1+r)\,S_0 - R \geq (1+r)\,S_0 - rS_0 = S_0$$

and then the statement is true for $k = 1$. Assuming the statement true for $k \geq 1$, we obtain:

$$S_{k+1} = (1+r)\,S_k - R \geq \qquad \boxed{\text{since } R \leq rS_0}$$

$$\geq (1+r)\,S_k - rS_0 \geq \qquad \boxed{\text{by induction}}$$

$$\geq (1+r)\,S_k - rS_k =$$

$$= S_k \ .$$

In conclusion, if $R \leq rS_0$, then $S_{k+1} \geq S_k$, that is, the debt does not decrease as time passes.

Solution 1.3 Y_{k+1} is a proportion of Y_k with coefficient $1 - p_Y$. Instead, to obtain X_{k+1} one must consider the proportion $(1 - p_X)\,X_k$ of material present at the previous time that does not decay, to which the contribution $p_Y Y_k$ of the first material must be added. Ultimately:

$$\begin{cases} Y_{k+1} = (1 - p_Y)\,Y_k \\ X_{k+1} = p_Y Y_k + (1 - p_X)\,X_k \end{cases}$$

For the solution of this simple case refer to Exercise 2.17. For the closed form solution of systems of linear difference equations in the general case, see Chap. 5.

Solution 1.4   The interest share at time $k + 1$, by definition, is $I_{k+1} = rD_k$. Then we can write

$$I_{k+1} - I_k = rD_k - rD_{k-1} = r\,(D_k - D_{k-1}) = -rC = -\frac{S}{n}r$$

that is

$$I_{k+1} = I_k - \frac{S}{n}r \ .$$

To determine the closed form expression of I_k we proceed iteratively:

$$I_{k+1} = I_k - \frac{S}{n}r = I_{k-1} - 2\frac{S}{n}r = I_{k-2} - 3\frac{S}{n}r = \cdots =$$

$$= I_1 - (k-1)\frac{S}{n}r = rS - (k-1)\frac{S}{n}r = rS\frac{n-k+1}{n} \ .$$

Solution 1.5   The interest share at time $k + 1$ is:

$$I_{k+1} = rD_k = r\,(D_{k-1} - C_k) =$$

$$= I_k - rC_k = \qquad \boxed{\text{because } C_k = R - I_k}$$

$$= (1+r)\,I_k - rR \ .$$

From the equality that expresses the constance of the installments $(C_{k+1} + I_{k+1} = C_k + I_k)$ it follows

$$C_{k+1} = C_k + (I_k - I_{k+1}) = C_k + rC_k = (1+r)\,C_k$$

from which $C_k = (1+r)^{k-1}\,C_1 = (1+r)^{k-1}\,(R - rS)$.

Solution 1.6 The amount S_{k+1} on the bank account after $k+1$ years is the sum of the amount S_k of the previous year and the matured interest rS_k minus the fixed expenses C:

$$S_{k+1} = S_k + rS_k - C.$$

Solution 1.7 $s = 100\left((C_{gg}/C_0)^{365/gg} - 1\right)$.

Solution 1.8 Since for $0 < r < 1$ it holds

$$m\,(r) = \frac{1}{n}\sum_{k=1}^{+\infty} nkr^k\,(1-r) = (1-r)\sum_{k=1}^{+\infty} kr^k = r\,(1-r)\sum_{k=1}^{+\infty} kr^{k-1} =$$

$$= (1-r)\,r\frac{d}{dx}\left(\frac{x}{1-x}\right)\Bigg|_{x=r} = \frac{r}{1-r}$$

and

$$d\,(r) = \max\left\{k \in \mathbb{N}:\ (1-r)^k > \frac{1}{2}\right\} = \max\left\{k \in \mathbb{N}:\ k \le -\frac{\log 2}{\log(1-r)}\right\} =$$

$$= \text{integer part of}\left(-\frac{\log 2}{\log(1-r)}\right)$$

we conclude that

$$m\,(r) \gtrless d\,(r) \quad \Leftrightarrow \quad \frac{r}{1-r} \gtrless \text{integer part of}\left(-\frac{\log 2}{\log(1-r)}\right).$$

Solution 1.9 Let Y_k be the number of operations we need to put back k blocks. If $k = 0$ obviously $Y_0 = 0$, while for $k = 1$ we have $Y_1 = 1$. If, instead, we want to put back $k + 2$ blocks in the basket, we have only two possibilities: at the first step one removes a block, having next Y_{k+1} different ways to conclude the operation, or, always at the first step, one removes two blocks, then having Y_k ways to operate. Definitively:

$$Y_{k+2} = Y_{k+1} + Y_k.$$

Notice that the recursive equation coincides with the one of Fibonacci numbers (Example 1.15).

Solution 1.10 Assume one must move $k + 1$ disks. In order to " free" the disk of greater radius one can think moving the remaining disks on the central rod (see Fig. 1.6): the minimum number of moves needed to obtain this result is Y_k because it coincides with the minimum number of moves needed to shift the first k disks onto the right-most rod. By one only move we can shift the disk of maximum radius from the left rod to the right one, and Y_k moves are needed to move from the central rod to the right one the remaining disks. We thus obtain:

$$Y_{k+1} = Y_k + 1 + Y_k = 2Y_k + 1.$$

We observe that the number of moves Y_{k+1} is obviously enough; that it is the minimum follows from the fact that Y_k is the minimum for k disks and the moving of the bigger disk is an operation required to complete the displacement of the tower.

Solution 1.11   If the disks are $k + 1$, by X_k moves one moves the first k disks on the right rod, next by a move one moves the last disk on the central rod; at this point, by X_k moves one moves the first k disks on the left rod, by a move one moves on the right rod the disk having bigger radius and finally by X_k moves one moves on the right rod the remaining disks. In conclusion:

$$X_{k+1} = 3X_k + 2.$$

Solution 1.12   Suppose we have divided each side of the triangle in $k+1$ parts: the number of triangles X_{k+1} is given by the number of triangles X_k that are obtained considering, starting from the vertex, the first k parts, to which must it be added the number $k + 1$ of triangles whose base lies on the base of the starting triangle and the k triangles having their vertex at the points that divide the base:

$$X_{k+1} = X_k + k + 1 + k = X_k + 2k + 1.$$

Solution 1.13   Fig. 8.2 contains the first iterations of the cobweb model (for $P_{k+1} = -0.9 P_k + 2$) obtained by the graphical method.

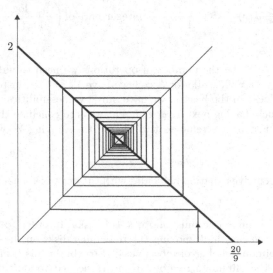

Fig. 8.2 Cobweb model

Solution 1.14   Bessel equation. Set $y(x) = x^h \sum\limits_{n=0}^{+\infty} A_n x^n$, with $h \geq 0$ and $A_0 \neq 0$, differentiating term by term and substituting in the given equation, we deduce the identically null value for the following quantity:

$$\sum_{n=0}^{+\infty}(n+h)(n+h-1)A_n x^{n+h} + \sum_{n=0}^{+\infty}(n+h)A_n x^{n+h} + \sum_{n=0}^{+\infty}A_n x^{n+h+2} - k^2 \sum_{n=0}^{+\infty}A_n x^{n+h}$$

Operating the change of indices $m = n+2$ into the third series and by collecting, we can write

$$\sum_{m=0}^{+\infty}\left[(m+h)^2 - k^2\right]A_m x^{m+h} + \sum_{m=2}^{+\infty}A_{m-2} x^{m+h} = 0.$$

At this point, equating the coefficients of the terms of equal degree, we get

$$\begin{cases} \left(h^2 - k^2\right)A_0 = 0 \\ \left((h+1)^2 - k^2\right)A_1 = 0 \\ \left((m+h)^2 - k^2\right)A_m + A_{m-2} = 0 \qquad \forall m \geq 2 \end{cases}$$

Since $A_0 \neq 0$, by the first relation we obtain $h = k$ and so, from the second equation, it follows $A_1 = 0$: then all the coefficients of odd order are zero while those of even order are identified starting from A_0 by the relation

$$A_{2n} = -\frac{A_{2n-2}}{4n(n+k)} \qquad n \geq 1.$$

For completeness we observe that, in addition to the solution found and to its multiples, there exist others defined only for $x \neq 0$.

Hermite equation. By setting $y(x) = \sum\limits_{n=0}^{+\infty} A_n x^n$, differentiating term by term and substituting in the given equation, we get

$$\sum_{n=2}^{+\infty} n(n-1)A_n x^{n-2} - 2\sum_{n=1}^{+\infty} nA_n x^n + 2k\sum_{n=0}^{+\infty} A_n x^n = 0.$$

Thanks to the indices change $m = n - 2$ in the first series, we rewrite it as

$$\sum_{m=0}^{+\infty}(m+1)(m+2)A_{m+2} x^m - 2\sum_{m=1}^{+\infty} mA_m x^m + 2k\sum_{m=0}^{+\infty} A_m x^m = 0$$

that is

$$\sum_{m=0}^{+\infty}\left[(m+1)(m+2)A_{m+2} + 2(k-m)A_m\right]x^m = 0.$$

By imposing the cancellation of the coefficients of the terms of each degree, we get the recursive relation:

$$A_{m+2} = \frac{2(m-k)A_m}{(m+1)(m+2)} \qquad m \in \mathbb{N}.$$

Laguerre equation. Proceeding as in the previous example, we get

$$\sum_{n=2}^{+\infty} n\,(n-1)\,A_n x^{n-1} + \sum_{n=1}^{+\infty} n A_n x^{n-1} - \sum_{n=1}^{+\infty} n A_n x^n + k \sum_{n=0}^{+\infty} A_n x^n = 0.$$

This time we have to "arrange" the indices of the first two series:

$$\sum_{n=1}^{+\infty} n\,(n+1)\,A_{n+1} x^n + \sum_{n=0}^{+\infty} (n+1)\,A_{n+1} x^n - \sum_{n=1}^{+\infty} n A_n x^n + k \sum_{n=0}^{+\infty} A_n x^n = 0.$$

One just has to match the coefficients of the terms of the same degree:

$$A_1 + k A_0 = 0, \qquad A_{n+1} = \frac{n-k}{(n+1)^2} A_n.$$

Legendre equation. With the same method:

$$\left(\sum_{n=1}^{+\infty} n A_n x^{n-1} - \sum_{n=1}^{+\infty} n A_n x^{n+1} \right)' + k\,(k+1) \sum_{n=0}^{+\infty} A_n x^n = 0$$

computing formally the derivative

$$\sum_{n=2}^{+\infty} n\,(n-1)\,A_n x^{n-2} - \sum_{n=1}^{+\infty} n\,(n+1)\,A_n x^n + k\,(k+1) \sum_{n=0}^{+\infty} A_n x^n = 0.$$

By the change of indices $n \to n+2$ in the first series, we obtain

$$\sum_{n=0}^{+\infty} (n+1)\,(n+2)\,A_{n+2} x^n - \sum_{n=1}^{+\infty} n\,(n+1)\,A_n x^n + k\,(k+1) \sum_{n=0}^{+\infty} A_n x^n = 0$$

and equating the coefficients of the terms of equal degree:

$$A_{n+2} = \frac{n\,(n+1) - k\,(k+1)}{(n+1)\,(n+2)} A_n \qquad n \in \mathbb{N}.$$

The explicitation of the coefficients in each case is shown in Exercise 2.38.

Solution 1.15   We use backward Euler with the discretization step $h > 0$. If Y_k approximates $y\,(hk)$, then

$$(1) \qquad (1 - h\,(k+1))\,Y_{k+1} = Y_k$$

$$(2) \qquad ah\,(Y_{k+1})^2 + (1 - ah)\,Y_{k+1} - Y_k = 0.$$

Solution 1.16   a)   Let $U_{j,k}$ be an approximation of $u\,(js, hk)$. Then to approximate the second derivatives of a generic function f, among the possible finite difference schemes, we use the **centered second differences**:

$$f''\,(\xi) = \frac{f\,(\xi+\delta) + f\,(\xi-\delta) - 2f\,(\xi)}{2\,\delta^2} = \frac{1}{2\delta} \left(\frac{f\,(\xi+\delta) - f\,(\xi)}{\delta} - \frac{f\,(\xi) - f\,(\xi-\delta)}{\delta} \right)$$

obtaining

$$U_{j,k} = \frac{s^2}{2\left(s^2 + h^2\right)}\left(U_{j,k+1} + U_{j,k-1}\right) + \frac{h^2}{2\left(s^2 + h^2\right)}\left(U_{j+1,k} + U_{j-1,k}\right).$$

Notice that if $s = h$ then $U_{j,k} = \frac{1}{4}\left(U_{j,k+1} + U_{j,k-1} + U_{j+1,k} + U_{j-1,k}\right)$, that is each value assumed by $U_{j,k}$ is the arithmetic mean of four adjacent values for each internal node. Therefore it is between the maximum and the minimum among them. It follows the **Discrete maximum principle** for the Laplace equation: if one solves the equation in a bounded open set with boundary data and a uniform grid ($s = h$), then $\max_{j,k} U_{j,k}$ is taken at the edge (the same goes for the minimum). In particular, *if the maximum (or the minimum) is taken in an internal node then $U_{j,k}$ is independent of k and j namely it is constant.*

b) Let $V_{j,k}$ be an approximation of $v\left(js, kh\right)$. Again using centered differences, we obtain:

$$2\left(h^2 - s^2\right) V_{j,k} = h^2\left(V_{j+1,k} + V_{j-1,k}\right) - s^2\left(V_{j,k+1} + V_{j,k-1}\right)$$

Solution 1.17   By making the substitutions (1.21) and putting $k = 2r/\sigma^2$, one obtains:

$$\begin{cases} w_\tau = w_{yy} + (k - 1)\, w_y - kw \\ w\left(y, 0\right) = \max\left(e^y - 1, 0\right). \end{cases}$$

Substituting (1.22), one gets:

$$\beta u + u_\tau = \alpha^2 u + 2\alpha u_y + u_{yy} + (k - 1)\left(\alpha u + u_y\right) - ku.$$

Imposing the conditions

$$\begin{cases} \beta = \alpha^2 + (k - 1)\,\alpha - k \\ 0 = 2\alpha + (k - 1) \end{cases}$$

and solving the system, we obtain the values of the parameters

$$\alpha = -\frac{1}{2}\left(k - 1\right) \qquad \beta = -\frac{1}{4}\left(k + 1\right)^2$$

with respect to which the problem of partial derivatives is written as

$$\begin{cases} u_t = u_{yy} & \text{in the half-plane:} \quad -\infty < y < +\infty, \ \tau > 0 \\[2mm] u\left(y, 0\right) = u_0\left(y\right) = \max\left(e^{\frac{1}{2}(k+1)y} - e^{\frac{1}{2}(k-1)y}, 0\right) \end{cases}$$

because

$$w\left(y, \tau\right) = e^{-\frac{1}{2}(k-1)y - \frac{1}{4}(k+1)^2 \tau}\, u\left(y, \tau\right).$$

Since it is known that $u\left(y, \tau\right) = \dfrac{1}{2\sqrt{\pi\tau}}\displaystyle\int_{-\infty}^{+\infty} u_0\left(s\right) e^{-(y-s)^2/4\tau}\, ds$

we conclude

$$\boxed{v\left(x, t\right) = xN\left(d_1\right) - Ee^{-r(T-t)}N\left(d_2\right)}$$

where

$$N\left(d\right) = \frac{1}{\sqrt{2\pi}} \int_{-\infty}^{d} e^{-s^2/2} ds$$

$$d_1 = \frac{\log\left(x/E\right) + \left(r + \frac{\sigma^2}{2}\right)(T - t)}{\sigma\sqrt{T - t}} \qquad d_2 = \frac{\log\left(x/E\right) + \left(r - \frac{\sigma^2}{2}\right)(T - t)}{\sigma\sqrt{T - t}}.$$

8.2 Solutions of the exercises in Chapter 2

Solution 2.1   We develop explicitly only points i) and ii).

(1) $X_k = -k, \ \forall k \in \mathbb{N}; \qquad \lim_k X_k = -\infty;$

(2) $X_k = -2\left(\dfrac{1}{3}\right)^k + 3, \ \forall k \in \mathbb{N}; \qquad \lim_k X_k = 3;$

(3) $X_k = -\dfrac{1}{12}4^k + \dfrac{1}{3}, \ \forall k \in \mathbb{N}; \qquad \lim_k X_k = -\infty;$

(4) $X_k = \dfrac{1}{2}\left(-\dfrac{1}{5}\right)^k + \dfrac{5}{2}, \ \forall k \in \mathbb{N}; \qquad \lim_k X_k = \dfrac{5}{2};$

(5) $X_k = \dfrac{3}{2}(-1)^k + \dfrac{1}{2}, \ \forall k \in \mathbb{N}; \qquad \nexists \lim_k X_k; \ X_k$ oscillates and takes only two values;

(6) $X_k = -\dfrac{1}{4}(-3)^k + \dfrac{5}{4}, \ \forall k \in \mathbb{N}; \qquad \nexists \lim_k X_k.$ The modulus $|X_k|$ diverges to $+\infty.$

Solution 2.2   Starting from (2.3) we obtain the following table:

$$X_1 = aX_0 + b$$
$$X_2 = aX_1 + b = a\left(aX_0 + b\right) + b = a^2 X_0 + b\left(1 + a\right)$$
$$X_3 = aX_2 + b = a\left(a^2 X_0 + b\left(1 + a\right)\right) = a^3 X_0 + b\left(1 + a + a^2\right)$$
$$\ldots = \ldots$$
$$X_k = a^k X_0 + b\left(1 + a + a^2 + \cdots + a^{k-1}\right).$$

Since (see previous exercise)

$$1 + a + a^2 + \cdots + a^{k-1} = \begin{cases} \dfrac{1 - a^k}{1 - a} & a \neq 1 \\ k & a = 1 \end{cases}$$

substituting the value of α, we obtain the solving formulas of Theorem 2.5.

Solution 2.3   The characteristic equation $\lambda^2 + a\lambda + b = 0$ admits the two roots

$$\lambda_{1,2} = \frac{-a \pm \sqrt{a^2 - 4b}}{2}.$$

If $b = a^2/4$, then the general solution is $X_k = (c_1 + c_2 k)\lambda_1^k$ with $\lambda_1 = -a/2$. Therefore, X_k has finite limit (equal to 0) for every initial datum if and only if $|\lambda_1| < 1$ namely if and only if $|a| < 2$.

If $b \neq a^2/4$, then the general solution is $X_k = c_1\lambda_1^k + c_2\lambda_2^k$ and X_k has finite limit for each initial datum if and only if $|\lambda_1| < 1$ and $|\lambda_2| < 1$. Therefore, to obtain convergence of all solutions:

- if $b > a^2/4$, then it must hold $a^2 + (4b - a^2) < 4$, that is $b < 1$;
- if $b < 0 < a^2/4$, then it must hold $\sqrt{a^2 - 4b} < \min\{2+a,\ 2-a\}$, that is
$$\begin{cases} a^2 < 4 \\ b > (|a| - 1) \end{cases};$$
- if $0 < b < a^2/4$, then it must hold $\left|\sqrt{a^2 - 4b} \pm a\right| < 2$, that is $\begin{cases} a^2 < 4 \\ (a-1) < b \end{cases}.$

Solution 2.4

(1) $X_k = c_1(-1)^k + c_2 3^k, \ \ k \in \mathbb{N}$;

(2) $X_k = c_1 \cos(\pi k/6) + c_2 \sin(\pi k/6), \ \ k \in \mathbb{N}$;

(3) $X_k = c_1 + c_2 \cos(\pi k/2) + c_3 \sin(\pi k/2), \ \ k \in \mathbb{N}$;

(4) $X_k = (c_1 + c_2 k)(-1)^k + c_3(-3)^k, \ \ k \in \mathbb{N}$;

(5) $X_k = c_1(-1)^k + c_2 3^k - 1/2, \ \ k \in \mathbb{N}$;

(6) $X_k = c_1 + c_2 \cos(\pi k/2) + c_3 \sin(\pi k/2) - k/2, \ \ k \in \mathbb{N}$.

Solution 2.5 $\alpha = 2,$ α is stable and attractive;
$X_k = 3(-2)^{-k-1} - 5(2)^{-k-1} + 2, \ \ k \in \mathbb{N}$.

Solution 2.6

(a) $X_k = c_1\left(\dfrac{-1-i}{2}\right)^k + c_2\left(\dfrac{-1+i}{2}\right)^k = \left(\dfrac{\sqrt{2}}{2}\right)^k\left(\tilde{c}_1 \cos\left(\dfrac{3}{4}\pi k\right) + \tilde{c}_2 \sin\left(\dfrac{3}{4}\pi k\right)\right)$;

(b) $\alpha = 0$ is stable and attractive (c) $X_k = \left(\sqrt{2}/2\right)^k (-2\sin(3\pi k/4)), \ \ k \in \mathbb{N}$.

Solution 2.7   (a) $X_k = c_1(2-i)^k + c_2(2+i)^k + 3, \ \ k \in \mathbb{N}$ (b) $\alpha = 3$ is not stable.

Solution 2.8

(a) $X_k = c_1\left(\dfrac{-1-i}{2}\right)^k + c_2\left(\dfrac{-1+i}{2}\right)^k - 2 =$

$= \left(\dfrac{\sqrt{2}}{2}\right)^k\left(\tilde{c}_1 \cos\left(\dfrac{3}{4}\pi k\right) + \tilde{c}_2 \sin\left(\dfrac{3}{4}\pi k\right)\right) - 2$;

(b) $\alpha = -2$ is stable and attractive.

Solution 2.9   We observe preliminarily that if $\mathcal{P}(\lambda) = 0$ is the characteristic equation associated to the considered difference equations, then:

$$1 + a_1 + a_0 = \mathcal{P}(1); \qquad 1 - a_1 + a_0 = \mathcal{P}(-1)$$

where a_0 is equal to the product of the two roots λ_1 and λ_2 (possibly complex) of $\mathcal{P}$: $a_0 = \lambda_1 \lambda_2$.
If the conditions on the coefficients apply, then we obtain:

$$-a_1 < 1 + a_0 < 2 \qquad \Rightarrow \qquad -\frac{a_1}{2} < 1$$

$$-a_1 > -1 - a_0 > -2 \qquad \Rightarrow \qquad -\frac{a_1}{2} > -1.$$

Therefore, $\mathcal{P}(1) > 0$ and $\mathcal{P}(-1) > 0$ hence the abscissa of the vertex of the parabola $\mathcal{P}(\lambda)$ belongs to the interval $(-1, 1)$: if the roots λ_1 and λ_2 are real, they necessarily belong to the interval $(-1, 1)$. If the roots are complex, as $\lambda_2 = \overline{\lambda}_1$, we obtain:

$$1 > a_0 = \lambda_1 \lambda_2 = \lambda_1 \overline{\lambda}_1 = |\lambda_1|^2$$

from which $|\lambda_1| = |\lambda_2| < 1$.
Conversely, if $\alpha = 0$ is attractive, then the roots λ_1 and λ_2 have modulus smaller than 1, hence the modulus of their product is smaller than 1.
If λ_1 and λ_2 are real, then $\lambda_1 \lambda_2 = a_0 < 1$ and necessarily $\mathcal{P}(-1) > 0$ and $\mathcal{P}(1) > 0$.
If λ_1 e λ_2 are complex conjugate, then

$$1 > |\lambda_1 \lambda_2| = |\lambda_1 \overline{\lambda}_1| = |\lambda_1|^2 = a_0$$

and, obviously, $\mathcal{P}(-1) > 0$, $\mathcal{P}(1) > 0$.

Solution 2.10 Putting $S_k = \sum_{k=1}^{n} k$, the exercise requires to write the explicit solution of the problem

$$S_{k+1} = S_k + (k+1), \qquad S_0 = 1.$$

It is a one-step nonhomogeneous linear equation. The general solution of the associated homogeneous equation is the set of all the constant trajectories. We look for a particular solution of the nonhomogeneous equation, of the type $k(ak + b)$: by substituting into the equation, we obtain

$$a(k+1)^2 + b(k+1) = ak^2 + bk + k + 1,$$

whose solution is $a = b = 1/2$. By imposing the initial condition, we obtain $c = 0$ from which the claim follows.

Alternatively, one could prove the formula by induction.
We also remember the constructive solution proposed by Gauss: by setting $S_k = 1 + 2 + \cdots + (k-1) + k = \sum_{n=1}^{k} n$, we can rewrite the terms of the summation as $S_k = k + (k-1) + \cdots + 3 + 2 + 1$ and add the two relations:

$$2S_k = (1+k) + (2+k-1) + \cdots + (k-1+2) + (k+1) =$$
$$= \underbrace{(k+1) + (k+1) + \cdots + (k+1)}_{k \text{ times}} = k(k+1).$$

Solution 2.11   As usual, if either the result is known or one wants to verify a conjecture about the outcome, one can make a check by induction.
However, we use a technique that is independent of the knowledge of the result.

Putting $S_k = \sum\limits_{n=1}^{k} n^2$, we can write:

$$\begin{cases} S_{k+1} - S_k = (k+1)^2 \\ S_0 = 0 \end{cases}$$

The general solution T of the associated homogeneous problem $T_{k+1} - T_k = 0$ is given by the constant sequences. We look for a particular solution X of the non-homogeneous problem of third-degree polynomial type (because the right-hand side is a two-degree polynomial, and the constant solves the homogeneous problem): $X_k = ak^3 + bk^2 + ck$. Substituting into the initial equation, we obtain:

$$a(k+1)^3 + b(k+1)^2 + c(k+1) = ak^3 + bk^2 + ck + k^2 + 2k + 1$$

$$(3a-1)k^2 + (3a+2b-2)k + (a+b+c-1) = 0$$

$$a = 1/3 \qquad b = 1/2 \qquad c = 1/6.$$

The general solution is

$$S_k = d + X_k = d + \frac{1}{3}k^3 + \frac{1}{2}k^2 + \frac{1}{6}k \qquad\qquad k \in \mathbb{N}$$

and, by the initial condition $S_0 = 0$, we deduce $d = 0$, wherefrom:

$$S_k = \frac{1}{6}k(2k+1)(k+1) \qquad\qquad k \in \mathbb{N}.$$

Solution 2.12 $S_k = \dfrac{1}{4}k^2(k+1)^2$.

Solution 2.13 We have to find the explicit solution of problem

$$P_{k+1} = P_k + z^{k+1}, \qquad P_0 = 1.$$

It is a one-step nonhomogeneous linear equation. If $z = 1$ then the solution of the problem is $P_k = k$.

Consider now $z \neq 1$. The general solution of the associated homogeneous equation is the set of all constant trajectories. We look for a particular solution of the nonhomogeneous equation with k–th term equal to az^{k+1}: substituting into the equation, we obtain

$$az^{k+2} = az^{k+1} + z^{k+1}, \qquad a = 1/(z-1).$$

The general solution of the nonhomogeneous equation is then $P_k = c + z^{k+1}/(z-1)$. By imposing the initial condition $P_0 = 1$ we obtain $c = 1/(1-z)$, from which

$$P_k = -\frac{1}{z-1} + \frac{1}{z-1}z^{k+1} = \frac{1-z^{k+1}}{1-z}.$$

If $z = 1$, then $P_k = k+1$. If $z \neq 1$, then $P_k = \dfrac{1-z^{k+1}}{1-z}$; this last relation can be proved by induction or by observing that

$$(1-z)P_k = \left(1 + z + z^2 + \cdots + z^k\right) - \left(z + z + z^2 + z^3 + \cdots + z^{k+1}\right) = 1 - z^{k+1}.$$

Dividing by $(1-z)$ we obtain the desired formula.

Solution 2.14 With the notation of Exercise 2.13, if $z = 1$, then $f_{h,k} = k + 1$. If $z \neq 1$, then $f_{h,k} = z^h P_k = z^h \dfrac{1 - z^{k+1}}{1 - z}$.

Solution 2.15 A necessary condition for the convergence of a series is that the general term is infinitesimal, so we do not have convergence when $|z| \geq 1$. If $z \in \mathbb{C}$ e $|z| < 1$, then the series absolutely converges and

$$\sum_{k=0}^{+\infty} z^k = \frac{1}{1 - z} \qquad \forall z \in \mathbb{C} : |z| < 1.$$

In fact, recalling the expression of the partial sums (see Exercise 2.13)

$$\sum_{k=0}^{+\infty} z^k = \lim_k \left(\sum_{n=0}^{k} z^n \right) = \lim_k \frac{1 - z^{k+1}}{1 - z} = \frac{1}{1 - z}.$$

Solution 2.16 $0.99999\ldots = \displaystyle\sum_{k=1}^{+\infty} 9 \cdot 10^{-k} = \frac{9}{10} \sum_{k=0}^{+\infty} 10^{-k} = \frac{9}{10} \frac{1}{1 - 1/10} = 1$.

Solution 2.17 With reference to the solution of Exercise 1.3, the equation in Y is decoupled and has solution $Y_k = (1 - p_Y)^k Y_0$. By substituting in the other equation, we obtain the first-order non-homogeneous linear equation:

$$X_{k+1} = (1 - p_X) X_k + p_Y (1 - p_Y)^k Y_0.$$

The general solution of the associated homogeneous equation is $c (1 - p_X)^k$. By using Table 2.1 to look for particular solutions, we conclude that:

- if $p_X \neq p_Y$, then $X_k = \left(X_0 - \dfrac{p_Y}{p_X - p_Y} Y_0 \right) (1 - p_X)^k + \dfrac{p_Y}{p_X - p_Y} Y_0 (1 - p_Y)^k$
  $k \in \mathbb{N}$;

- if $p_X = p_Y \neq 1$, then looking for particular solutions of the type $ak (1 - p_X)^k$, we obtain
  $$X_k = X_0 (1 - p_X)^k + \frac{p_X}{1 - p_X} Y_0 k (1 - p_X)^k \quad k \in \mathbb{N};$$

- if $p_X = p_Y = 1$, then $X_k = Y_k = 0$, $\forall k > 1$.

Solution 2.18

(1) $X_k = c_1 (-1)^k + c_2 3^k + \dfrac{1}{12} (-3)^k, \quad k \in \mathbb{N}$;

(2) $X_k = c_1 \cos \left(\dfrac{\pi}{6} k \right) + c_2 \sin \left(\dfrac{\pi}{6} k \right) - k + \dfrac{18 - 10\sqrt{3}}{3}, \qquad k \in \mathbb{N}$;

(3) $X_k = c_1 + c_2 \cos \left(\dfrac{\pi}{2} k \right) + c_3 \sin \left(\dfrac{\pi}{2} k \right) + \dfrac{1}{4} \sin \left(\dfrac{\pi}{2} k \right) + \dfrac{1}{4} \cos \left(\dfrac{\pi}{2} k \right), \qquad k \in \mathbb{N}$;

(4) $X_k = \left(c_1 + c_2 k + \dfrac{1}{4} k^2 \right) (-1)^k + c_3 (-3)^k, \quad k \in \mathbb{N}$.

Solution 2.19   We use the $\mathcal{Z}$-transform. Setting $\mathcal{Z}\{X\} = x(z)$, by Table in Appendix G, we get

$$\mathcal{Z}\{k\} = \frac{z}{(z-1)^2}$$

$$\mathcal{Z}\{X_{k+1}\} = z(x(z) - X_0) = zx(z)$$

$$\mathcal{Z}\{X_{k+2}\} = z^2 x(z) - z^2 X_0 - zX_1 = z^2 x(z) - z.$$

By substituting into the equation

$$(z^2 - z - 2)\, x(z) = \frac{z}{(z-1)^2} + z$$

one obtains

$$x(z) = \frac{z}{(z-2)(z+1)(z-1)^2} + \frac{z}{(z-2)(z+1)}.$$

Thus

$$\mathcal{Z}^{-1}\left\{\frac{z}{(z-2)(z+1)}\right\} = \mathcal{Z}^{-1}\left\{z\left(\frac{1/3}{z-2} - \frac{1/3}{z+1}\right)\right\} = \frac{1}{3}\left(2^k - (-1)^k\right)$$

$$\mathcal{Z}^{-1}\left\{\frac{z}{(z-2)(z+1)(z-1)^2}\right\} = \operatorname{Res}(f,2) + \operatorname{Res}(f,-1) + \operatorname{Res}(f,1) =$$

$$= \frac{1}{3}2^k - \frac{1}{12}(-1)^k + \frac{d}{dz}\left.\frac{z^k}{(z-2)(z+1)}\right|_{z=1} =$$

$$= \frac{1}{3}2^k - \frac{1}{12}(-1)^k - \frac{k}{2} - \frac{1}{4}.$$

Ultimately:

$$X_k = \frac{2^{k+1}}{3} - \frac{5}{12}(-1)^k - \frac{k}{2} - \frac{1}{4} = \frac{1}{12}\left(2^{k+3} + 5(-1)^{k+1} - 6k - 3\right) \qquad k \in \mathbb{N}.$$

Solution 2.20   By using the $\mathcal{Z}$-transform, setting $x(z) = \mathcal{Z}\{X\}$ and transforming the equation

$$z^2 x(z) - 5zx(z) + 6x(z) = 1 \qquad \Rightarrow \qquad (z^2 - 5z + 6)\, x(z) = 1$$

one gets

$$x(z) = \mathcal{Z}\{X\} = \frac{1}{(z-2)(z-3)}.$$

Thus:

$$X_0 = \operatorname{Res}(g_0, z=0) + \operatorname{Res}(g_0, z=2) + \operatorname{Res}(g_0, z=3) = \frac{1}{6} - \frac{1}{2} + \frac{1}{3} = 0$$

$$X_k = \operatorname{Res}(g_k, z=2) + \operatorname{Res}(g_k, z=3) = 3^{k-1} - 2^{k-1} \qquad k \geq 1.$$

Alternatively $\mathcal{Z}^{-1}\left\{\dfrac{1}{(z-2)(z-3)}\right\}$ can be calculated by the discrete convolution:

$$\mathcal{Z}^{-1}\left\{\frac{1}{(z-2)(z-3)}\right\} = \mathcal{Z}^{-1}\left\{\frac{1}{z-2}\right\} * \mathcal{Z}^{-1}\left\{\frac{1}{z-3}\right\} = Y * Z$$

having set

$$\begin{cases} Y_k = 2^{k-1} & \forall k \geq 1 \\ Y_0 = 0 \end{cases} \qquad \begin{cases} Z_k = 3^{k-1} & \forall k \geq 1 \\ Z_0 = 0 \end{cases}$$

We obtain $X_0 = Y_0 Z_0 = 0$ and

$$X_k = \sum_{j=0}^{k} Y_{n-j} Z_j = \sum_{j=1}^{k-1} 3^{j-1} 2^{n-j-1} = \frac{3^{k-1} - 2^{k-1}}{3 - 2} = 3^{k-1} - 2^{k-1} \qquad \forall k \geq 1$$

having used the identity

$$a^m - b^m = (a - b)\left(a^{m-1} + a^{m-2}b + \cdots + ab^{m-2} + b^{m-1}\right)$$

in the last equality.

Solution 2.21 By taking the $\mathcal{Z}$-transform of both sides of the given equation, we get:

$$zx(z) - z - 2x(z) = \frac{3}{z^4}$$

and, solving in x,

$$x(z) = \frac{z}{z - 2} + \frac{3}{z^4(z - 2)} = \frac{z}{z - 2} + 3z^{-5}\frac{z}{z - 2}.$$

Computing the inverse transform of this expression by Table in Appendix G, we deduce

$$X_k = \left(\mathcal{Z}^{-1}\{x\}\right)_k = 2^k + 3 \cdot 2^{k-5} U_{k+5} = \begin{cases} 2^k & 0 \leq k \leq 4 \\ 2^k + 3 \cdot 2^{k-5} & k \geq 5 \end{cases}$$

Solution 2.22 It is a Volterra equation.
By replacing $k + 1$ by k into the expression of X_k and dividing by 2, we obtain

$$X_{k+1} = 1 + 16 \sum_{s=0}^{k} (k - s) X_s = 1 + 16\, k * X_k.$$

By applying the $\mathcal{Z}$-transform to both members, we get

$$zx(z) - z = \frac{z}{z - 1} + 16\frac{z}{(z - 1)^2} x(z)$$

$$x(z) = \frac{z(z - 1)}{(z - 5)(z + 3)} = z\left(\frac{1}{2(z - 5)} + \frac{1}{2(z + 3)}\right) = \frac{1}{2}\frac{z}{z - 5} + \frac{1}{2}\frac{z}{z + 3}$$

and then $X_k = \frac{1}{2}5^k + \frac{1}{2}(-3)^k$.

Solution 2.23 We first compute $\mathcal{Z}\{Y\}$: if $|z| < 1$, then $\mathcal{Z}\{Y\} = \dfrac{-3! z^3}{(z - 1)^4}$ for

$$\sum_{k=0}^{+\infty} \frac{k(k + 1)(k + 2)}{z^k} = (-z)^3 \left(\frac{d}{dz}\right)^3 \sum_{k=0}^{+\infty} \frac{1}{z^k} = (-z)^3 \left(\frac{d}{dz}\right)^3 \frac{1}{1 - z} = 3!\frac{-z^3}{(z - 1)^4}.$$

By the identity $k^3 = k(k+1)(k+2) - 3k^2 - 2k$ and by

$$\mathcal{Z}\{k\} = \frac{z}{(z-1)^2} \qquad \mathcal{Z}\{k^2\} = \frac{z(z+1)}{(z-1)^3}$$

we get by linearity

$$\mathcal{Z}\{X\} = \mathcal{Z}\{Y\} - 3\mathcal{Z}\{k^2\} - 2\mathcal{Z}\{k\} = \frac{-6z^3}{(z-1)^4} - 3\frac{z(z+1)}{(z-1)^3} - 2\frac{z}{(z-1)^2} =$$

$$= \frac{-11z^3 + 4z^2 + z}{(z-1)^4}.$$

Solution 2.24 Integrating by parts, for each b, $b > 0$, and for each $k \in \mathbb{N}$, one has:

$$\int_0^b e^{-x}x^k\,dx = \left[-e^{-x}x^k\right]_0^b + k\int_0^b e^{-x}x^{k-1}\,dx = -e^{-b}b^k + k\int_0^b e^{-x}x^{k-1}\,dx.$$

Taking the limit as $b \to +\infty$, one deduces

$$\begin{cases} I_k = kI_{k-1} \\ I_0 = 1 \end{cases}$$

whose explicit solution is $I_k = \left(\prod_{s=1}^{k} s\right)I_0 = k!$

Solution 2.25

(1) $X_1 = 0 \cdot X_0 = 0$ and then $X_k = 0$ for each k (and for each initial datum X_0);

(2) $X_k = \dfrac{3k-2}{3k+4}\dfrac{3k-5}{3k+1}\cdots\dfrac{4}{11}\dfrac{1}{7}X_0 = X_0\prod_{j=0}^{k-1}\dfrac{3j+1}{3j+7}$;

(3) $X_k = e^{3(k-1)}X_{k-1} = e^{3(k-1)}e^{3(k-2)}\ldots e^3 e^0 X_0 = X_0\exp\left(3\sum_{j=0}^{k-1} j\right) =$

$$= X_0\exp\left(\frac{3}{2}k(k-1)\right);$$

(4) $X_k = e^{\cos 2(k-1)}e^{\cos 2(k-2)}\ldots e^{\cos 2}e^{\cos 0}X_0 = X_0\exp\left(\sum_{j=0}^{k-1}\cos 2j\right).$

Solution 2.26 The given equation is first-order, linear nonhomogeneous with variables coefficients (Definition 2.63).
If k starts from 0 (with generic initial datum $X_0 \in \mathbb{R}$), then $X_1 = 1$ and by (2.24) one obtains, for every $k \geq 2$ (the restriction applies to the calculations, but the formula is true for $k \geq 1$):

$$X_k = 1 + \sum_{h=0}^{k-2}(k-1)(k-2)\cdots(h+1) = 1 + (k-1)!\sum_{h=0}^{k-2}\frac{1}{h!} = (k-1)!\sum_{h=0}^{k-1}\frac{1}{h!}.$$

So, after a few iterations the general term X_k grows very rapidly and is well approximated (with a negligible relative error) by $e(k-1)!$

If one starts from $k = 1$ (with generic initial datum $X_1 \in \mathbb{R}$), the closed form solution is obtained from the simplified formula provided by Remark 2.59, by setting $A_k = k$ and $B_K = 1$ $\forall k \geq 2$ (the restriction applies to the calculations, but the formula is true for $k \geq 1$):

$$X_k = (k-1)! X_1 + \sum_{h=1}^{k-1} (k-1)(k-2)\cdots(h+1) =$$

$$= (k-1)! X_1 + \sum_{h=1}^{k-1} \frac{(k-1)(k-2)\cdots(h+1)\,h!}{h!} =$$

$$= (k-1)! \left(X_1 + \sum_{h=1}^{k-1} \frac{1}{h!} \right) \sim (e + X_1)(k-1)! \, .$$

An alternative (more simple) way to obtain the general solution starting from X_0 for $k = 0$, is to change variable by putting $Z_k = X_k/(k-1)!$ $\forall k \geq 1$: in this way we obtain the recursive relation $Z_{k+1} = Z_k + 1/k!$ $\forall k \geq 2$, having immediate solution ($Z_k = \sum_{h=0}^{k-1} 1/h!$) which yields X_k (one has to make the obvious changes if one starts from X_1 when $k = 1$).

The results are consistent with Theorems 2.64 and 2.66: the one-step difference equation with nonconstant coefficients has infinitely many solutions, the set of solutions is a one-dimensional vector space. Consistently with Remark 2.59, the problem initialized to $k = 0$ presents memory loss from $k = 1$: the solutions differ only in the initial value X_0, but for each of them $X_1 = 1$ and all coincide for $k \geq 1$.

Furthermore, with reference to Theorem 2.66 i) and ii) are valid but not iii) (here $n = 1$ and the determinant of the 1x1 matrix coincides with its unique element): in fact, the solutions of the homogeneous problem $Y_{k+1} = kY_k$ with initial datum Y_0 in $k = 0$ verify $Y_k = 0$ $\forall k \geq 1$; this does not contradict the theorem, but shows that the assumption $A_k^0 \neq 0$ $\forall k$ can not be removed (in the case we are considering, $A_0^0 = 0$).

Solution 2.27   We have $x \cdot x = 1 \cdot (1 - x)$ that is $x^2 + x - 1 = 0$ which admits the two solutions $x_{1,2} = (-1 \pm \sqrt{5})/2$: only the positive one is acceptable. The length of the golden section of a segment measures $(\sqrt{5} - 1)/2$.

Solution 2.28   (1) X_k solves a problem like (2.28) with $a = 0$ and $b = c = d = 1$. We operate a substitution as in (2.30), choosing $\alpha = x = (\sqrt{5} - 1)/2$.

Setting $Y_k = \left(X_k - \dfrac{\sqrt{5} - 1}{2} \right)^{-1}$, we write (2.32) as

$$Y_{k+1} = \frac{\dfrac{\sqrt{5}-1}{2} + 1}{\dfrac{1-\sqrt{5}}{2}} Y_k + \frac{1}{\dfrac{\sqrt{5}-1}{2}} = \frac{\sqrt{5}+1}{1-\sqrt{5}} Y_k + \frac{2}{\sqrt{5}-1}$$

and by Theorem 2.5, we obtain

$$Y_k = \left(X_0 - \frac{1}{\sqrt{5}} \right) \left(\frac{1+\sqrt{5}}{1-\sqrt{5}} \right)^k + \frac{1}{\sqrt{5}}$$

and then

$$X_k = \frac{\sqrt{5}-1}{2} + \left(\left(X_0 - \frac{1}{\sqrt{5}} \right) \left(\frac{1+\sqrt{5}}{1-\sqrt{5}} \right)^k + \frac{1}{\sqrt{5}} \right)^{-1}.$$

(2) From $\lim\limits_k \left| \dfrac{1+\sqrt{5}}{1-\sqrt{5}} \right|^k = +\infty$ it follows $\lim\limits_k X_k = \dfrac{\sqrt{5}-1}{2} = x$.

Solution 2.29 We express X_{k+1} as a function of X_k: $X_{k+1} = (4X_k + 9)/(X_k + 2)$.
If we set $X_k = \dfrac{Z_{k+1}}{Z_k} - 2$, the given equation can be rewritten as

$$Z_{k+2} + 2Z_{k+1} + Z_k = 0$$

which has general solution $Z_k = c_1 (-1)^k + c_2 k (-1)^k$. Consequently, the solutions
of the starting equation are

$$X_k = \frac{-3 - c_3 (3k+1)}{1 + c_3 k}$$

where $c_3 = c_2/c_1$ and c_3 arbitrary constant. Notice that to this representation it
must be added the solution $X_k = -\dfrac{k+1}{k} - 2$ corresponding to the case $c_1 = 0$.

We have solved a particular case of **Riccati equation**: given the sequences P, Q,
and R, determine X verifying the relation

$$X_{k+1}X_k + P_k X_{k+1} + Q_k X_k + R_k = 0$$

that, through the change of variables $X_k = \dfrac{Z_{k+1}}{Z_k} - P_k$, is transformed into a linear
equation in Z:

$$Z_{k+2} + (Q_k - P_{k+1}) Z_{k+1} + (R_k - P_k Q_k) Z_k = 0.$$

Solution 2.30 (a) $X_k = 2^k - 1$;   (b) $X_k = 3^k - 1$.

Solution 2.31 $X_k = 3^{-k} (\lambda + 3) - 3$, $k \in \mathbb{N}$. Therefore, if $k \to +\infty$, then $X_k \to -3$, whatever the value of λ. Graphically we have:

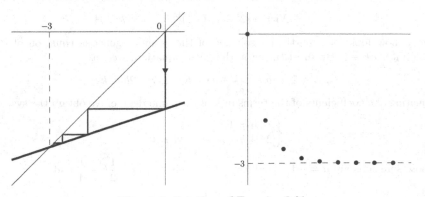

Fig. 8.3 Solution of Exercise 2.31

Solution 2.32 $X_k = (-1)^k (\lambda - 1.5) + 1.5$ $k \in \mathbb{N}$. So, if $k \to +\infty$, does not exist the limit of X_k unless it is $\lambda = 1.5$; in this case the solution is constant $(X_k = 1.5 \ \forall k)$, otherwise the solution is 2 periodic. If $\lambda = 0.5$ then graphically we have

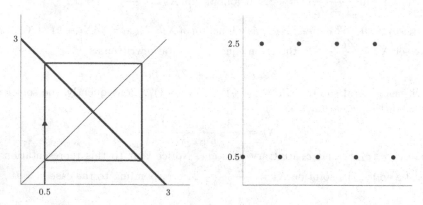

Fig. 8.4 Solution of Exercise 2.32

Solution 2.33

1) $X_k = c_1 (-1)^k + c_2 \cos \left(\dfrac{\pi}{3} k \right) + c_3 \sin \left(\dfrac{\pi}{3} k \right),$ $\forall k \in \mathbb{N};$

2) $X_k = c_1 + c_2 (-1)^k + c_3 \cos \left(\dfrac{\pi}{2} k \right) + c_4 \sin \left(\dfrac{\pi}{2} k \right),$ $\forall k \in \mathbb{N};$

3) $X_k = c_1 \cos \left(\dfrac{\pi}{4} k \right) + c_2 \sin \left(\dfrac{\pi}{4} k \right) + c_3 \cos \left(\dfrac{3\pi}{4} k \right) + c_4 \sin \left(\dfrac{3\pi}{4} k \right),$ $\forall k \in \mathbb{N};$

4) $X_k = 2^k (c_1 + c_2 k) \cos \left(\dfrac{\pi}{2} k \right) + 2^k (c_3 + c_4 k) \sin \left(\dfrac{\pi}{2} k \right),$ $\forall k \in \mathbb{N}.$

Solution 2.34   The characteristic equation $\lambda^2 - \lambda - 2 = 0$ admits the simple roots $\lambda = -1$ and $\lambda = 2$. Therefore the general solution of the homogeneous problem is

$$X_k = c_1 2^k + c_2 (-1)^k \qquad k \in \mathbb{N}.$$

Let us now look for a particular solution of the nonhomogeneous equation of the type $\overline{X}_k = ak + b$. By substituting in the given equation, we get:

$$a (k + 2) + b - a (k + 1) - b - 2ak - 2b = k.$$

Equating the coefficients of the terms in k of the same degree, we obtain the system

$$\begin{cases} -2a = 1 \\ 2a + b - a - b - 2b = 0 \end{cases}$$

whose solutions are $a = -1/2$ and $b = -1/4$. Then $\overline{X}_k = -\dfrac{1}{2} k - \dfrac{1}{4}$ and

$$X_k = c_1 2^k + c_2 (-1)^k - \dfrac{1}{2} k - \dfrac{1}{4} \qquad k \in \mathbb{N}.$$

Finally, by imposing the initial data in the general solution, we deduce $c_1 = 2/3$ and $c_2 = -5/12$. The required solution is then:

$$X_k = \frac{1}{3}2^{k+1} - \frac{5}{12}(-1)^k - \frac{1}{2}k - \frac{1}{4} \qquad k \in \mathbb{N}.$$

Solution 2.35 We use the $\mathcal{Z}$-transform. By setting $x(z) = \mathcal{Z}\{X\}$, we transform the equation

$$(z^2 - 5z + 6)\, x(z) = 1$$

$$x(z) = \frac{1}{(z-3)(z-2)}$$

$$X_k = \operatorname{res}\left(\frac{z^{k+1}}{(z-2)(z-3)}, 0\right) + \operatorname{res}\left(\frac{z^{k-1}}{(z-2)(z-3)}, 3\right) = -2^{k-1} + 3^{k-1} \qquad k \geq 1.$$

Notice that

$$X_0 = \operatorname{res}\left(\frac{1}{z(z-2)(z-3)}, 0\right) + \operatorname{res}\left(\frac{1}{z(z-2)(z-3)}, 2\right) +$$

$$+ \operatorname{res}\left(\frac{1}{z(z-2)(z-3)}, 3\right) = \frac{1}{6} - \frac{1}{2} - \frac{1}{3} = 0.$$

Alternatively, from

$$\frac{1}{(z-3)(z-2)} = \frac{1}{z-3} - \frac{1}{z-2}$$

and from

$$\mathcal{Z}^{-1}\left\{\frac{1}{z-a}\right\} = \tau_1\{a^k\}$$

we get

$$x(z) = \left(\mathcal{Z}^{-1}\left\{\frac{1}{z-3} - \frac{1}{z-2}\right\}\right)_k = \begin{cases} 0 & \text{if } k = 0 \\ 3^{k-1} - 2^{k-1} & \text{if } k \geq 1 \end{cases}$$

Solution 2.36 In the particular case $k = a$ and $s = a + b$ one solves also the problem of Example 1.17.
Moreover, one observes that the possible different stories of a ruin are infinite and therefore it is impractical to calculate the corresponding probabilities and then to sum them up.
Let be $q = 1 - p$. The ruin probability R_{k+1} starting the game with $k+1$ Euro can be seen as the sum of the probabilities of victory in the first round followed by ruin and the probability of loss in the first round followed by ruin:

$$R_{k+1} = pR_{k+2} + qR_k$$

that for $p \neq 0$ is equivalent to

$$R_{k+2} = \frac{1}{p}R_{k+1} - \frac{q}{p}R_k.$$

The characteristic equation

$$\lambda^2 - \frac{1}{p}\lambda + \frac{q}{p} = 0$$

admits the roots $\lambda_1 = 1$ and $\lambda_2 = q/p$. By setting $r = q/p$

- if $p \neq q$, then $R_k = c_1 + c_2 r^k$ and taking into account that $R_0 = 1$ and $R_s = 0$, we obtain

$$\boxed{R_k = \frac{r^k - r^s}{1 - r^s} \qquad \text{if } p \neq \frac{1}{2}}$$

- if $p = q = 1/2$, then $R_k = c_1 + c_2 k$ and taking into account that $R_0 = 1$ and $R_s = 0$, we obtain

$$\boxed{R_k = \frac{s - k}{s} \qquad \text{if } p = \frac{1}{2}}$$

For instance, $p = 0.48$, $k = 10$, $s = 20$, falls within the case $p \neq 1/2$ and shows how, even for p slightly less than 0.5 is $R_{10} > 0.5$ ($R_{10} = 0.690066$). So, in this case, it is far more likely to be ruined that to preserve and/or increase capital.

If calculations are repeated with $k = 100$ and $s = 200$, the result was an even more dramatic evidence: $R_{100} > 0.999$.

The example $p = 0.5$, $k = 10$, $s = 20$, falls in the case $p = 1/2$: we obtain $R_k = 0.5$. The situation is much better than the previous, although it is possible to be ruined in a case out of two! Notice that even $p = 0.5$, $k = 100$, $s = 200$ lead to $R_k = 0.5$.

So, it is better not to bet repeatedly when probabilities are even slightly contrary. If one is forced to bet having adverse probability ($p < 1/2$), then the best way to hope to double the capital is to bet everything at once!

In summary, if one wants to double the capital ($s = 2k$), one has three cases:

- if $p < q$, then $r > 1$ and $R_k = \dfrac{r^k - 1}{r^k - r^{-k}} \nearrow 1$;

- if $p = q$, then $r = 1$ and $R_k = \dfrac{1}{2}$, $\forall k$;

- if $p > q$, then $r < 1$ and $R_k = \dfrac{r^k - 1}{r^k - r^{-k}} \searrow 0$.

We observe that the function $k \mapsto R_k$ does not describe a history of subsequent bets, but it corresponds to the dependence on the initial sum k ($0 \leq k \leq s$) of the probability R_k of ruin alternatively to the achievement of the sum s.

Solution 2.37 We study the two models separately.

(1) Setting $P_N = P_E$, by the equality $Q_{k+1}^d = Q_{k+1}^o$ we obtain the equation

$$P_{k+1} = -\frac{d}{b}(1 - \delta) P_k + \frac{a+c}{b} - \frac{d}{b} \delta P_E.$$

Notice that the equation admits the unique equilibrium $\alpha = P_E$. The closed form solution is

$$P_k = \left(-\frac{d}{b}(1 - \alpha) \right)^k (P_0 - P_E) + P_E.$$

Since

$$\left| -\frac{d}{b}(1 - \alpha) \right| < \left| -\frac{d}{b} \right| \qquad \alpha \in (0, 1)$$

it follows that:

- if there is convergence in the cobweb model, there is also convergence in the model with normal price;
- if there are oscillations of constant amplitude in the cobweb model, then they become damped oscillations;
- if there are significant oscillations in the cobweb model, anything can happen in the model with normal price.

(2) By $Q^o_{k+1} = -c + dP^e_{k+1}$ and the equilibrium condition $Q^o_{k+1} = Q^d_{k+1}$ one obtains

$$P^e_{k+1} = \frac{Q^o_{k+1} + c}{d} = \frac{a + c - bP_{k+1}}{d}$$

and by substituting into

$$P^e_{k+1} - (1 - \beta) P^e_k = \beta P_k \qquad \beta \in (0, 1)$$

one gets

$$P_{k+1} = \left(1 - \beta\left(1 + \frac{d}{b}\right)\right) P_k + \beta\frac{a + c}{b}.$$

The unique equilibrium is P_E, then it is:

$$P_k = \left(1 - \beta\left(1 + \frac{d}{b}\right)\right)^k (P_0 - P_E) + P_E .$$

The stability condition is

$$\left|1 - \beta\left(1 + \frac{d}{b}\right)\right| < 1 \quad \Leftrightarrow \quad \beta\left(1 + \frac{d}{b}\right) < 2 \quad \Leftrightarrow \quad \frac{d}{b} < \frac{2}{\beta} - 1.$$

Since $\frac{2}{\beta} - 1 > 1$, the last stability condition is looser than that of the cobweb model (Example 2.8).

Solution 2.38 With the notations of Exercise 1.14, we use Theorem 2.56 (with X_n equal to A_n, A_{2n} or A_{2n+1} depending on the case) to determine the coefficients of the series expansion of a particular solution y of the proposed differential equations.

Bessel: The coefficients of odd index are null. From $A_{2n} = -\dfrac{A_{2n-2}}{4n(n + k)}$ it follows

$$A_{2n} = \frac{(-1)^n \, k!}{n! \, (n + k)! 2^{2n}} A_0$$

Dividing all the coefficients A_n by the same constant $k!A_0$ we obtain the particular solution J_k, called *Bessel function of the first kind of order k*:

$$J_k(x) = \sum_{n=0}^{+\infty} \frac{(-1)^n}{n!(n + k)!} \left(\frac{x}{2}\right)^{2n+k}$$

For the general solution of the Bessel equation see [9, 15]).

Hermite: We start from $A_{n+2} = \dfrac{2(n - k) A_n}{(n + 1)(n + 2)}$.
If k is even we consider $A_0 = 1$ and $A_1 = 0$: it follows

$$A_{2n+1} = 0 \,\, \forall n\,, \quad A_{2n} = 0 \,\, \forall n > k/2\,, \quad A_{2n} = (-1)^n \, \frac{2^n \, (k - 2n + 2)!!}{(2n)!} \,\, \text{if } 0 \le n \le k/2,$$

where $m!! = m\,(m-2)\,(m-4)\ldots$. If k is odd, we consider $A_0 = 0$ and $A_1 = 1$: it follows

$$A_{2n} = 0 \;\; \forall n, \quad A_{2n+1} = 0 \;\; \forall n > (k-1)/2,$$

$$A_{2n+1} = (-1)^n \frac{2^n\,(k-2n+1)!!}{(2n+1)!} \;\; \text{if} \;\; 0 \leq n \leq (k-1)/2.$$

In both cases the considered solutions are polynomials H_k of degree less than or equal to k, called *Hermite polynomials of order* k.

Laguerre: from $A_{n+1} = \dfrac{n-k}{(n+1)^2} A_n$ it follows

$$A_n = \frac{(-1)^n}{n!} \binom{k}{n} A_0 \quad n = 0, 1, \ldots, k, \qquad A_n = 0 \;\; \forall n > k,$$

from which we get the particular solution p_k (*Laguerre polynomial*):

$$p_k(x) = \sum_{n=0}^{+\infty} \frac{(-1)^n\,k!}{(n!)^2(k-n)!}\, x^n =$$

$$= (-1)^k \left(\frac{1}{k!}x^k - \frac{k}{1!(k-1)!}x^{k-1} + \cdots + (-1)^{k-1}kx + (-1)^k \right).$$

Legendre: from $A_{n+2} = \dfrac{n\,(n+1) - k\,(k+1)}{(n+1)\,(n+2)} A_n = \dfrac{(n-k)\,(n+k+1)}{(n+1)\,(n+2)} A_n$.

If k is even we consider $A_0 = 1$ and $A_1 = 0$: it follows

$$A_{2n+1} = 0 \;\; \forall n, \quad A_{2n} = 0 \;\; \forall n > k/2,$$

$$A_{2n} = \frac{(-1)^n}{(2n)!} \frac{(k-2n+1)!!}{(k-2n)!!} \frac{k!!}{(k+1)!!} \;\; \text{if} \;\; 0 \leq n \leq k/2.$$

Obviously in all the coefficients the common factor $k!!/(k+1)!!$ can be omitted.
If k is odd we consider $A_0 = 0$ and $A_1 = 1$: it follows

$$A_{2n} = 0 \;\; \forall n, \quad A_{2n+1} = 0 \;\; \forall n > (k-1)/2,$$

$$A_{2n+1} = \frac{(-1)^n}{(2n+1)!} \frac{(k+2n)!!}{(k-2n-1)!!} \frac{(k+1)!!}{k!!} \;\; \text{if} \;\; 0 \leq n \leq (k-1)/2.$$

Obviously in all the coefficients the common factor $(k+1)!!/k!!$ can be omitted
In both cases the considered solutions are polynomials P_k of degree less than or equal to k, called *Legendre polynomials of order* k.

Solution 2.39 The **present value** V_A of a payment of amount (o **nominal value**) V_N after n years at a constant compound interest rate r is given by the formula

$$V_A = V_N\,(1+r)^{-n}.$$

The choice of r is not simple and a correct determination is important for the effectiveness of the model. In each case the appropriate values exist. For instance, to evaluate the opportunity to make an investment r can be chosen equal to the government bonds yield with a duration compatible with the considered investment. Going back to the exercise:

$$A_k = \frac{1}{1+r} + \frac{1}{(1+r)^2} + \cdots + \frac{1}{(1+r)^k} = \frac{1}{1+r}\frac{1-(1+r)^{-k}}{1-(1+r)^{-1}} = \frac{1-(1+r)^{-k}}{r}.$$

Then

$$A_k = \frac{1 - v^k}{r} \qquad v = (1+r)^{-1}$$

If A_k is the present value at the compound rate r of an annuity of k unit install-ments at dates $1, \ldots, k$, we obtain A_{k+1} adding to A_k the present value of the unit installment drawn at time $k + 1$:

$$A_{k+1} = A_k + (1 + r)^{-k+1} = A_k + v^{k+1}.$$

Applying the formula of Remark 2.61 the result determined above can be found. If $r = r_k$, then

$$A_k = \sum_{n=1}^{k} \frac{1}{(1 + r_k)^n} \qquad A_{k+1} = A_k + (1 + r_k)^{-k+1}.$$

Solution 2.40   Using the formula of the previous exercise, we obtain the relation $M = Q\dfrac{1 - v^k}{r}$ from which

$$Q = \frac{Mr}{1 - v^k} \qquad v = (1+r)^{-1}$$

Solution 2.41   a) Since $\tau_k > 0$, it follows $r_k \in [1.5 , 15]$. The rate 1.5% is called **floor** of the bond.
b) One must satisfy the equation

$$\frac{P_k}{r_k} = \frac{100}{\tau_{l_0}}$$

Consequently, under our simplifying assumptions, the price should be

$$P_k = 100 \max \{1.5 , 15 - 2\tau_k\} / \tau_k .$$

We observe the high **volatility** of the bond: if $\tau_k = 7.25$ then P_k reduces to 20.68 (in base 100) that is, one suffers a great loss; but, if $\tau_k \to 0$ then the price tends to $+\infty$. It is clear that the values of P_k vary greatly even if τ_k varies little (**leverage effect**).
c) $P_k + 100\,(r_1 + r_2 + \cdots + r_{k-1})$. Selling the security to maturity, one gets the value $100\,(1 + r_1 + \cdots + r_{10})$.
d) $100\,(1 + 5 + 5.1 + 5.2 + 4.9 + 4 + 3 + 2.5 + 2.6 + 4 + 7.5)$.

Solution 2.42   Since the terms of absolutely convergent series can be rearranged, we have:

$$\mathcal{Z}\{X * Y\} = \sum_{k=0}^{+\infty} (X * Y)_k \frac{1}{z^k} = \sum_{k=0}^{+\infty} \sum_{h=0}^{k} X_h Y_{k-h} \frac{1}{z^k} = \sum_{k=0}^{+\infty} \sum_{h=0}^{k} \frac{X_h}{z^h} \frac{Y_{k-h}}{z^{k-h}} =$$

$$= \sum_{s=0}^{+\infty} \sum_{h=0}^{+\infty} \frac{X_h}{z^h} \frac{Y_s}{z^s} = \left(\sum_{h=0}^{+\infty} \frac{X_h}{z^h}\right) \left(\sum_{s=0}^{+\infty} \frac{Y_s}{z^s}\right) = \mathcal{Z}\{X\}\,\mathcal{Z}\{Y\} .$$

Solution 2.43   From Newton's binomial formula it follows:

$$(m+1)^k = \sum_{s=0}^{k} \binom{k}{s} m^{k-s} \qquad m \in \mathbb{N}.$$

By replacing subsequently in this formula m with the first n positive integers and summing these relations, one obtains:

$$\sum_{m=1}^{n} (m+1)^k = \sum_{m=1}^{n} \sum_{s=0}^{k} \binom{k}{s} m^{k-s} = \sum_{s=0}^{k} \binom{k}{s} \sum_{m=1}^{n} m^{k-s} = \sum_{s=0}^{k} \binom{k}{s} S_{k-s}$$

which can be rewritten as

$$S_k - 1 + (n+1)^k = \binom{k}{0} S_k + \sum_{s=1}^{k} \binom{k}{s} S_{k-s}$$

which in turn yields

$$S_{k-1} = (n+1)^k - 1 \sum_{s=2}^{k} \binom{k}{s} S_{k-s}.$$

Solution 2.44   If $n = 2$, then

$$\det \begin{bmatrix} 1 & \mu_1 \\ 1 & \mu_2 \end{bmatrix} = \mu_2 - \mu_1$$

If $n = 3$, then by replacing μ_1 with μ in the matrix

$$\begin{bmatrix} 1 & \mu_1 & \mu_1^2 \\ 1 & \mu_2 & \mu_2^2 \\ 1 & \mu_3 & \mu_3^2 \end{bmatrix}$$

and computing the determinant, we obtain a polynomial $P(\mu)$ of degree two. By substituting μ also to μ_2 or μ_3 the determinant vanishes, and therefore μ_2 and μ_3 are roots of $P(\mu)$. This means that the equality $P(\mu) = a(\mu - \mu_2)(\mu - \mu_3)$ holds, where a is the coefficients of the term μ^2 of P. Therefore a coincides with the algebraic complement of μ_1^2 and consequently

$$\det \begin{bmatrix} 1 & \mu_1 & \mu_1^2 \\ 1 & \mu_2 & \mu_2^2 \\ 1 & \mu_3 & \mu_3^2 \end{bmatrix} = (\mu_3 - \mu_2)(\mu_1 - \mu_2)(\mu_1 - \mu_3).$$

The formula for the case of dimension n is obtained analogously, assuming that one knows the one of $n-1$–dimensional case.

Solution 2.45 $\mathcal{Z}\left\{\dfrac{1}{k!}\right\} = \sum_{k=0}^{+\infty} \dfrac{1}{z^k k!} = e^{1/z}.$

8.3 Solutions of the exercises in Chapter 3

Solution 3.3 $\left\{\dfrac{2}{7}, \dfrac{4}{7}, \dfrac{6}{7}\right\}$, $\left\{\dfrac{2}{9}, \dfrac{4}{9}, \dfrac{8}{9}\right\}$.

Solution 3.4 The equilibria are solutions of $x^2 = x$ that is $\alpha_1 = 0$ and $\alpha_2 = 1$. To determine possible cycles, we note that the k-th iterate of the map is x^{2k} and then one has to solve

$$x^{2k} = x$$

which has the solutions $\alpha_1 = 0$ and $\alpha_2 = 1$ We conclude that there are no periodic orbits.

Solution 3.5 The equation

$$x \in [-3, 3] : \quad \sqrt{9 - x^2} = x$$

is equivalent to

$$\begin{cases} x \geq 0 \\ 2x^2 = 9 \end{cases} \quad \Rightarrow \quad \alpha = \frac{3}{2}\sqrt{2}.$$

To determine possible 2 periodic orbits, we solve

$$\sqrt{9 - \left(\sqrt{9 - x^2}\right)^2} = x \quad \Rightarrow \quad \sqrt{x^2} = x \quad \Rightarrow \quad |x| = x.$$

Every $x \geq 0$ is a solution and then (except for α) belongs to a 2 periodic orbit. Finally, it is sufficient to notice that, if $x \in [-3, 0)$, then $f(x) \geq 0$ and therefore these points belong to eventually 2 periodic orbits or $(-3\sqrt{2}/2)$ to eventually stationary orbits.

Solution 3.6 The equilibria are solution of

$$\frac{|r| - r}{2} = x \quad \Rightarrow \quad |x| = 3x \quad \Rightarrow \quad \alpha = 0.$$

As if $X_0 > 0$ then $f(X_0) = 0$ these points belong to eventually stationary orbits. Furthermore, if $X_0 < 0$ then $f(X_0) > 0$ and therefore these points belong to eventually stationary orbits too. In conclusion, there are no periodic orbits.

Solution 3.7 Let $\{\alpha_0, \alpha_1\}$ be a cycle of order 2 for the DDS $\{I, f\}$. Since by definition $\alpha_0 \neq \alpha_1$, we can assume $\alpha_0 < \alpha_1$ Then:

$$\alpha_0 < \alpha_1 = f(\alpha_0); \qquad \alpha_1 > \alpha_0 = f(\alpha_1).$$

By the continuity of f and $\alpha_0, \alpha_1 \in I$ it follows the existence of $c \in (\alpha_0, \alpha_1)$ such that $f(c) = c$.

Solution 3.8 We observe that

$$X_1 - X_0 = \sqrt{X_0 + 2} - X_0 = \frac{X_0 + 2 - X_0^2}{\sqrt{X_0 + 2} + X_0} \geq 0 \quad \Leftrightarrow \quad -2 \leq X_0 \leq 2.$$

We prove that $\{X_k\}$ is strictly increasing for $\lambda \in [-2, 2)$. To this aim, after noticing that the sequence has positive terms, we verify that if $X_0 \in [-2, 2)$ then $X_k < 2$ for each $k \in \mathbb{N}$. In fact, $X_0 \in [-2, 2)$; furthermore, if we assume, for $k \geq 1$, (induction hypothesis) that $X_k < 2$, we get:

$$X_{k+1} = \sqrt{X_k + 2} < \sqrt{2 + 2} = 2.$$

We show now that, if $X_0 \in [-2, 2)$, then $\{X_k\}$ is an increasing sequence. For each $k \geq 1$, we have:

$$X_{k+1} - X_k = \sqrt{X_k + 2} - X_k = \boxed{\text{by multiplying and dividing by } \sqrt{X_k+2}+X_k}$$

$$= \frac{X_k + 2 - X_k^2}{\sqrt{X_k + 2} + X_k} = \boxed{\text{by factoring}}$$

$$= \frac{-(X_k + 1)(X_k - 2)}{\sqrt{X_k + 2} + X_k} \geq \boxed{\text{as } X_k < 2}$$

$$\geq 0.$$

It follows that, for each $\lambda \in [-2, 2)$, the DDS $\{[-2, 2), \sqrt{x + 2}\}$ identifies a sequence converging to $\alpha = 2$ (compare with the discussion of Example 3.26).
We invite the reader to analyze the case $\lambda > 2$, checking that we get a sequence (decreasing and) converging to $\alpha = 2$.

Solution 3.9   Let $\{\alpha_0, \alpha_1\}$ be the 2 periodic orbit, with $\alpha_0 < \alpha_1$.

$$\alpha_0 < \alpha_1 = f(\alpha_0) \qquad \alpha_1 > \alpha_0 = f(\alpha_0).$$

By applying the theorem of zeros to the continuous function $x - f(x)$ in the interval $[\alpha_0, \alpha_1] \subset I$, it follows the existence of $\alpha \in (\alpha_0, \alpha_1)$ such that $\alpha = f(\alpha)$.

Solution 3.11 $X_1 = -(X_0)^3 = -X_0^3$, $X_2 = -(-X_0^3)^3 = (-1)^2 (X_0)^{3^2} = X_0^9$.
By iterating: $X_k = (-1)^k (X_0)^{3^k} \qquad k \in \mathbb{N}$.

Solution 3.12   First we draw the graph of the function $f : \mathbb{R} \to \mathbb{R}$, $f(x) = (x + |x|)/4$. The unique equilibrium point is $\alpha = 0$. If $X_0 < 0$, then $X_1 = f(X_0) = 0$ and therefore $X_k = 0$ for every $k \in \mathbb{N}$. If $X_0 > 0$, the graphical analysis suggests that the sequence $\{X_k\}$ is decreasing and converging to 0: the confirmation can be obtained by induction.

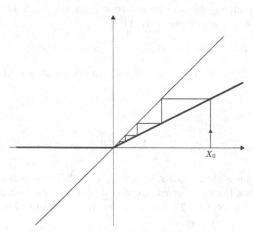

Fig. 8.5 Graph of $f(x) = (x + |x|)/4$

Solution 3.13   The dynamical system presents two equilibria: $\alpha_1 = -\pi$ and $\alpha_2 = \pi$. By the graphical analysis we deduce that for any initial datum in $(-\pi, \pi)$ the corresponding sequence $\{X_k\}$ is increasing and converges to π.

Solution 3.14   We prove the algorithms in the most general case $f : I \to I$, $I \subseteq \mathbb{R}$ closed interval.

Algorithm I. Consider $X_0 \leq X_1$. We show by induction that X is increasing. If $k = 1$ the relation is true by assumption. If (induction assumption) $X_{k-1} \leq X_k$, since f is increasing, we obtain $f(X_{k-1}) \leq f(X_k)$ which equals $X_k \leq X_{k+1}$. One proceeds in the same way if $X_0 \geq X_1$.

Now, let f be increasing and continuous on the closed interval I, and $X_0 < X_1$. If I has l.u.b. $b \in \mathbb{R}$, then, by Theorem 3.23, f admits at least one fixed point in $[X_0, b]$. The set of fixed points of f on $[X_0, b]$ is bounded (as $[X_0, b]$ is bounded) and nonempty (as f admits at least one fixed point). Thus, the set of fixed points of f has g.l.b., say $\overline{\alpha}$, which is also a minimum. In fact, if $\{\alpha_n\}$ is a sequence of fixed points such that $\alpha_n \to \overline{\alpha}$ as $n \to +\infty$, by the continuity of f it follows $f(\alpha_n) \to f(\overline{\alpha})$. Since $f(\alpha_n) = \alpha_n$, $\forall n$, this implies $f(\overline{\alpha}) = \overline{\alpha}$ that is $\overline{\alpha}$ belongs to the set of fixed points. We verify now that $X_k \leq \overline{\alpha}$ for each k. If $k = 0$, the relation is true because $\overline{\alpha} \in (X_0, b]$. If (induction hypothesis) $X_k \leq \overline{\alpha}$, as f is increasing, one has

$$X_{k+1} = f(X_k) \leq f(\overline{\alpha}) = \overline{\alpha}.$$

In conclusion, X is an increasing and bounded from above sequence: therefore, it admits finite limit $L \leq \overline{\alpha}$. Since f is continuous, by Theorem 3.25 L is a fixed point for f and since it can not be less than $\overline{\alpha}$ one gets $L = \overline{\alpha}$.

Finally, if I is unbounded from above and f does not admit fixed points "to the right" of X_0, X must be an unlimited sequence, because otherwise its limit would be a fixed point of f.

Using the same technique one proves the case $X_1 < X_0$.

Algorithm II. The function $g = f \circ f = f^2$ is increasing and $X_{2k} = f(X_{2k-1})$ implies

$$X_{2k+1} = f(X_{2k}) = g(X_{2k-1}).$$

We conclude that the sequence $\{X_{2k+1}\}$ is recursively defined by the increasing function g and conclusions follow from Algorithm I. One proceeds similarly for $\{X_{2k}\}$. Let be now $X_1 \leq X_2$. Suppose that, for $k \geq 1$, is $X_{2k-1} \leq X_{2k}$. We deduce:

$$X_{2k+1} = g(X_{2k-1}) \leq \boxed{\text{as } g \text{ is increasing and } X_{2k-1} \leq X_{2k}}$$
$$\leq g(X_{2k}) = X_{2k+2}.$$

By the induction principle it follows $X_{2k-1} \leq X_{2k}$ for each $k \geq 1$.

Solution 3.16   The function $f : \mathbb{R} \to \mathbb{R}$ defined by $f(x) = |x - 1|/2$ admits a unique fixed point $\alpha = 1/3$, that solves

$$x \in \mathbb{R}: \qquad \frac{|x-1|}{2} = x.$$

Since f is a contraction (with constant $1/2$), Theorem 3.32 leads to the conclusion that whatever the initial datum $X_0 \in \mathbb{R}$, the corresponding sequence $X_k = f(X_{k-1})$ converges to $1/3$. By the graphical analysis one can deduce that the convergence is not monotone.

Solution 3.18 In the various cases the equilibria verify:

1. $\alpha = 1, 0, -1$; $f'(x) = \frac{1}{2}(3x^2 + 1)$; $f'(0) = \frac{1}{2}$ (l.a.s.); $f'(\pm 1) = 2$ (repelling);

2. $\alpha = \frac{1}{2}\sqrt{5} - \frac{1}{2}, -\frac{1}{2}\sqrt{5} - \frac{1}{2}, 0$; $f'(x) = 3x^2 + 2x$; $f'(0) = 0$ (l.a.s.); $f'\left(\frac{1}{2}\sqrt{5} - \frac{1}{2}\right) = \frac{7}{2} - \frac{1}{2}\sqrt{5} > 1$, $f'\left(-\frac{1}{2}\sqrt{5} - \frac{1}{2}\right) = \frac{1}{2}\sqrt{5} + \frac{7}{2} > 1$ (repelling);

3. $\alpha = 0, 1$; $f'(x) = (1-x)e^{1-x}$; $f'(0) = e$ (repelling); $f'(1) = 0$ (l.a.s.);

4. $\alpha = -1, 0$; $f'(x) = 3x^2 + 2x + 1$; $f'(0) = 1$; $f''(x) = 6x + 2$; $f''(0) = 2$ (s.r. and i.l.a.s.); $f'(-1) = 2$ (repelling);

5. $\alpha = 0$; $f'(x) = e^x$; $f'(0) = 1$; $f''(x) = e^x$; $f''(0) = 1$ (s.r. and i.l.a.s.);

6. $\alpha = 0$, $f'(x) = 1 - 3x^2$, $f'(0) = 1$, $f''(x) = -6x$, $f''(0) = 0$, $f'''(x) = -6 < 0$ (l.a.s.);

7. $\alpha = 0$; $f'(x) = 3x^2 + 1$; $f'(0) = 1$; $f''(x) = 6x$; $f''(0) = 0$; $f''(0) = 6 > 0$ (unstable);

8. $\alpha = 0, 2$; $f'(x) = 2x - 1$; $f'(0) = -1$; $2f'''(0) + 3(f''(x))^2 = 4 > 0$ (l.a.s.); ; $f'(2) = 3$ (repelling)

9. $\alpha = \sqrt{2}, -\sqrt{2}, 0$; $f'(x) = 3x^2 - 1$; $f'(0) = -1$, $f'(\pm\sqrt{2}) = 5$ (repelling); $f''(x) = 6x$; $f'''(x) = 6$; $2f'''(0) + 3(f''(0))^2 = 30 > 0$ (l.a.s.).

Solution 3.19 For the first strategy (fixed quantity):
$f(x) = (1+a)x - ax^2 - b$, the equilibria are $\alpha_{1,2} = \left(1 \pm \sqrt{1 - 4b/a}\right)/2$, and

- if $0 < b < a/4$, there are two positive equilibria: $\alpha_1 > \alpha_2 > 0$, with α_1 stable and attractive and α_2 repelling;
- if $b = a/4$, then $\alpha_1 = \alpha_2 = 1/2$ that it is unstable;
- if $b > a/4$, there are no equilibria.

Summing up, the maximum sustainable fished with the first strategy is $a/4$, which, however, corresponds to an unstable situation.
With the second strategy (fixed proportion), $f(x) = (1 + a - r)x - ax^2$, the equilibria are $\alpha_2 = 0$, $\alpha_1 = 1 - r/a$, $f'(\alpha_1) = 1 - a + r$, α_1 stable if $0 < r < a$ (the condition $r > a - 2$ is surely satisfied $a < 2$, as $r \in (0, 1)$).
Therefore, with the second strategy

$$P_{\max}(a) = \max_{[a-2,a] \cap [0,1]} r\alpha_1 = \begin{cases} a/4 & 0 \le a \le 2 \\ 1 - 1/a & a > 2 \end{cases}.$$

We observe that the maximum fished quantity $P_{\max}(a)$ is an increasing function of a. Moreover, if $a > 2$ the first strategy may be convenient, provided one does not exceed... .

Solution 3.21   The possible equilibria are solution of the equation

$$x \in \mathbb{R}: \qquad (1-a)x = b.$$

If $a = 1$ and $b = 0$, each $c \in \mathbb{R}$ is an equilibrium; if $a = 1$ and $b \ne 0$, there are no equilibrium points; finally, if $a \ne 1$, there is only one equilibrium point: $c = b/(1-a)$.
As regards stability, if $a = 1$ and $b = 0$, each equilibrium point is stable but not

attractive; if $a \neq 1$, since $g'(x) = a$, by Theorem 3.57, each equilibrium is globally asymptotically stable if $|a| < 1$, repelling if $|a| > 1$. Compare the obtained results with the graphs and comments in paragraph 2.1.

Solution 3.22   We use the Algorithm II.

The function $f(x) = 1 - \dfrac{4}{\pi} \arctan x$ is continuous and strictly decreasing, with

$$-1 < 1 - \frac{4}{\pi} \arctan x < 1 \qquad\qquad \forall x \in \mathbb{R}.$$

If $Y_0 = 4$, then $Y_1 = 1 - \dfrac{4}{\pi} \arctan 4 \neq Y_0$, and $Y_2 < Y_0$. One must then determine any fixed points of

$$f^2(x) = 1 - \frac{4}{\pi} \arctan \left(1 - \frac{4}{\pi} \arctan x \right).$$

We observe that $f^2(0) = 0$ e $f^2(1) = 1$. Furthermore:

$$(f^2)'(x) = \frac{16}{\pi^2} \frac{1}{(1+x^2)\left(1 + \left(1 - \dfrac{4}{\pi}\arctan x\right)^2\right)}$$

and then $(f^2)(x) < 1$ for each $x \geq 1$: to the right of $x = 1$ there are no other fixed points of f^2. Then, by Algorithm II, one deduces $Y_{2k} \searrow 1$ and $Y_{2k+1} \nearrow f(1) = 0$.

Solution 3.23   Plotting the graph of the function $f : \mathbb{R} \to \mathbb{R}$ defined by $f(x) = \left(x^2 + 1\right)/3$ we realize that there are no equilibrium points: whatever $a > 0$, the sequence is increasing and unbounded, then it diverges to $+\infty$.
In fact, $\forall a \geq 0$, we have:

$$Y_{k+1} - Y_k = \frac{Y_k^2 + 4}{3} - Y_k = \frac{1}{3}\left(Y_k^2 - 3Y_k + 4\right) \geq 0 \qquad \forall k \in \mathbb{N}$$

since the discriminant of the quadratic polynomial in parentheses is negative. Moreover, since

$$4Y_k \leq Y_k^2 + 4$$

the difference between two consecutive terms grows with k

$$Y_{k+1} - Y_k = \frac{1}{3}\left(Y_k^2 - 3Y_k + 4\right) \geq \frac{1}{3}Y_k$$

therefore the sequence is not bounded from above.

Solution 3.24   Plotting the graph of the function $f : \mathbb{R} \to \mathbb{R}$ defined by $f(x) = \left(x^2 + 6\right)/5$, we realize that there are two intersection points in $x = 2$ and $x = 3$, between its graph and the bisector of the first and third quadrant. This means that there are two equilibrium points:

$$\begin{array}{lll} \text{se } a = 2 & Y_k = 2 & \forall k \in \mathbb{N} \\ \text{se } a = 3 & Y_k = 3 & \forall k \in \mathbb{N}. \end{array}$$

By a graphical analysis we deduce that the first equilibrium is attractive, the second one is repelling and, more precisely:

- if $|a| < 3$, then $\lim_k Y_k = 2$;
- if $|a| > 3$, then $\lim_k Y_k = +\infty$.

We look for an analytical confirmation of these results. Thanks to Theorem 3.57, since

$$f'(2) = \frac{4}{5} < 1 \qquad\qquad f'(3) = \frac{6}{5} > 1$$

we deduce that $\alpha = 2$ is a locally asymptotically stable equilibrium, while $\alpha = 3$ is repelling. Moreover, since f, for $x > 0$, is strictly increasing, thanks to Algorithm I we deduce:

- since $Y_1 < Y_0 = a$ for each $a \in (2,3)$, then $\lim_k Y_k = 2$;
- since $Y_1 > Y_0$ for each $a \in (0,2)$, then $\lim_k Y_k = 2$;
- since $Y_1 > Y_0$ for each $a \in (3,+\infty)$, then $Y_k \nearrow +\infty$.

The results when $x < 0$ are obtained by symmetry from the previous ones.

Solution 3.25   First, we draw the graph of the function $f : \mathbb{R} \to \mathbb{R}$, $f(x) = 2x \exp(-x/2)$. To this end we note that

$$\lim_{x \to -\infty} f(x) = -\infty \qquad\qquad \lim_{x \to +\infty} f(x) = 0^+;$$

$$f'(x) = e^{-x/2}(2-x) \qquad\qquad f''(x) = e^{-x/2}\left(\frac{x}{2} - 2\right)$$

and then

$$f'(x) \geq 0 \qquad \Leftrightarrow \qquad x \leq 2$$
$$f''(x) \geq 0 \qquad \Leftrightarrow \qquad x \geq 4.$$

To determine any fixed points, we observe that $f(0) = 0$ and $f'(0) = 2$. This means that, for $x > 0$, the graph of the identity function intersects the graph of f at one

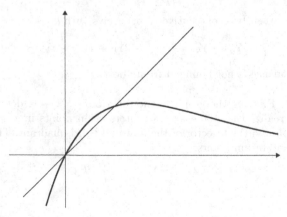

Fig. 8.6 Graph of $2xe^{-x/2}$

point only, say $\bar{x}$. Moreover, since

$$f(1) = \frac{2}{\sqrt{e}} > 1 = f(1) \qquad\qquad f(2) = \frac{4}{e} < 2 = f(2)$$

we have $1 < \bar{x} < 2$.

From above, by Algorithm I, the sequence Y converges to $\bar{x}$, as $0 < a < \bar{x}$ or $\bar{x} < a \leq 2$. If $a > 2$, as necessarily

$$Y_1 = f(Y_0) < 2$$

one falls in the previous case and then again there is convergence to the value $\bar{x}$.

Solution 3.26 For each $Y_0 = a > \sqrt{b}$ the sequence computed recursively by

$$Y_{k+1} = \frac{1}{2}\left(Y_k + \frac{b}{Y_k}\right) \qquad \forall k \in \mathbb{N}$$

is well defined. In fact, by induction, one proves that $Y_k > \sqrt{b}$, $\forall k \in \mathbb{N}$. For $k = 1$, the statement is true by assumption. Assume now $Y_k > \sqrt{b}$ (induction hypothesis). We have:

$$Y_{k+1} - \sqrt{b} = \frac{1}{2}\left(Y_k + \frac{b}{Y_k}\right) - \sqrt{b} =$$

$$= \frac{1}{2Y_k}\left(Y_k^2 - 2\sqrt{b}Y_k + b\right) = \boxed{\text{square of the binomial}}$$

$$= \frac{1}{2Y_k}\left(Y_k - \sqrt{b}\right)^2 > \boxed{\text{induction assumption}}$$

$$> 0.$$

The sequence Y is strictly decreasing. In fact, for every $k \geq 1$,

$$Y_{k+1} - Y_k = \frac{1}{2}\left(Y_k + \frac{b}{Y_k}\right) - Y_k = \frac{b - Y_k^2}{Y_k} = \frac{\left(\sqrt{b} - Y_k\right)\left(\sqrt{b} + Y_k\right)}{Y_k} < 0$$

as shown in the previous point. Then, Y is strictly decreasing, bounded from below, and therefore admits limit $L \in \mathbb{R}$. Moreover, L must be a fixed point for g, and then:

$$L = \frac{1}{2}\left(L + \frac{b}{L}\right) \qquad \Leftrightarrow \qquad L = \sqrt{\alpha}.$$

From what we have just obtained, if $Y_0 = \sqrt{b}$ the sequence Y is constant: $Y_k = \sqrt{b}$, $\forall k \in \mathbb{N}$.

We evaluate the distance ϵ_{k+1} from the equilibrium in step $k + 1$ as a function of the corresponding distance ϵ_k in step k:

$$0 \leq \epsilon_{k+1} = Y_{k+1} - \sqrt{b} = \boxed{\text{see above}}$$

$$= \frac{\left(Y_k - \sqrt{b}\right)^2}{2Y_k} = \boxed{\text{by definition}}$$

$$= \frac{\epsilon_k^2}{2Y_k} < \boxed{\text{if } Y_k > \sqrt{b} > 0}$$

$$< \frac{\epsilon_k^2}{2\sqrt{b}}.$$

Putting $\beta = 2\sqrt{b}$, we get

$$\epsilon_{k+1} < \beta \left(\frac{\epsilon_k}{\beta}\right)^2 < \beta \left(\frac{\epsilon_{k-1}}{\beta}\right)^4 < \cdots < \beta \left(\frac{\epsilon_1}{\beta}\right)^{2^k} \qquad \forall k \geq 1.$$

Therefore one has a good algorithm to calculate the square root of the number α given that the recursive expression $Y_{k+1} = f(Y_k)$ is simple and the error estimate ϵ_k quickly converges to zero.

For instance, if $b = 3$ and $Y_0 = 2$, it is $\beta < 4$ and

$$\frac{\epsilon_1}{\beta} = \frac{2 - \sqrt{3}}{2\sqrt{3}} = \frac{1}{2\sqrt{3}(2 + \sqrt{3})} = \frac{1}{6 + 4\sqrt{3}} < \frac{1}{10}.$$

Accordingly

$$\epsilon_2 < 4 \cdot 10^{-4} \qquad \epsilon_3 < 4 \cdot 10^{-8}$$

and then

$$\sqrt{3} \simeq Y_3 = \frac{1}{2}\left(Y_2 + \frac{3}{Y_2}\right) = \frac{1}{2}\left(\frac{97}{56} + \frac{3 \cdot 56}{97}\right) = 1.7321 .$$

Solution 3.27   We consider the DDS $\{(-6, +\infty), f\}$ where $f(x) = \sqrt{6 + x}$.

We put $X_{k+1} = \sqrt{6 + X_k}$ having fixed $X_0 > -6$. Then there is only one equilibrium $\alpha = 3$, because 3 is the unique solution of $0 < \alpha = \sqrt{6 + \alpha}$.

α is locally asymptotically stable because $f'(3) = 1/6$. More precisely, it is globally attractive because if $X_0 = 3$, then $X_k = 3$ for each k, otherwise $X_k \geq 0$ for each $k \geq 1$ and

$$x > 3 \quad \Leftrightarrow \quad 3 < f(x) < x$$

$$0 < x < 3 \quad \Leftrightarrow \quad x < f(x) < 3$$

Then each trajectory $\{X_k\}$ is strictly monotone and converges to 3.

Solution 3.28   Suppose $\max_{\mathbb{R}} f = \beta < 0$. Then:

$$Y_1 = Y_0 + f(Y_0) < Y_0 + \beta$$
$$Y_2 = Y_1 + f(Y_1) < Y_1 + \beta < Y_0 + 2\beta$$
$$\cdots = \cdots$$
$$Y_{k+1} = Y_k + f(Y_k) < Y_k + \beta < Y_0 + (k+1)\beta.$$

By the comparison theorem on limits, $\lim_k (k+1)\beta = -\infty$ implies $\lim_k Y_k = -\infty$. One can proceed similarly to prove the second part of the thesis.

Solution 3.29 $(1 - s)$ is the probability that the bacterium (or a descendant) is extinguished.

$(1 - s)^2$ is the probability that both the possible direct successors become extinct. By comparing two successive generations, we get the equation in s (p is given!):

$$s = p\left(1 - (1 - s)^2\right). \tag{8.1}$$

In fact, the bacterium in order not to become extinct should reproduce itself (with probability p) and must not be extinguished both the heirs. Equation (8.1) has two solutions: $s = 0$ and $s = 2 - 1/p$.

If $p \le 1/2$, then only $s = 0$ is acceptable (because probability is never negative), therefore, the bacterium is extinguished.

If $p > 1/2$, then it is necessary to choose the correct value of s between the two solutions: let S_k be the probability that the bacterium is able to reproduce itself for at least k generations. Then

$$\begin{cases} S_{k+1} = p\left(1 - (1 - S_k)^2\right) \\ S_1 = p \end{cases}$$

since a bacterium comes to the $(k + 1)$-th generation if and only if it splits and at least one of its two descendants reproduces k times. The graphical analysis shows that if $p > 0$, then $\{S_k\}$ is monotonically convergent to $2 - 1/p$.

Concluding, for completeness, if $p > 1/2$ the bacterium has probability $s = 2 - 1/p$ to be "immortal" (it is sure only if $p = 1$), because the probability to be immortal is equal to $\lim_{k \to +\infty} S_k$.

To prove the monotonicity, we study the DDS $\{\mathbb{R}^+, f\}$ with $f(x) = 2px - px^2$ when p varies in $[0, 1]$.

If $p = 0$ the dynamic is trivial.

If $p \in (0, 1]$, the graph of f is a (concave) parabola. We compute the possible equilibria:

$$-px^2 + 2px = x \quad \Leftrightarrow \quad x\left[px - (2p - 1)\right] = 0 \quad \Leftrightarrow \quad x = 0 \text{ or } x = 2 - 1/p.$$

Since $f'(x) = 2p(1 - x)$, taking into account that it must be $x \ge 0$, we conclude that:

- if $p \in (0, 1/2)$, then there is only the null equilibrium that it is stable as $f'(0) = 2p < 1$;
- if $p = 1/2$, then the unique equilibrium is 0 that is stable in $[0, 1]$ because it is upper semi-stable in $\mathbb{R}$;
- if $p \in (1/2, 1]$, then there are two equilibria $\alpha_0 = 0$ and $\alpha_1 = 2 - 1/p$; α_0 is unstable as $f'(\alpha_0) = 2p > 1$, and α_1 is locally asymptotically stable since $f'(\alpha_1) = 2(1 - p) < 1$, and its basin is $(0, +\infty)$. Then, starting from $S_1 = p > \alpha_1$, S_k decreasing converges to α_1.

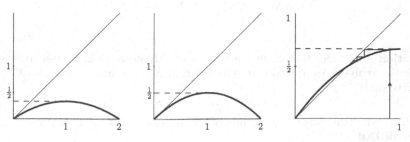

Fig. 8.7 Cobwebs of Exercise 3.29: $0 \le p < 1/2$, $p = 1/2$, $1/2 < p < 1$

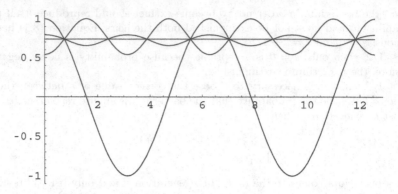

Fig. 8.8 Graphs of f, f^2, f^5, f^{20} where $f(x) = \cos x$

Solution 3.32   We draw the graph of $f(x) = \cos x$ and of the identity function, obtaining the existence of a unique equilibrium α of the given DDS.
Necessarily $\alpha \in (0,1)$ and by $f'(x) = -\sin x$ we deduce $|f'(\alpha)| = |\sin \alpha| < 1$. So, by Theorem 3.57, α is a stable equilibrium.
We observe now that $X_1 \in [-1,1]$ and $X_2 \in [0, \cos 1]$ for any $X_0 \in \mathbb{R}$. Furthermore

$$-1 < -\sin 1 \le f'(x) \le 0 \qquad \forall x \in [0, \cos 1].$$

Therefore, if $k \ge 3$, then $X_k - \alpha$ and $X_{k+1} - \alpha$ have opposite signs and $|X_{k+1} - \alpha| \le c|X_k - \alpha|$ having set $c = \sin 1$. Hence $\{X_k\}$ converges to α by oscillating. Solving the equation $\alpha = \cos \alpha$ numerically one can obtain the desired value.

Solution 3.33 Since the map $f(x) = x + x^3$ is strictly increasing, we use Algorithm 1. We consider an initial datum X_0 and compute $X_1 = X_0 + X_0^3$. We have:

$$X_1 = X_0 \quad \Leftrightarrow \quad X_0 = 0$$

then $\alpha = 0$ is the unique equilibrium of the DDS. Furthermore:

$$X_1 > X_0 \quad \Leftrightarrow \quad X_0 > 0$$

and then, by Algorithm 1, we have:

$$\forall X_0 > 0 \quad \lim_k f^k(X_0) = +\infty$$

$$\forall X_0 < 0 \quad \lim_k f^k(X_0) = -\infty.$$

Solution 3.34 Since the considered map is strictly decreasing, we use Algorithm 2 for the analysis. We consider an initial datum $X_0 > 0$ and given $X_1 = X_0^{-2}$ we observe that

$$X_1 = X_0 \quad \Leftrightarrow \quad X_0^3 = 1 \quad \Leftrightarrow \quad X_0 = 1.$$

Then $\alpha = 1$ is the unique equilibrium of the system. For $X_0 \ne 1$, as $X_2 = X_0^4$, we conclude that

$$X_2 = X_0 \quad \Leftrightarrow \quad X_0^3 = X_0 \quad \Leftrightarrow \quad X_0 = 1$$

that is there are no 2 cycle. Since

$$X_2 > X_0 \quad \Leftrightarrow \quad X_0^3 > X_0 \quad \Leftrightarrow \quad X_0 > 1$$

by Algorithm 2,

$$\lim_k X_{2k} = +\infty; \quad \lim_k X_{2k+1} = 0^+ .$$

If $X_0 \in (0,1)$, then

$$\lim_k X_{2k} = 0^+; \quad \lim_k X_{2k+1} = +\infty .$$

Solution 3.35 The function f_β is continuous and increasing for each β. We are looking for any of its fixed points:

$$\{x \geq 0 : \beta + \sqrt{x} = x\} = \{x \geq 0 : \sqrt{x} = x - \beta\} = \{x \geq \max\{0, \beta\} : x = (x - \beta)^2\} =$$
$$= \{x \geq \max\{0, \beta\} : x^2 - (2\beta + 1)x + \beta^2 = 0\} .$$

The discriminant $\Delta = (2\beta + 1)^2 - 4\beta^2 = 1 + 4\beta$ is nonnegative if and only if $\beta \geq -1/4$. In conclusion:

- if $\beta < -1/4$ there are no equilibria;
- if $\beta = -1/4$ there is a unique equilibrium: $\alpha = \beta + 1/2 = 1/4$;
- if $\beta > -1/4$ we have to solve the system

$$\begin{cases} x \geq 0 \\ x \geq \beta \\ x_1 = \beta + \dfrac{1 + \sqrt{1 + 4\beta}}{2} \ \text{ or } \ x_2 = \beta + \dfrac{1 - \sqrt{1 + 4\beta}}{2} \end{cases} .$$

Obviously x_1 is always acceptable while x_2 is acceptable only when

$$1 - \sqrt{1 + 4\beta} \geq 0 \quad \Leftrightarrow \quad \beta \leq 0.$$

We study now the different identified cases. If $\beta < -1/4$, after a finite number of steps, the iteration must stop as there is a $\overline{k}$ such that $X_{\overline{k}} < 0$. Then, in this case, the sequence is not defined.
If $\beta = -1/4$, then $X_1 < X_0$ for each $X_0 \neq 1/4$ and, by Algorithm I, if $X_0 < 1/4$ then $X_k \to \alpha$ while for $0 \leq X_0 < 1/4$ the system stops after a finite number of steps.
If $-1/4 < \beta \leq 0$ there exist two equilibria, $\alpha_1 = \beta + \dfrac{1 + \sqrt{1 + 4\beta}}{2}$ and $\alpha_2 = \beta + \dfrac{1 - \sqrt{1 + 4\beta}}{2}$ with $0 < \alpha_2 < \alpha_1$. Again according to Algorithm I, we can conclude that if $X_0 > \alpha_1$ then $X_0 \to \alpha_1$; if $\alpha_2 < X_0 < \alpha_1$, as $X_1 > X_0$ and $\alpha_1 > X_0$, we conclude again that $X_k \to \alpha_2$; finally, starting to the left of α_2, for $\beta \neq 0$, the system stops after a finite number of steps.
If $\beta > 0$ there is a unique equilibrium $\alpha = \beta + \dfrac{1 + \sqrt{1 + 4\beta}}{2}$. If $X_0 > \alpha$, then $X_1 < X_0$: the Algorithm I tells us that $X_k \to \alpha$ and the same thing happens for $X_0 < \alpha$ as $X_1 > X_0$.

Solution 3.36 The equation $T^3(x) = x$ has 8 solutions (see Fig. 8.9): two are equilibria, to the others 6 correspond two 3 periodic orbits (see Exercise 3.3). The two 3 periodic orbits are both unstable as deduced by adapting the argument of The-

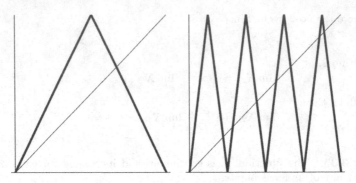

Fig. 8.9 Graphs of T and T^3

orem 3.73. One of these periodic orbits is $\{2/7,\ 4/7,\ 6/7\}$. Moreover $T\,(1/7) = 2/7$, then $1/7$ falls is into the basin of the 3 cycle and the trajectory that starts from $1/7$ is eventually periodic:

$$\frac{1}{7},\ \frac{2}{7},\ \frac{4}{7},\ \frac{6}{7},\ \frac{2}{7},\ \frac{4}{7},\ \frac{6}{7},\ \frac{2}{7},\ \cdots$$

Solution 3.37 By contradiction, if there were $\alpha_0,\ \alpha_1 \in I$ such that $\alpha_0 \neq \alpha_1$ e $\alpha_1 = f(\alpha_0),\ \alpha_1 = f(\alpha_0)$, we would obtain:

$$0 \neq \int_{\alpha_0}^{\alpha_1} (1 + f'(x))\,dx = \alpha_1 - \alpha_0 + f(\alpha_1) - f(\alpha_0) = 0\,.$$

Solution 3.38 The characteristic equation $\lambda^2 - \lambda - 1 = 0$ admits the two solutions $\lambda_{1,2} = \left(1 \pm \sqrt{5}\right)/2$ so the general solution of the equation is

$$X_k = c_1 \left(\frac{1 + \sqrt{5}}{2}\right)^k + c_2 \left(\frac{1 - \sqrt{5}}{2}\right)^k.$$

By substituting X_1 and X_2 one determines the constants c_1 and c_2 and the particular solution

$$X_k = \frac{1}{\sqrt{5}}\,\frac{2\alpha - 1 + \sqrt{5}}{1 + \sqrt{5}}\left(\frac{1 + \sqrt{5}}{2}\right)^k - \frac{1}{\sqrt{5}}\,\frac{2\alpha - 1 - \sqrt{5}}{1 - \sqrt{5}}\left(\frac{1 - \sqrt{5}}{2}\right)^k.$$

Notice that if $\alpha = 1$, X_k coincides for $k \geq 1$ with the sequence of Fibonacci numbers.
(b) One has $Y_{k+1} = f\,(Y_k)$ where $f : \mathbb{R}\backslash\{-1\} \to \mathbb{R}$, $f\,(x) = (1 + x)^{-1}$ is a non linear function. The sequence Y has positive terms because if $x > 0$, then $f\,(x) > 0$.
A possible finite limit must solve $\{x \geq 0 :\ f(x) = x\}$ whose solution is $l = (-1 + \sqrt{5})/2$. Indeed it is $\lim_k Y_k = l$. In fact, since f is strictly decreasing in the interval $(0, +\infty)$ and $f\,(l) = l$, then follow the implications

$$x > l \quad \Rightarrow \quad f\,(x) < l; \qquad\qquad 0 < x < l \quad \Rightarrow \quad f\,(x) > l$$

and then $Y_{2k+1} < \dfrac{-1 + \sqrt{5}}{2} < Y_{2k}$ for each $k \geq 1$.

Moreover, the sequence $\{Y_{2k}\}$ is strictly decreasing and $\{Y_{2k+1}\}$ is strictly increasing. In fact, set $g(x) = f^2(x) = f(f(x)) = 1 - (2+x)^{-1}$, it holds $0 < g'(x) < 1/2$ for each $x > 0$, from which, by applying Lagrange Theorem to $g \in C^1(0, +\infty)$

$$|Y_{2k+2} - l| = |f(f(Y_{2k})) - f(f(l))| = |g(Y_{2k}) - g(l)| =$$
$$= |g'(\xi)| \cdot |Y_{2k} - l| < \frac{1}{2}|Y_{2k} - l| < \frac{|1 - l|}{2^{k+1}} \qquad k \geq 0$$

and similarly

$$|Y_{2k+1} - l| < \frac{1}{2}|Y_{2k-1} - l| < \frac{l - 1/2}{2^k} \qquad k \geq 0.$$

Concluding, Y_k quickly converges to $(-1 + \sqrt{5})/2$ oscillating around such a value.
(c) Y_k can be expressed by the continued fraction

$$Y_k = \cfrac{1}{1 + \cfrac{1}{1 + \cdots}}.$$

The calculation of the first values of the sequence

$$Y_2 = 1, \quad Y_3 = \frac{1}{2}, \quad Y_4 = \frac{2}{3}, \quad Y_5 = \frac{3}{5}, \quad Y_6 = \frac{5}{8}, \ldots$$

leads to conjecture the expression

$$Y_k = \frac{F_{k-1}}{F_k} = 2\frac{(1 + \sqrt{5})^{k-1} \quad (1 - \sqrt{5})^{k-1}}{(1 + \sqrt{5})^k - (1 + \sqrt{5})^k} \qquad \forall k \geq 2$$

where $\{F_k\}$ is the sequence of Fibonacci numbers. Let us prove it by induction. If $k = 2$ then $Y_2 = 1 = F_1/F_2$. Assuming the relation true for a given $k \geq 2$, one has

$$Y_{k+1} = \frac{1}{1 + Y_k} = \frac{F_k}{F_k + F_{k-1}} = \frac{F_k}{F_{k+1}}.$$

Notice that

$$\lim_k \frac{F_{k-1}}{F_k} = \lim_k Y_k = \frac{\sqrt{5} - 1}{2}.$$

(d) The DDS presents only two equilibria $\alpha_1 = (\sqrt{5} - 1)/2$ and $\alpha_2 = (-\sqrt{5} - 1)/2$. α_1 is stable and attractive because $f \in C^1(\mathbb{R} \setminus \{-1\})$ and $|f'(\alpha_1)| < 1$; α_2 is unstable because $|f'(\alpha_2)| > 1$.
If $\beta > -1$, then we can repeat the analysis of step (b), concluding that $Z_k > 0$ for every $k \geq 2$. In this case Z_k quickly tends to l for any initial datum, oscillating with respect to l.
The case $\beta \leq -1$ is more delicate: in fact, the sequence $\{Z_k\}$ is well-defined for each $k \geq 2$ if and only if $\beta \neq -1$ and each value of Z_k computed from β is different from -1. To this aim, we consider the inverse function f^{-1} of f: $f^{-1}(t) = \frac{1-t}{t}$ and we denote by W_k the sequence

$$\begin{cases} W_{k+1} = f^{-1}(W_k) \\ W_1 = -1 \end{cases}$$

whose first terms are $\{-1, -2, -3/2, -5/3, -8/5, -13/8, \ldots\}$.

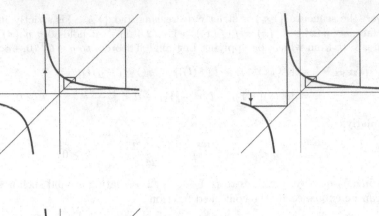

Fig. 8.10 Cobwebs concerning Exercise 3.28

By induction one verifies that $W_k = -F_{k+1}/F_k$ for each $k \geq 1$:

$$W_1 = -1 = -\frac{F_2}{F_1}, \qquad W_{k+1} = \frac{1 - W_k}{W_k} = -\frac{F_k + F_{k+1}}{F_{k+1}} = -\frac{F_{k+2}}{F_{k+1}}.$$

Notice that $-2 \leq W_k \leq -1$, for each k.

So the DDS is well-defined for every initial datum $\alpha \in \mathbb{R}$ different from $-F_{k+1}/F_k$, $\forall k \in \mathbb{N}$, where F_k is the k-th Fibonacci number.

If $\beta < -2$ is an initial datum (automatically eligible), then $Z_k > -1$ for each $k \geq 3$ and for the foregoing analysis Z_k converges to l oscillating.

If $-2 < \beta < 1$ is an eligible initial datum (that is $\beta \neq -F_{k+1}/F_k$ for each k) then, as in point (b), it can be shown that Z_k oscillates around α_2, passing from the region $\{x \in \mathbb{R} : -2 < x < \alpha_2\}$ to $\{x \in \mathbb{R} : \alpha_2 < x < -1\}$ and viceversa in such a way that $|Z_k - \alpha_2|$ is strictly increasing, until $X_{\bar{k}} < -2$, after which $Z_{\bar{k}+1} > -1$ and the trajectory turns out to be attracted too from equilibrium α_1 in the manner described in (b).

Solution 3.39   Assuming $X_0 > 0$ and $X_1 > 0$, one has $X_k > 0$ for each $k \in \mathbb{N}$. Then, the assigned problem is equivalent to

$$\log X_{k+2} = 5 \log X_{k+1} - 6 \log X_k \qquad k \in \mathbb{N}.$$

This is a second order linear difference equation (in the new unknown $Z_k = \log X_k$) obtained by taking the logarithm of both members. The characteristic equation

$\lambda^2 - 5\lambda + 6 = 0$ admits the two roots $\lambda_1 = 2$ and $\lambda_2 = 3$, and so the general solution is

$$Z_k = c_1 2^k + c_2 3^k \qquad k \in \mathbb{N}.$$

which yields

$$X_k = \exp\left\{c_1 2^k + c_2 3^k\right\} \qquad k \in \mathbb{N}.$$

We observe that if $X_0 = 0$ or $X_1 = 0$, the sequence is not defined. Moreover, if $X_0 \neq 0 \neq X_1$, then the sequence has the same sign of X_1, $\forall k \geq 1$.

8.4 Solutions of the exercises in Chapter 4

Solution 4.1   For the existence of the 3 cycle $\{1, 2, 3\}$ one must check $X_0 = 1$, $f(X_0) = X_1 = 2$, $f(X_1) = X_2 = 3$ and $f(X_2) = X_3 = 1$, that is, the system

$$\begin{cases} a + b + c = 2 \\ 4a + 2b + c = 3 \\ 9a + 3b + c = 1 \end{cases}$$

must admit solution. It results that this system admits the unique solution: $a = -3/2$, $b = 11/2$, $c = -2$.

Solution 4.5 The equation $x \in [0, 1] : h_a(x) = x$ admits the solutions $\alpha = 0$ and $\alpha_a = (a - 1)/a$ that are the fixed points of the DDS. The points of the 2 periodic orbit are solutions of

$$x \in [0, 1] : \quad h_a^2(x) = x, \ h_a(x) \neq x.$$

Since

$$h_a^2(x) = a[ax(1-x)][1 - ax(1-x)] = a^2 x(1-x)(1 - ax + ax^2)$$

one gets

$$h_a^2(x) - x = x[a^2(1 - ax + ax^2 - x + ax^2 - ax^3) - 1] =$$
$$= x[-a^3 x^3 + 2a^3 x^2 - 2a^2(a+1)x + a^2 - 1] =$$
$$= -a^3 x \left[x^3 - 2x^2 + \frac{1+a}{a}x - \frac{a^2 - 1}{a^3}\right].$$

Knowing that α_a is a fixed point for h_a, we factorize dividing by $(x - \alpha_a)$, thus obtaining

$$h_a^2(x) - x =$$
$$= -a^3 x \left(x - \frac{a-1}{a}\right)\left(x^2 - \frac{a+1}{a}x + \frac{a+1}{a^2}\right) =$$
$$= -a^3 x \left(x - \frac{a-1}{a}\right)\left(x - \frac{a + 1 + \sqrt{a^2 - 2a - 3}}{2a}\right)\left(x - \frac{a + 1 - \sqrt{a^2 - 2a - 3}}{2a}\right)$$

So, if $a > 3$ we find that

$$\beta_a = \frac{a + 1 + \sqrt{(a+1)(a-3)}}{2a}, \qquad \gamma_a = \frac{a + 1 - \sqrt{(a+1)(a-3)}}{2a}$$

are the points of the 2 periodic orbit. Notice that $\beta_3 = \gamma_3$ and $\beta_a > \gamma_a$ when $a > 3$. In particular:

$$a_2 = 1 + \sqrt{6} = 3.449489\ldots$$
$$\beta_{a_2} = 1 + \sqrt{3} - \sqrt{2} = 1.3178\ldots$$
$$\gamma_{a_2} = 1 + \sqrt{2} - \sqrt{3} = 0.68216\ldots$$

The calculations directly prove that $\{\beta_a, \gamma_a\}$ is the unique 2 periodic orbit of $\{\mathbb{R}, h_a\}$.

Solution 4.6 Since we explicitly know the values β_a and γ_a by the previous exercise, we can compute $h'_a(\beta_a)$ and $h'_a(\gamma_a)$. Since $h'_a(x) = a(1 - 2x)$, one gets

$$h'_a(\beta_a) = a\left(1 - 2\frac{a + 1 + \sqrt{(a+1)(a-3)}}{2a}\right) = -1 - \sqrt{(a+1)(a-3)}$$

$$h'_a(\gamma_a) = a\left(1 - 2\frac{a + 1 - \sqrt{(a+1)(a-3)}}{2a}\right) = -1 + \sqrt{(a+1)(a-3)}$$

and then

$$h'_a(\beta_a)\, h'_a(\gamma_a) = 1 - (a+1)(a-3).$$

So

$$|h'_a(\beta_a)\, h'_a(\gamma_a)| < 1 \quad \Leftrightarrow \quad |1 - (a+1)(a-3)| < 1$$
$$\Leftrightarrow \quad -1 < 1 - (a+1)(a-3) < 1$$
$$\Leftrightarrow \quad -2 < -(a+1)(a-3) < 0$$

$$\boxed{\text{if } a > 3} \quad \Leftrightarrow \quad -2 < -(a+1)(a-3)$$
$$\Leftrightarrow \quad a^2 - 2a - 5 < 0$$
$$\Leftrightarrow \quad a < 1 + \sqrt{6}.$$

The conclusion follows by Theorem 3.69.
Notice that in the interval $3 < a < 1 + \sqrt{6}$ we have $h'_a(\beta_a) < -1$ and $0 < h'_a(\gamma_a) < 1$.

Solution 4.7 The relations proposed by the hint ensure, thanks to Theorem 3.65, that β_a and γ_a are stable and locally attractive equilibria for $g = f^2$. Lemma 3.68 allows to prove the statement in the hint.

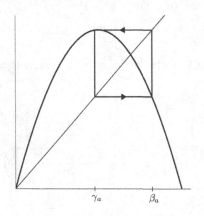 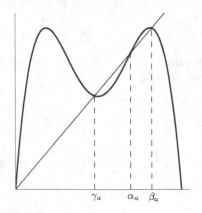

Fig. 8.11 Graphs of $h_{3.1}$ and $h_{3.1}^2$

Take now $f = h_{a_2}$ and $g = h_{a_2}^2$. We calculate

$$\left(h_a^2\right)' = h_a'(h_a)\, h_a'$$

$$\left(h_a^2\right)'' = h_a''(h_a)\left(h_a'\right)^2 + h_a'(h_a)\, h_a''$$

$$\left(h_a^2\right)''' = h_a'''(h_a)\left(h_a'\right)^3 + 3h_a''(h_a)\, h_a''h_a' + h_a'(h_a)\, h_a'''.$$

Considering

$$h_a'(x) = a\,(1 - 2x), \qquad h_a''(x) = -2a, \qquad h_a'''(x) = 0$$

and

$$h_a(\gamma_a) = \beta_a, \qquad h_a(\beta_a) = \gamma_a, \qquad h_a^2(\gamma_a) = \gamma_a, \qquad h_a^2(\beta_a) = \beta_a$$

$$h_a'(\gamma_a)\, h_a'(\beta_a) = -a^2 + 2a + 4$$

we obtain $g'(\gamma_{a_2}) = g'(\beta_{a_2}) = -1$. Using the explicit expressions of γ_a and β_a obtained in Exercise 4.5,

$$g'''(\gamma_a) = -12a^2\left(1 - \sqrt{a^2 - 2a - 3}\right) = 12a^2\left(\sqrt{2} - 1\right) > 0$$

from which $2g'''(\gamma_{a_2}) + 3\left(g''(\gamma_{a_2})\right)^2 > 0$

$$g''(\gamma_a) = -2a\left(a^2\,(1 - 2\gamma_a)^2 + a\,(1 - 2\beta_a)\right) =$$

$$= -2a^2\left(a\left(1 - \frac{a + 1 - \sqrt{a^2 - 2a - 3}}{a}\right)^2 + a\left(1 - \frac{a + 1 - \sqrt{a^2 - 2a - 3}}{a}\right)\right)$$

$$g''(\gamma_{a_2}) = 2\left(1 + \sqrt{6}\right)\left(\left(1 + \sqrt{3}\right)\sqrt{6} - 2\right)$$

$$g'''(\beta_a) = 3\,(-2a)\,(-2a)\,a\,(1 - 2\beta_{a_2}) = 12a^3\left(\frac{-1 - \sqrt{a^2 - 2a - 3}}{a}\right)$$

$$g'''(\beta_{a_2}) = -12\left(7 + 2\sqrt{6}\right)\left(1 + \sqrt{2}\right)$$

$$3\left(g''(\beta_{a_2})\right)^2 = 12\left(7 + 2\sqrt{6}\right)\left(28 + 12\sqrt{3} - 4\sqrt{6} - 12\sqrt{2}\right) >$$

$$> 12\left(7 + 2\sqrt{6}\right)\left(2 + +2\sqrt{2}\right) = -2g'''(\beta_{a_2}).$$

Solution 4.8   We suggest that one starts from $X_0 = 1/2$ (in order to use the Fatou Theorem) and calculates sixty iterations, verifying that $\{X_{2k}\}$ and $\{X_{2k+1}\}$ are stabilized with a certain number of decimal and then calculate $h'(X_{60})\, h'(X_{61})$. Note that anyway, due to stability, all the initial values different from 0, α_a, 1 are in the attraction basin of the periodic orbit.

The uniqueness of the 2 cycle follows from the fact that $h_a^2(x) = x$ is a fourth degree equation, and therefore has at most four solutions, two of which are equilibria.

Solution 4.10   1) α_a is a super-attractive equilibrium if it solves $a\,(1 - 2\alpha_a) = 0$ with $1 < a \leq 3$. So, $b_0 = 2$ and $\alpha_2 = 1/2$ is the super-attractive equilibrium.
2) By the Fatou Theorem $1/2$ is into the attraction basin of the orbit $\{\gamma_a, \beta_a\}$. By solving w.r.t. $a \in (a_1, a_2)$ the equation $h_a^2(1/2) = 1/2$, we obtain the value b_1 of a for which $1/2$ belongs to the 2 cycle. For this value, the 2 cycle is super-attractive

thanks to Theorem 3.69:

$$\left(h_a^2\right)'(\gamma_a) = \left(h_a^2\right)'(\beta_a) = (h_a)'(\gamma_a)(h_a)'(\beta_a) \ .$$

One has to solve

$$\begin{cases} a^3 - 4a^2 + 8 = 0 \\ 3 < a \leq 1 + \sqrt{6} \end{cases} .$$

Alternatively, exploiting the explicit knowledge of γ_a and β_a (see Exercise 4.5), one solves w.r.t. $a \in (a_1, a_2)$ the equation $\gamma_a = 1/2$. One obtains $b_1 = 1 + \sqrt{5}$, $\gamma_{b_1} = 1/2$, $\beta_{b_1} = \left(1 + \sqrt{5}\right)/4$.

For the determination of the values b_k one can proceed in a similar way. However, the resolution w.r.t. $a \in (a_k, a_{k+1})$ of the equation $(h_a)^{2^k}(1/2) = 1/2$ is delicate. However the super-attractiveness of the 2^k cycle corresponding to $a = b_k$ is assured by Theorem 3.73.

Solution 4.11   The points of a 3 periodic orbit solve the equation

$$x \in [0, 1] : \quad h_a \left(h_a \left(h_a \left(x\right)\right)\right) = x \quad e \quad h_a \left(x\right) \neq x.$$

For a close to a_ω, the graph of $(h_a)^3$ presents the following qualitative behavior: it is symmetric with respect to $x = 1/2$ that it is a relative minimum point; there are 4 mode (relative maxima), the complex roots of $(h_a)^3 (x) = x$ are 8, the possible real roots are all in $[0, 1]$ (Theorem 4.1).

If $a < a_\omega$ the real roots of $(h_a)^3 (x) = x$ are only two (the equilibria of h_a), therefore there are no 3 periodic orbits.

If $a = a_\omega$, then the distinct real roots of $(h_a)^3 (x) = x$ are 5 including three doubles due to the points of tangency. Since 3 is a prime number, the double roots necessarily correspond to a 3 periodic orbit.

If $a > a_\omega$, there are 8 distinct real roots. Necessarily, those that are not fixed points of h_a are two 3 periodic orbits, one of which is stable, the other one unstable, as can be graphically verified by observing the slopes at the points of crossing the bisecting line (see Fig. 8.12).

So, the value $a = a_\omega$ is of bifurcation for h_a, different from those classified in Section 4.3.

Taking advantage of the information in the bifurcation diagram of Fig. 4.13 and 4.14, the elements of the 3 cycle are numerically obtained, with $a = 3, 84 : \{0, 1494, 0, 4880, 0, 9594\}$ approximated to 4 decimal places. It occurs numerically that such a 3 cycle

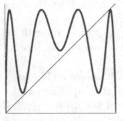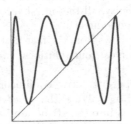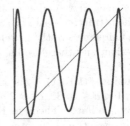

Fig. 8.12 Graphs of $(h_a)^3$ with a near a_ω for: $a < a_\omega$, $a = a_\omega$, $a > a_\omega$

is stable (Theorem 3.73):

$$h'_{3.84}(0.1494)\, h'_{3.84}(0.4880)\, h'_{3.84}(0.9594) = -0.87552\,.$$

Solution 4.12 Solving $f_a(x) = x$ one finds 3 equilibria for $a > 1$ and 1 for $a \le 1$; moreover, there is a single 2 periodic orbit (α_1, α_2) for each $a > 0$, where α_j are the solutions different from 0 of $f_a(x) = -x$.

Solution 4.13 A possible description is given by the following discrete dynamical system $\{[0, 2\pi), (\theta + \pi/30) \bmod 2\pi\}$ where the angular variable θ is measured in radians. All the orbits of the DDS are periodic (with period 60) and therefore the periodic orbits are dense. The DDS is not topologically transitive since if $\theta_0 \ne \psi + h\pi/30$, $\forall h \in \{0, 1, \ldots, 59\}$, then the trajectories with initial datum θ_0 have positive minimum distance from ψ. Finally, there is sensitive dependence on initial data because for every $\varepsilon > 0$ if $\theta_0 < \pi/30 < \varphi_0$ and $\varphi_0 - \theta_0 < \varepsilon$ then $\theta_{60} - \varphi_{60} > 59\pi/30$, having denoted by θ_k and φ_k the trajectory of initial point θ_0 and φ_0, respectively. Concluding, the dynamic is not chaotic.

However, the sensitive dependence on the initial data is somewhat surprising. In fact, the function describing the DDS is not continuous. Nevertheless, it is possible to describe in a more satisfactory way the movement of the lancet modeling it with another DDS (the choice of a model is as important as its analysis). We consider the set $S = \{z \in \mathbb{C} : |z| = 1\}$, that is the unit circle of the complex plane (remember that all its points are of the kind $e^{i\theta}$ with $\theta \in \mathbb{R}$) and the function $f : S \to S$ defined by $f\left(e^{i\theta}\right) = e^{i(\theta + \pi/30)}$, that is continuous with respect to the distance between complex numbers. Then, the DDS $\{S, f\}$ describes more satisfactorily the motion of the seconds hand (see Example 4.21 and Section 4.7). Even the DDS $\{S, f\}$ has dense periodic orbits and it is not topologically transitive, therefore it has no chaotic dynamics. However, unlike the previous one, it does not have sensitive dependence on the initial data:

$$|\theta_k - \varphi_k| = |\theta_0 - \varphi_0| \qquad \forall \theta_0, \varphi_0 \in S \text{ and } \forall k \in \mathbb{N}.$$

Solution 4.14 Unlike the previous exercise, in this case the significant variable is the angular one $\theta \in \mathbb{R}$, considered modulo 2π with reference to the position in the quadrant and the DDS is $\{A, g\}$ with $A = \left\{h\dfrac{\pi}{30}, \quad h = -59, \ldots, +55\right\}$ and $g : A \to A$ such that, starting from $X_0 = 0$, is $X_k = g^k(0)$ with

$$X_k = -k\pi/30 \qquad \text{if } k = 0, 1, \ldots, 59$$
$$X_{60} = \pi/6,\ X_{61} = 4\pi/30, \ldots,\ X_{65} = 0,\ X_{66} = -\pi/30,$$
$$X_{67} = -2\pi/30, \ldots,\ X_{119} = -54\pi/30\quad X_{120} = \pi/3,\ X_{121} = 9\pi/30, \ldots$$

The other trajectories can be deduced from that of the initial data 0. X_k describes a periodic motion with period 720. A is a finite set, all the orbits are periodic and then are dense. All the trajectories run all over A, therefore the system is topologically transitive. Finally, there can be no sensitive dependence on the initial data for any DDS whose domain has a finite number of points.

Solution 4.15 We explicit the general term: $X_k = 10^k X_0$, $\forall X_0 \in \mathbb{R}$. Then, the DDS is not topologically transitive and it has no periodic orbits. In particular, the dynamic is not chaotic.

However, it has sensitive dependence on the initial data, because a small error in the measurement of X_0 is enormously amplified:

$$\text{if } X_0 - \tilde{X}_0 = \varepsilon > 0 \quad \text{then} \quad X_k - \tilde{X}_k = 10^k \varepsilon .$$

In other words, since each iteration moves the decimal point to the right of a position in the decimal representation of X_0, to calculate all the values of X_k it is necessary to know X_0 with infinite precision, that is, to have *all* the digits of its decimal representation.

Solution 4.16 $E = \{x \in \mathbb{R} : \; x = h2^n; \; h, n \in \mathbb{N}, \; h \le 2^n\}, \; k_0(h2^n) = n.$

Solution 4.18   The Cantor set is of the kind $\dfrac{1}{3}$-middle, then

$$S_1(x) = x/3 \qquad S_2(x) = 1 + (x-1)/3 \qquad \rho_i = 1/3, \quad i = 1, 2.$$

The dimension s of $\mathcal{C}$ verifies the equation $2\left(\dfrac{1}{3}\right)^s = 1$, from which

$$\dim_{\mathcal{H}}(\mathcal{C}) = \dim_{\mathcal{B}}(\mathcal{C}) = (\ln 2)/(\ln 3) .$$

For the set $\mathcal{C}_t$ of kind Cantor t middle we have

$$S_1(x) = \frac{1-t}{2} x \qquad S_2(x) = 1 + \frac{1-t}{2}(x-1) \qquad \rho_i = \frac{1-t}{2}, \quad i = 1, 2.$$

We obtain the equation $2\left(\dfrac{1-t}{2}\right)^s = 1$, from which

$$\dim_{\mathcal{H}}(\mathcal{C}_t) = \dim_{\mathcal{B}}(\mathcal{C}_t) = \frac{\ln 2}{\ln 2 - \ln(1-t)}.$$

Solution 4.19   From the previous exercise $\dim_{\mathcal{H}} A = 1/2$. By using as covering the squares involved in the iterative construction of $A \times A$ to the k-th iteration one obtains the inequality $\mathcal{H}_\delta^1(A \times A) \le 4^k 4^{-k}\sqrt{2} = \sqrt{2}$. On the other hand, the projections of these same squares on the perpendicular lines of equation $y = 2x$ and $y = -x/2$ are segments of length $3/\sqrt{2}$. It follows

$$\frac{3}{\sqrt{5}} \le \mathcal{H}_\delta^1(A \times A) \le \sqrt{2} \qquad \forall \delta > 0$$

$$\frac{3}{\sqrt{5}} \le \mathcal{H}^1(A \times A) \le \sqrt{2} .$$

Therefore the Hausdorff dimension of $A \times A$ is 1. Despite having integer dimension, it is anything but an elementary set and in no way corresponds to the intuitive idea of one-dimensional line.

Solution 4.20 $\dim_{\mathcal{H}}(Q) = \dim_{\mathcal{B}}(Q) = 2\ln 2/\ln 3.$

Solution 4.21 $\dim_{\mathcal{H}}(W) = \dim_{\mathcal{B}}(W) = 1$. The example shows that there exist subsets of the plane with Hausdorff dimension 1 that do not possess any of the intuitive properties of an elementary curve.

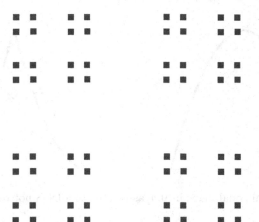

Fig. 8.13 Product of Cantor sets: $\mathcal{C}_{1/2} \times \mathcal{C}_{1/2}$ (third iteration)

Solution 4.22   From $\varphi(x) = x - b/(1-a)$ follow $\varphi^{-1}(x) = x + b/(1-a)$ and $\varphi \circ f \circ \varphi^{-1} = g$.

Solution 4.23   Among the four DDS $\{\mathbb{R}, x/2\}$, $\{\mathbb{R}, 2x\}$, $\{\mathbb{R}, x\}$ and $\{\mathbb{R}, x+1\}$ there is no pair of DDS topologically conjugate. The statement can be deduced from Theorem 4.42.

Solution 4.24   If $f : I \to I$, $g : J \to J$ e $\varphi : I \to J$ is the topological conjugation, with $\tilde{t} = \varphi(\tilde{x})$, one has:

$$g\left(\tilde{t}\right) = g\left(\varphi\left(\tilde{x}\right)\right) = \boxed{\varphi \text{ is a topological conjugation}}$$

$$= \varphi\left(f\left(\tilde{x}\right)\right) = \boxed{\tilde{x} \text{ is a fixed point}}$$

$$= \varphi\left(\tilde{x}\right) = \tilde{t}.$$

Solution 4.25   If $t \in \varphi(H)$ there exists $x \in H$ such that $t = \varphi(x)$. For these values is, by hypothesis,

$$\varphi^{-1}(t) = x < f(x) = f\left(\varphi^{-1}(t)\right).$$

By applying φ to both sides of $\varphi^{-1}(t) < f\left(\varphi^{-1}(t)\right)$ and taking into account the monotonicity of φ

$$t = \varphi\left(\varphi^{-1}(t)\right) < \varphi\left(f\left(\varphi^{-1}(t)\right)\right) = g(t).$$

Notice that for φ decreasing, by the same technique, one can prove $t > g(t)$.

Solution 4.26   There are several ways to prove it.
I way: it is sufficient to note (see Fig. 8.14) that h_4 and g have a different number of fixed points (respectively 2 and 3).
II way: for the previous exercise if there was an increasing topological conjugation φ then $h_4(x) - x$ and $g(x) - x$ should have the same sign in a right neighborhood of

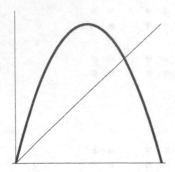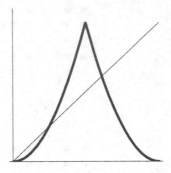

Fig. 8.14 Two unimodal maps in $[0,1]$ associated to a DDS not topologically conjugate

0, a false conclusion; if it existed φ decreasing, then $h_4(x) - x$ and $g(x) - x$ should have the opposite sign near 0, that is false.

The two dynamics are actually different: in particular 0 is repelling for h_4, while it is locally attractive for g.

Solution 4.27 Let f and φ be increasing. Then, for each t_1, $t_2 \in \varphi(H)$ with $t_1 < t_2$ there exist x_1, $x_2 \in H$ such that $x_1 < x_2$ and $\varphi_1(x_1) = t_1$, $\varphi_2(x_2) = t_2$,

$$g(t_1) = \varphi\left(f\left(\varphi^{-1}(t_1)\right)\right) = \varphi(f(x_1)) < \varphi(f(x_2)) = \varphi\left(f\left(\varphi^{-1}(t_2)\right)\right) = g(t_2).$$

The other three cases are treated in the same way (remember that φ is always monotone).

Solution 4.28 If I is open there exists a continuous $\varphi : I \to \mathbb{R}$ with continuous inverse. Theorem 4.4 applies to $\{\mathbb{R}, g\}$ where $g = \varphi \circ f \circ \varphi^{-1}$ and by Theorem 4.42 the dynamic of $\{\mathbb{R}, g\}$ has the same properties of the dynamic of $\{I, f\}$.

Solution 4.29 The Cauchy problem has a unique solution

$$u(t) = \left(1 + \frac{1 - u_0}{u_0} e^{b(t_0 - t)}\right)^{-1},$$

that is a strictly increasing function in $[0, +\infty)$, of which there is the limit $\lim_{t \to +\infty} u(t) = 1$.

Euler method
Given the discretization parameter $h > 0$ we obtain the recursive scheme

$$U_0 = u_0 \in (0,1), \qquad U_{k+1} = (1 + bh)\, U_k - b\, h\, (U_k)^2, \quad k \geq 1.$$

The function $f(x) = (1 + bh)x - bhx^2$ vanishes at 0 and $x = 1 + (bh)^{-1}$ and has maximum equal to $(1 + bh)^2 / (4bh)$, therefore f transform the interval $\left[0, 1 + (bh)^{-1}\right]$ in a proper subset for each $h \in (0, 3/b)$. We analyze in detail the solution produced by the numerical scheme in the case $0 < h < 3/b$.

The DDS $\{[0, 1 + (bh)^{-1}], f\}$ and $\{[0, 1], h_a\}$, where $a = 1 + bh$, are topologically conjugate, by the function φ, with $\varphi(x) = \dfrac{bh}{1+bh}x$, $\varphi^{-1}(x) = \dfrac{1+bh}{bh}x$, that is

$$\varphi \circ f \circ \varphi^{-1}(x) = h_{1+bh}(x) .$$

By varying h in $(0, 3/b)$, the parameter of the logistic $a = 1 + bh$ varies in $(1, 4)$. So there is always the equilibrium α_a of the logistic, that is globally asymptotically stable if $0 < h \leq 2/b$ (super-attractive if $h = 1/b$) to which the trajectories approach with exponential speed and with the correct monotonicity if and only if $0 < h \leq 1/b$. We observe explicitly that the equilibrium of $\{(0, 1 + (bh)^{-1}), f\}$, corresponding in the topological conjugation to α_a, is $x = 1$ (that is it coincides with the limit, as t tends to $+\infty$, of the continuous solution u). Furthermore, if $0 < h \leq b^{-1}$ the value 1 is the monotone limit too, as k tends to $+\infty$, of the sequence U_k corresponding to the discrete solution.

We summarize the main qualitative information that can be drawn from the topological conjugation with the logistics discrete dynamics which is described in Chap. 4. The discretization with the Euler scheme gives a a good description of the solution of the problem under consideration, if and only if $0 < h \leq b^{-1}$.

If $b^{-1} < h \leq 2b^{-1}$, then the numerical solution is given by a sequence that converges to 1 in an oscillating manner.

If $h > 2b^{-1}$, then 1 ceases to be attractive for the s.d.d. associated with f and, as h, grows periodic oscillations appear in all periods integers (numerical oscillations that are no reflected in the exact solution), until h is approaching to $3b^{-1}$, the numeric solution has a chaotic behavior (the trajectories have an erratic behavior in the interval $(0, 1)$) so it is far from describing the monotonic behavior of the exact solution.

If $h > b^{-1}$ to the erratic trajectories others diverging to $-\infty$ are added.

Backward Euler method Given the discretization parameter $h > 0$ one obtains the recursive scheme

$$V_0 = u_0 \in (0, 1) , \qquad bh\,(V_{k+1})^2 + (1 - bh)\,V_{k+1} - V_k = 0, \quad k \geq 1.$$

At every step a quadratic equation in V_{k+1} with known V_k must be solved. To make the choice between the two algebraic solutions, we remember that the (increasing) solution of the differential equation must take positive values. Furthermore, the discriminant is greater than zero for each $h > 0$, and there are always a variation and a permanence (that ensure a positive and a negative root). We choose the unique positive root:

$$V_{k+1} = g(V_k)$$

$$g(x) := \frac{1}{2bh}\left(-1 + bh + \sqrt{(1 - bh)^2 + 4bh\,x}\right) .$$

The function g transforms the interval $[0, 1]$ into itself. The DDS $\{[0, 1], g\}$ has a unique equilibrium $(x = 1)$ (in addition to 0 that is repelling). It is $g \in C^1$, $g'(1) = (1 + bh)^{-1}$, $0 < g'(1) < 1$ for every $h > 0$. Therefore (Theorem 3.51) $x = 1$ is stable and attractive (its basin is $(0, 1]$) and, for each $h > 0$, all the trajectories generated by the backward Euler numerical scheme converge to 1 (in a monotone way) as t tends to $+\infty$. This behavior is in good agreement with the exact solution of the differential equation.

So the backward Euler scheme, even though it needs the solution of a nonlinear equation, is more satisfactory because it provides a good approximation of the exact solution (V_k is an approximation of $u(kh)$), without special restrictions on the discretization step h.

The fact that one is able to determine explicitly for h a range of values in which the discretization is satisfactory even if carried out with the Euler method, depends on the particular choice of the differential equation of which are well-known the exact solution and the dynamics associated with the corresponding discretization. All of this is not known for general nonlinear differential equations (otherwise it would be useless to make numerical approximations ...).

In the general case of an autonomous ordinary differential equation $u' = F(u)$ with initial datum u_0 and F' bounded ($|F'| \leq L$), the backward Euler method leads to the following situation:

$$V_{k+1} - h\,F(V_{k+1}) - V_k = 0$$

$$\frac{d}{du}(Id - h\,F) \neq 0 \qquad \text{if} \quad h < L^{-1}.$$

Therefore $Id - hF$ is locally invertible. Let $g = (Id - hF)^{-1}$ be the local inverse. Then $\alpha = g(\alpha)$ if and only if $F(\alpha) = 0$; that is α is an equilibrium ($\alpha = g(\alpha)$) of the DDS if and only if $u(t) \equiv \alpha$ is the solution of the Cauchy problem with $u_0 = \alpha$. We set

$$V_{k+1} = g(V_k) \qquad k \geq 1.$$

From $g' = (1 - hF')^{-1} \circ g$ it follows

$$g'(\alpha) = \frac{1}{1 - hF'(g(\alpha))} = \frac{1}{1 - hF'(\alpha)}.$$

Concluding, if $F'(\alpha) < 0$, then $0 < g'(\alpha) < 1$, α is stable for the DDS associated to g and $V_k \nearrow \alpha$. This means that the backward Euler method is stable and accurate for each $h \in (0, L^{-1})$.

Solution 4.30 $e^x - 3$ is a C^∞ and convex function. $e^1 = e < 3 < e^{5/4}$, then applying Theorems 4.48 and 4.49 there is a unique zero $\alpha = \ln 3 = 1.09861...$ in $(1, 5/4)$ which is simple and, therefore, super-attractive for the Newton method. Chosen $X_0 = 5/4$ we have $\max_{[1,5/4]} g'' = e^{5/4}$, $\min_{[1,5/4]} g' = e$, so the following error estimation holds

$$0 < X_{k+1} - \alpha < \sqrt[4]{e}\,(X_k - \alpha)^2/2$$

and from $X_1 - X_0 < 1/4$ it follows $0 < X_{k+1} - \alpha < 4^{-2k} < 10^{-3}$ if $k = 3$. Iterating $N_g(x) = x - 1 + 3e^{-x}$ one obtains $1.098 < \alpha < 1.099$.

Solution 4.31   If $|z| > |c| + 1$, $q(z) = z^2 + c$, then:

$$|q(z)| = |z^2 + c| \geq \boxed{\text{triangle inequality}}$$
$$\geq |z^2| - |c| \geq$$
$$\geq (|c| + 1)^2 - |c| = |c|^2 + |c| + 1.$$

Then

$$|q^2(z)| = |q(q(z))| = ||q(z)|^2 + c| \geq$$
$$\geq |q(z)|^2 - |c| \geq (|c|^2 + |c| + 1)^2 - |c| =$$
$$= |c|^4 + 2|c|^3 + 3|c|^2 + |c| + 1 \geq$$
$$\geq 3|c|^2 + |c| + 1.$$

By induction, one obtains

$$\left| q^k(z) \right| \geq \left(2^k - 1 \right) |c|^2 + |c| + 1.$$

Since the right-hand side of the inequality tends to $+\infty$ if k diverges, the statement follows.

Solution 4.32 Let α be an equilibrium of $\{I, f\}$, $\varphi \in C^1(I)$ a scalar function of a variable with values in J with inverse $C^1(J)$ and $g = \varphi \circ f \circ \varphi^{-1}$. Then $(\varphi^{-1})'(\varphi(\alpha)) = (\varphi'(\alpha))^{-1}$, $\varphi(\alpha)$ is an equilibrium of $\{J, g\}$ and

$$g'(\varphi(\alpha)) = (\varphi \circ f \circ \varphi^{-1})'(\varphi(\alpha)) =$$
$$= \varphi'(f(\varphi^{-1}(\varphi(\alpha)))) f'(\varphi^{-1}(\varphi(\alpha))) (\varphi^{-1})'(\varphi(\alpha)) =$$
$$= \varphi'(f(\alpha)) f'(\alpha) (\varphi^{-1})'(\varphi(\alpha)) =$$
$$= f'(\alpha).$$

Solution 4.33 Let $\{\alpha_1, \ldots, \alpha_s\}$ be a periodic orbit of $\{I, f\}$, $\varphi \in C^1(I)$ a scalar function of one variable with values in J with inverse $C^1(J)$ and $g = \varphi \circ f \circ \varphi^{-1}$. Then $\{\varphi(\alpha_1), \ldots, \varphi(\alpha_s)\}$ is a periodic orbit of $\{J, g\}$, even $\{I, f^s\}$ and $\{J, g^s\}$ are topologically conjugate through φ, therefore the multiplier $(f^s(\alpha_1))'$ of the equilibrium α_1 of $\{I, f^s\}$ is preserved (Exercise 4.32), but $(f^s(\alpha_1))' = \prod_{j=1}^s f'(\alpha_j)$ is the multiplier of the orbit too.

8.5 Solutions of the exercises in Chapter 5

Solution 5.1   1) The eigenvectors of $\mathbb{M}$ are $\lambda_1 = 1$ and $\lambda_2 = 2$ to which there correspond, respectively, the (linearly independents) eigenvectors

$$\mathbf{V}^1 = \begin{bmatrix} 1 & 0 \end{bmatrix}^T \qquad\qquad \mathbf{V}^2 = \begin{bmatrix} 3 & 1 \end{bmatrix}^T.$$

The general solution of the system is then

$$\mathbf{X}_k = c_1 \begin{bmatrix} 1 \\ 0 \end{bmatrix} + c_2 2^k \begin{bmatrix} 3 \\ 1 \end{bmatrix} = \begin{bmatrix} c_1 + 3c_2 2^k \\ c_2 2^k \end{bmatrix} \qquad k \in \mathbb{N}.$$

By imposing the initial condition, one obtains the values of the constants: $c_1 = -5$, $c_2 = 2$.

Notice that given the triangular shape of the matrix $\mathbb{M}$, one could proceed more quickly by solving the second equation of the system and subsequently the first, after replacing the solution found in the first step.

2) $\mathbf{X}_k = \begin{bmatrix} 4 \\ 4 \\ 1 \end{bmatrix} + 5\,(-1)^k \begin{bmatrix} 0 \\ -1 \\ 1 \end{bmatrix} - 6\,(-3)^k \begin{bmatrix} 0 \\ 0 \\ 1 \end{bmatrix} \qquad k \in \mathbb{N}\,.$

3) $\mathbf{X}_k = 3\,(-1)^k \begin{bmatrix} 0 \\ 1 \\ 0 \end{bmatrix} - (-\sqrt{2})^k \begin{bmatrix} \sqrt{2} - 1 \\ 0 \\ 1 \end{bmatrix} \qquad k \in \mathbb{N}\,.$

Solution 5.3   We recall the identity $\det \mathbb{L} = \det \mathbb{L}^T$. Formula (D.2) of Appendix D, called **Laplace expansion of the determinant** with respect to the i-th row, applied to the matrix $\mathbb{L}^T$ assures us that the determinant can be calculated similarly by developing with respect to any column. By applying this expansion since the last column, we obtain

$$\det \mathbb{L} = (-1)^{1+n}\, \varphi_n \det \operatorname{diag}\{\sigma_1, \sigma_2, \ldots, \sigma_{n-1}\}\,.$$

Solution 5.4   For the Perron-Frobenius Theorem and Remark 5.20 the matrix $\mathbb{A}$ has a strictly positive dominant eigenvalue $\lambda_{\mathbb{A}} = 1/2$ (since the matrix $\mathbb{A}$ has rank 1 all the other eigenvalues are null). Since $\mathbb{A} \gg \mathbb{O}$, $\mathbf{B} \gg \mathbf{0}$ Theorem 5.28 ensures that the vector DDS has a unique stable and attractive equilibrium given by $\mathbf{A} = (\mathbb{I} - \mathbb{A})^{-1}\mathbf{B}$. The general solution is given by $\mathbf{X}_k = (1/n2^k)\,\mathbb{U}\,(\mathbf{X}_0 - \mathbf{A}) + \mathbf{A}$ where $\mathbb{U}$ is the matrix whose elements are all equal to 1.

Solution 5.5   a) By applying Remark 5.20 to the weakly positive matrix $\mathbb{M}$ we deduce the inequalities $7 \le \lambda_{\mathbb{M}} \le 9$.
b)   With reference to Examples 5.34 and 5.36, it holds $\mathbb{M} = t\mathbb{B}$ with t is a real number to be determined. Solving the system:

$$\begin{cases} t - 2\alpha t = 5 \\ t\alpha = 2 \end{cases}$$

we obtain $\alpha = 2/9$ e $t = 9$. From $0 < 2/9 < 1/4$ it follows that the matrix $\mathbb{M}$ is positive definite and

$$\lambda_k\,(\mathbb{M}) = 9\lambda_k\,(\mathbb{B}) = 9 - 8\left(\sin \frac{k\pi}{2\,(N+1)}\right)^2 \qquad 1 \le k \le N\,.$$

In particular $1 \le \lambda_k\,(\mathbb{M}) \le 9$ for each k, whatever the size of $\mathbb{M}$.

Solution 5.7   Since $\mathbb{M} = \mathbb{S} + \mathbb{T}$ and $\mathbb{S}\mathbb{T} = \mathbb{T}\mathbb{S}$, using the Newton's binomial formula (that is true if the summands commute), one can write

$$\mathbb{M}^k = (\mathbb{S} + \mathbb{T})^k = \sum_{j=0}^{k} \binom{k}{j} \mathbb{S}^j \mathbb{T}^{k-j};$$

but $\mathbb{T}^2 = \mathbb{O}$ and then $\mathbb{T}^k = \mathbb{O}$ for each $k \ge 2$, from which

$$\mathbb{M}^k = \mathbb{S}^k \mathbb{T}^0 + k\mathbb{S}^{k-1}\mathbb{T} = \mathbb{S}^k + k\mathbb{S}^{k-1}\mathbb{T}.$$

Similarly we can prove relation (5.7) used in the proof of Theorem 5.8.

Solution 5.8 λ solves $\det(\mathbb{A} - \lambda\mathbb{I}) = 0$, that is a root of polynomial with real coefficient. Recalling that $\mathbb{A}$ is real

$$0 = \det(\mathbb{A} - \lambda\mathbb{I}) = \det(\overline{\mathbb{A} - \lambda\mathbb{I}}) = \det(\mathbb{A} - \overline{\lambda}\mathbb{I})$$

that is even $\overline{\lambda}$ is an eigenvalue of $\mathbb{A}$. Since $\mathbb{A}$ is real, $\mathbb{A}\mathbf{V} = \lambda\mathbf{V}$ with $\mathbf{V} \neq 0$ implies

$$\mathbf{V} \notin \mathbb{R}^n \quad \text{e} \quad \mathbb{A}\overline{\mathbf{V}} = \overline{\mathbb{A}\mathbf{V}} = \overline{\lambda\mathbf{V}} = \overline{\lambda}\,\overline{\mathbf{V}}$$

that is $\overline{\mathbf{V}} \neq 0$ is an eigenvector associated to the eigenvalue $\overline{\lambda}$ and $\mathbf{V}, \overline{\mathbf{V}} \in \mathbb{C}^n \backslash \mathbb{R}^n$. Putting

$$\mathbf{V}^1 = \frac{1}{2i}(\mathbf{V} - \overline{\mathbf{V}}) = \operatorname{Im}(\mathbf{V}) \qquad \mathbf{V}^2 = \frac{1}{2}(\mathbf{V} + \overline{\mathbf{V}}) = \operatorname{Re}(\mathbf{V})$$

then $\mathbf{V}^1, \mathbf{V}^2 \in \mathbb{R}^n$ and

$$\mathbb{A}\mathbf{V}^1 = \frac{1}{2i}(\lambda\mathbf{V} - \overline{\lambda}\overline{\mathbf{V}}) = \operatorname{Im}(\lambda\mathbf{V}) = b\operatorname{Re}(\mathbf{V}) + a\operatorname{Im}(\mathbf{V}) = a\mathbf{V}^1 + b\mathbf{V}^2$$

$$\mathbb{A}\mathbf{V}^2 = \frac{1}{2}(\lambda\mathbf{V} + \overline{\lambda}\overline{\mathbf{V}}) = \operatorname{Re}(\lambda\mathbf{V}) = a\operatorname{Re}(\mathbf{V}) - b\operatorname{Im}(\mathbf{V}) = -b\mathbf{V}^1 + a\mathbf{V}^2$$

namely, rewriting $\mathbb{A}$ w.r.t. the basis $\{\mathbf{V}^1, \mathbf{V}^2\}$, putting $\mathbb{U}^{-1} = [\mathbf{V}^1\ \mathbf{V}^2]$, one obtains

$$\mathbb{B} = \mathbb{U}\mathbb{A}\mathbb{U}^{-1} = \begin{bmatrix} a & -b \\ b & a \end{bmatrix}.$$

In fact the columns of the matrix associated to a linear application are the images (with respect to the same basis) of the vectors of the basis.
If $\lambda = 0$, then $a = b = 0$ and $\mathbb{B} = \mathbb{O}$.
If $0 \neq \lambda = a + ib = \rho e^{i\theta}$ with $\rho > 0$, then:

$$\mathbb{B} = \begin{bmatrix} a & -b \\ b & a \end{bmatrix} = \begin{bmatrix} \rho & 0 \\ 0 & \rho \end{bmatrix} \begin{bmatrix} \cos\theta & -\sin\theta \\ \sin\theta & \cos\theta \end{bmatrix} = \rho \begin{bmatrix} \cos\theta & -\sin\theta \\ \sin\theta & \cos\theta \end{bmatrix}.$$

Solution 5.9 $\mathbb{B}$ is a matrix that verifies the assumptions of Theorem 5.13 (if σ_1, ..., σ_{n-1} and φ_1, ..., φ_{n-1} are strictly positive). By Remark 5.22, $\det\mathbb{L} = 0$ and $\det\mathbb{B} \neq 0$. Therefore, the dominant eigenvalue $\lambda_{\mathbb{B}} > 0$ and the dominant eigenvector $\mathbf{V}^{\mathbb{B}}$ of $\mathbb{B}$ are such that even for $\mathbb{L}$ (which adds to the eigenvalues of $\mathbb{B}$ the only zero eigenvalue because $\det\mathbb{L} = 0$):

$$\lambda_{\mathbb{L}} = \lambda_{\mathbb{B}} \qquad \mathbf{V}^{\mathbb{L}} = \begin{bmatrix} \mathbf{V}^{\mathbb{B}} \\ t \end{bmatrix} \qquad t \in \mathbb{R}.$$

It remains to prove that $t > 0$ (to obtain $\mathbf{V}^{\mathbb{L}} \gg 0$): explicitly from $\mathbb{L}\mathbf{V}^{\mathbb{L}} = \lambda\mathbf{V}^{\mathbb{L}}$ it follows $t = \lambda_{\mathbb{L}}^{-1}\sigma_{n-1}\mathbf{V}_{n-1}^{\mathbb{B}} > 0$. By studying the dynamic of the system $\mathbf{X}_{k+1} = \mathbb{B}\mathbf{X}_k$ Theorem 5.18 ensures that the age classes distribution that is observed in the long run is the one provided by the vector $\mathbf{V}^{\mathbb{L}}$.

Solution 5.10 The element in the first row and $n - 2$–th column of $\mathbb{L}^2$ is equal to $\sigma_{n-2} \times \varphi_{n-1} = \varphi_{n-1} > 0$. By induction one verifies that, for any increase of the power, the strictly positive elements of the first row increase of at last one towards the left, and those who already were, retain such property, until obtaining $\mathbb{L}^n$ with the entire first row strictly positive, except at most the least element. We are therefore reduced to the case of the previous exercise: $\mathbb{L}^{n-2}$ verifies the assumptions introduced in Exercise 5.9. Then Theorems 5.13, 5.14 and 5.18 ensures that there exists a dominant eigenvalue $\lambda_L > 0$ with a unique associated dominant eigenvector $\mathbf{V}^L > \mathbf{0}$ and the distribution in age classes that is observed in the long run is the one provided by the vector $\mathbf{V}^L$.

Solution 5.11   By using Remark 5.20 by rows of $\mathbb{A}$ and by columns of $\mathbb{B}$ one obtains:

$$\lambda_\mathbb{A} = 10 \qquad\qquad 5 \le \lambda_\mathbb{B} \le 6 .$$

Solution 5.12 $\mathbf{X}_k = \left[2^k \ 0 \ (-1)^{k+1} \right]^T , \ k \in \mathbb{N}.$

8.6 Solutions of the exercises in Chapter 6

Solutions 6.1 and 6.2 $\mathbb{M}_1$ is absorbing and not regular, $\mathbb{M}_2$ is regular and not absorbing, $\mathbb{M}_3$ is neither regular nor absorbing.

Solution 6.3   We have to determine a vector $\mathbf{U} = \left[U_1 \ 1 - U_1 \right]^T$, where U_1 belongs to $[0, 1]$, such that $\mathbf{U} = \mathbb{M}\mathbf{U}$, namely U_1 is a solution of the system

$$\begin{cases} aU_1 + b\,(1 - U_1) = U_1 \\ (1 - a)\,U_1 + (1 - b)\,(1 - U_1) = 1 - U_1 \end{cases} \quad\Rightarrow\quad U_1 = \frac{b}{1 + b - a}.$$

Since $b > 0$ and $1 - a > 0$, it follows $U_1 \in (0, 1)$ and then $\mathbf{U}$ is a strictly positive stochastic vector.

Solution 6.4   The three eigenvectors of $\mathbb{M}$ are distinct:

$$\lambda_1 = 1; \qquad \lambda_2 = \frac{1}{2}; \qquad \lambda_3 = \frac{1}{4}.$$

Thanks to the representation (6.10) of $\mathbf{P}_k = \mathbb{M}^k \mathbf{P}_0$ and by observing that the h-th column of $\mathbb{M}^k$ is equal to $\mathbb{M}^k \mathbf{e}_0$, we know that the term $m_{23}^{(k)}$ can be expressed in the form

$$m_{23}^{(k)} = a + b \left(\frac{1}{2} \right)^k + c \left(\frac{1}{4} \right)^k$$

where a, b, c are constants to be determined. Since $m_{23}^{(0)} = 0$, $m_{23}^{(1)} = 1/2$ and, as one can easily check by computing $\mathbb{M}^2$, $m_{23}^{(2)} = 3/8$, one obtains the linear system

$$\begin{cases} a + b + c = 0 \\ a + \dfrac{1}{2}b + \dfrac{1}{4}c = \dfrac{1}{2} \\ a + \dfrac{1}{4}b + \dfrac{1}{16}c = \dfrac{3}{8} \end{cases}$$

whose solutions are $a = 0$, $b = 2$, $c = -2$.

Solution 6.5   We define the following events:

* $E_1 =$ "the switches A and D are closed";
* $E_2 =$ "the switches A, C, E are closed";
* $E_3 =$ "the switches B, C, D are closed";
* $E_4 =$ "the switches B and E are closed".

The reliability is then given by: $P(E_1 \cup E_2 \cup E_3 \cup E_4)$. From the independence hypothesis is obtained:

$$P(E_1) = ad \quad P(E_2) = ace \quad P(E_3) = bcd \quad P(E_4) = be$$

$$P(E_1 \cap E_2) = acde \quad P(E_1 \cap E_3) = abcd \quad P(E_1 \cap E_4) = abde$$

$$P(E_2 \cap E_3) = abcde \quad P(E_2 \cap E_4) = abce \quad P(E_3 \cap E_4) = bcde$$

$$P(E_1 \cap E_2 \cap E_3) = P(E_1 \cap E_2 \cap E_4) = P(E_1 \cap E_3 \cap E_4) = P(E_2 \cap E_3 \cap E_4) =$$
$$= P(E_1 \cap E_2 \cap E_3 \cap E_4) = abcde.$$

We can thus conclude, thanks to the inclusion-exclusion principle (see point 7 of Appendix C), that the reliability of the Wheatstone bridge is:

$$P(E_1 \cup E_2 \cup E_3 \cup E_4) =$$
$$= ad + ace + bcd + be - acde - abcd - abde - abce - bcde + 2abcde \,.$$

Solution 6.6   A permutation matrix has only one element equal to 1 in each row and each column and all others zero. If the element t_{ij} of T is 1, then for each vector **V**, the i-th components of **TV** is equal to the j-th component of **V**: the effect of the multiplication by **V** is a permutation of the components of **V**. In particular, T is also a matrix of orthogonal change of coordinates: $\mathbb{T}^{-1} = \mathbb{T}^T$.

Solution 6.7   Set $\mathbb{H} = \mathbb{TMT}^{-1} = \mathbb{TMT}^T$. Then, from $m_{ij} = [\mathbb{M}]_{ij} \geq 0$, $t_{ij} = [\mathbb{T}]_{ij} \geq 0$ for every i and j, it follows $h_{ij} = [\mathbb{H}]_{ij} \geq 0$ for each i and j. Moreover, from $\sum_{i=1}^{n} t_{ij} = \sum_{i=1}^{n} m_{ij} = 1$ for each j, it follows

$$\sum_{i=1}^{n} h_{ij} = \sum_{i,s,k=1}^{n} t_{is} m_{sk} t_{jk} = 1 \qquad \forall j \,.$$

In other words, the change of coordinates induced by the permutation matrix $\mathbb{T}$ corresponds to setting $h_{ij} = m_{t(i),t(j)}$ where $t(s)$ is the column position in the s-th row of $\mathbb{T}$ where there is 1.

Solution 6.8   Defined the events

* $H_i =$ "a piece chosen at random has been produced by the machinery i";
* $A =$ "a piece chosen at random is faulty",

to determine the solution we obtain from the tree–diagram of Fig. 8.15 the probabilities

$$P(H_i) = \frac{r_i}{r} \qquad\qquad P(A|H_i) = \frac{d_i}{r_i} \qquad i = 1, 2, 3 \,.$$

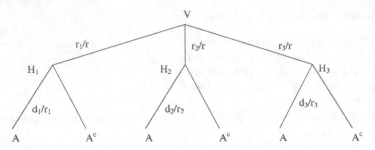

Fig. 8.15 Tree–diagram concerning Exercise 6.8

By the Bayes law, we thus get the probability that the defected piece comes from machinery i:

$$P\left(H_i|A\right) = \frac{d_i}{d_1 + d_2 + d_3} \qquad i = 1, 2, 3$$

which is the ratio between the number of faulty parts produced by it and the total number of faulty parts.

Solution 6.9   This time the coin is not fair...

Solution 6.10   If $a > b = 0.5$ and $c > 0$, then C has a greater possibility of A also if $a = 1$ (namely A is a infallible shooter). In fact, if $a = 1$ then B is doomed and the duel is to be concluded in a few rounds. Precisely:

$$P(A \text{ wins}) = 0.5\left(1 - c\right)^2 , \quad P(C \text{ wins}) = 0.5(1 + c)$$

$$P(A \text{ die}) = 0.5 + 1.5c - c^2 , \quad P(B \text{ die}) = 1, \quad P(A \text{ all die}) = c\left(1 - c\right).$$

Instead, if $a < 1$, then set c, b must be reduced until the win probabilities A exceed those of C. This is possible only for small values of c: with reference to Example 6.30, it should be imposed

$$\begin{cases} P\left(E_2\right) > P\left(E_3\right) \\ P\left(E_2\right) > P\left(E_4\right) \end{cases} \text{namely} \begin{cases} P_\infty^2 > P_\infty^3 \\ P_\infty^2 > P_\infty^1 . \end{cases}$$

Solution 6.13   The permutation matrix $\mathbb{T} = \begin{bmatrix} 0\,0\,1 \\ 1\,0\,0 \\ 0\,1\,0 \end{bmatrix}$ has the eigenvalues $\lambda_1 = 1$, $\lambda_{2,3} = \left(-1 \pm i\sqrt{3}\right)/2$. All the eigenvalues have modulus equal to 1. The eigenvalue λ_1 is not dominant, λ_2 and λ_3 are not real.

Notice that the Markov chain associated to $\mathbb{T}$ presents the 3 cycle

$$\left\{\left[1\,0\,0\right]^T, \left[0\,1\,0\right]^T, \left[0\,0\,1\right]^T\right\} .$$

Moreover, the real Jordan form of $\mathbb{T}$ in suitable coordinates is given by

$\mathbb{J} = \begin{bmatrix} 1 & 0 & 0 \\ 0 & -1/2 & -\sqrt{3}/2 \\ 0 & \sqrt{3}/2 & -1/2 \end{bmatrix}$ and shows that a generic change of coordinates does not retain the property of being a stochastic matrix.

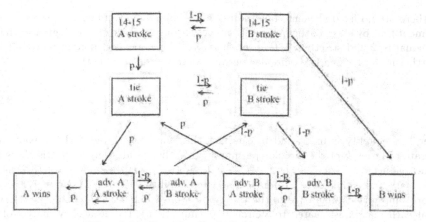

Fig. 8.16 Set of volleyball from score $14 - 15$

Solution 6.14   Fig. 8.16 shows the graph relative to the set of volleyball, where the situations of 15 deuce and vantaggio pari are considered equivalent. The other answers can be easily deduced from the graph.

Solution 6.15   If C shoots A, his survival probabilities are $59/189$. If C shoots B, his survival probabilities are $50/189$ (it should be noted that when the round of B comes, he must shoot A if A is alive, when the round of A comes, these must shoot B if B is alive). Therefore C should shoot at the air: this magnanimous gesture gives him a survivor probability of $25/63$.

8.7 Solutions of the exercises in Chapter 7

Solution 7.2 Given n elements:

- the number of their permutations is $n!$, as one verifies by induction on the number of the elements;
- their cyclic permutations are n, that corresponds to the number of possible distinct choices of t_{1j}, related to the cyclic permutations different from the identity $n - 1$;
- their circular permutations are $(n - 1)!$. In fact, distinguishing the "chairs for the guests", there are exactly $n!$ ways to accommodate them on the chairs (permutations); but there are accommodations on chairs equivalent up to a rotation; possible rotations are n (cyclic permutations); therefore the circular permutations are $(n - 1)! = n!/n$.

Solution 7.3   (a) The identity $\mathbb{I}_n$ is obviously reducible for each $n > 1$.
(b) Among the $n!$ permutation matrices of order $n > 1$, all and only the $(n-1)!$ permutation matrices of order n associated to a circular permutation are irreducible. In fact, if there is an element left fixed by the permutation, then the associated matrix is reducible.

If there are no fixed elements but the permutation is not circular, then there is an element that by effect of the permutation goes through a cycle of length greater than or equal to 2 and strictly less than n, therefore, the associated matrix is reducible. Furthermore, if $\mathbb{T}$ is a fixed circular permutation of order $n > 1$, then

$$(\mathbb{I}_n + \mathbb{T})^{n-1} = \sum_{k=0}^{n-1} \binom{n-1}{k} \mathbb{T}^k \gg \mathbb{O} \tag{8.2}$$

The last equality is motivated by the fact that for every i and j there exists an iterate of order k of the circular permutation $\mathbb{T}$ that sends j in i, that is there is k depending on i and j such that $t_{ij}^{(k)} = 1$. Since $\mathbb{T} \geq 0$ the validity of (8.2) is a necessary and sufficient condition for the irreducibility of $\mathbb{T}$ (see Theorem 7.29).

More simply (with reference to the theory developed in Sect. 7.2), one can observe that each matrix associated to a circular permutation is the adjacency matrix of a strongly connected graph, hence it is irreducible.

Solution 7.6 If in Definition 7.4 the dimension of the zero sub-matrix of $\widetilde{\mathbb{A}}$ is $(n - r) \times r$, with $0 < r < n$, then:

$$(n - r) \times r - (n - 1) = (r - 1)(n - r - 1) \geq 0.$$

Solution 7.16 We compute the characteristic polynomial $\mathcal{P}(\lambda) = \det(T - \lambda \mathbb{I}_n)$ by the formula

$$\det \mathbb{B} = \sum_{\sigma} \varepsilon(\sigma) \prod_{i=1}^{n} b_{i,\sigma(i)}$$

where the sum is extended to all permutations σ of n elements and $\varepsilon(\sigma)$ is the signature of these permutations.

If $\mathbb{T} = \mathbb{I}_n$ then $\mathcal{P}(\lambda) = (1 - \lambda)^n$. In particular, the statement follows for $n = 1$.

Let it now be $n > 1$. If $\mathbb{T}$ is irreducible, then it is also the adjacency matrix of a strongly connected graph; therefore it is the permutation matrix associated with a circular permutation, therefore it verifies $\det \mathbb{T} = (-1)^{n-1}$, from which $\mathcal{P}(\lambda) = (-1)^{n-1} + (-\lambda)^n = (\lambda^n - 1)(-1)^n$, i.e. the polynomial looked for, up to the sign. In the other cases ($\mathbb{T}$ reducible), up to a permutation $\mathbb{S}$ of the coordinates, the matrix has an irreducible square block structure on the diagonal, bordered by zeros:

$$A = STS^{-1} = \begin{bmatrix} \mathbb{A}_1 & & & & \\ & \mathbb{A}_2 & & & \\ & & \mathbb{A}_3 & & \\ & & & \ddots & \\ & & & & \mathbb{A}_m \end{bmatrix} \qquad \dim \mathbb{A}_j = k_j .$$

The explicit construction is obtained by considering all the cycles of period k_j contained in the permutation σ.

Solution 7.17 If the matrix is irreducible, then the statement is true. If the matrix is reducible, then it can be represented by a block of zeros in a NE position. The corresponding square block of SE must be stochastic. One applies the above reasoning to that block until one finds, by the finiteness of the matrix, a irreducible sub-matrix.

Solution 7.18 From the analysis of the corresponding graph G_M it follows that it is strongly connected, because there exists the cycle $\{1, 2, 3, 4, 1\}$, then the matrix M is irreducible.

Solution 7.20 Since the survival rates are all strictly positive, in the graph associated to the matrix there are all the arcs (v_i, v_{i+1}) for $i = 1, \ldots, n - 1$. Accordingly, if $\varphi_n > 0$, then there exists also the arc (v_n, v_1) and the graph is strongly connected.

Solution 7.21 If two consecutive classes are fertile, e.g. there exists $i \in \{1, \ldots, n-1\}$ such that $\varphi_i > 0$, $\varphi_{i+1} > 0$, then there are two closed paths in the graph associated to the matrix having length respectively i and $i + 1$ (whose GCD is 1). Therefore the matrix is aperiodic.
The condition is not necessary: in fact, the GCD of the length of the cycles is 1 only if $\varphi_1 > 0$. We observe that not even considering "consecutive" in a cyclic sense would make the condition necessary: for instance, $\varphi_i > 0$, $\varphi_k > 0$ with j and k coprime ensure the presence of cycles that make the matrix primitive.

Solution 7.23. No, it fails.
Precisely, if you analyze the problem with the adjacency matrix of the graph

$$A = \begin{bmatrix} 0 & 1 & 0 & 0 & 1 \\ 1 & 0 & 0 & 1 & 0 \\ 0 & 1 & 0 & 0 & 0 \\ 1 & 1 & 1 & 0 & 0 \\ 1 & 0 & 0 & 0 & 0 \end{bmatrix} \tag{8.3}$$

one obtains $\lambda_A = 1.905166$, $x_A = [0.396483 \ 0.5472576 \ 0.287249 \ 0.646133 \ 0.208109]$. If one analyzes the problem with the matrix made stochastic by normalizing the columns:

$$M = \begin{bmatrix} 0 & 1/3 & 0 & 0 & 1 \\ 1/3 & 0 & 0 & 1 & 0 \\ 0 & 1/3 & 0 & 0 & 0 \\ 1/3 & 1/3 & 1 & 0 & 0 \\ 1/3 & 0 & 0 & 0 & 0 \end{bmatrix} \tag{8.4}$$

one obtains $\lambda_M = 1$, $x_M = [0.176470 \ 0.352941 \ 0.117647 \ 0.294117 \ 0.058823]$.
Notice that the operation has slightly improved the authority index (much less with the use of M) and allowed to pass (only) page P_3 which had the same index: the example illustrates as this operation does not allow to bypass very significant differences in terms of links.

Appendix A

Sums and series

1) $\sum_{k=0}^{n} A_k = A_0 + A_1 + \cdots + A_n$

2) $\sum_{k=0}^{n} A_k = \sum_{k=r}^{n+r} A_{k-r}$

3) $\sum_{k=s}^{n} \sum_{j=t}^{m} A_{k,j} = \sum_{j=t}^{m} \sum_{k=s}^{n} A_{j,k}$

4) $\sum_{k=0}^{n} cA_k = c \sum_{k=0}^{n} A_k$

5) $\sum_{k=0}^{n} (A_k + B_k) = \sum_{k=0}^{n} A_k + \sum_{k=0}^{n} B_k$

6) Summation by parts

$$\sum_{k=0}^{n} A_k B_k = B_n \left(\sum_{k=0}^{n} A_k \right) - \sum_{s=0}^{n-1} (B_{s+1} - B_s) \left(\sum_{k=0}^{s} A_k \right)$$

7) $\sum_{k=0}^{n} x^k = \dfrac{1 - x^{n+1}}{1 - x} \qquad \forall x \neq 1$

8) $\sum_{k=1}^{n} k = \dfrac{1}{2} n (n + 1)$

9) $\sum_{k=1}^{n} k^2 = \dfrac{1}{6} n (n + 1) (2n + 1)$

10) $\sum_{k=1}^{n} k^3 = \dfrac{1}{4} n^2 (n + 1)^2$

11) $\sum_{k=1}^{n} k^4 = \dfrac{1}{30} n (n + 1) (2n + 1) (3n^2 + 3n - 1)$

12) $\sum_{k=1}^{n} k^5 = \dfrac{1}{12} n^2 (n + 1)^2 (2n^2 + 2n - 1)$

E. Salinelli, F. Tomarelli: *Discrete Dynamical Models.*
UNITEXT – La Matematica per il 3+2 76
DOI 10.1007/978-3-319-02291-8_A, © Springer International Publishing Switzerland 2014

13) $\displaystyle\sum_{k=1}^{n} \frac{1}{k} = \gamma + \frac{\Gamma'(n+1)}{n!} = \gamma_n + \ln n$

14) $\displaystyle\sum_{k=1}^{n} \frac{1}{k^2} = \frac{\pi^2}{6} - \left(\frac{\Gamma'}{\Gamma}\right)'(n+1)$

15) $\displaystyle\sum_{k=1}^{n} \frac{1}{k^3} = \left(\frac{\Gamma'}{\Gamma}\right)''(n+1) + \zeta(3)$

where: $\Gamma(x) = \int_0^{+\infty} e^{-t} t^{x-1} dt$, $\forall x > 0$ **Euler Gamma function**

$\gamma = -\Gamma'(1) = 0.57721566490153286060...$ **Euler-Mascheroni constant**

$\gamma_n \in [0, 1]$, $\gamma_n \nearrow \gamma$

$\zeta(x) = \displaystyle\sum_{k=1}^{+\infty} k^{-x}$, $\forall x > 1$ **Riemann Zeta function**

16)   **Cauchy-Schwarz inequality:**

$$\left(\sum_{j=1}^{n} A_j B_j\right)^2 \leq \left(\sum_{j=1}^{n} A_j^2\right)\left(\sum_{j=1}^{n} B_j^2\right) \qquad \forall A_j, B_j \in \mathbb{R}$$

17)   **Lagrange identity:** $\forall A_j, B_j \in \mathbb{R}$

$$\left(\sum_{j=1}^{n} A_j^2\right)\left(\sum_{j=1}^{n} B_j^2\right) - \left(\sum_{j=1}^{n} A_j B_j\right)^2 = \frac{1}{2}\sum_{j=1}^{n}\sum_{k=1}^{n}(A_j B_k - A_k B_j)^2$$

18) $\displaystyle\sum_{k=0}^{+\infty} A_k = \lim_{n \to +\infty}\left(\sum_{k=0}^{n} A_k\right)$ if the limit exists and is finite

19) $\displaystyle\sum_{k=1}^{+\infty} \frac{1}{k} = +\infty$

20) $\displaystyle\sum_{k=1}^{+\infty} \frac{1}{k^2} = \frac{\pi^2}{6}$

21) $\displaystyle\sum_{k=0}^{+\infty} x^k = \frac{1}{1-x}$ $|x| < 1$

22) $\displaystyle\sum_{k=0}^{+\infty} kx^k = \frac{x}{(1-x)^2}$ $|x| < 1$

23) $\displaystyle\sum_{k=0}^{+\infty} k(k-1)x^k = \frac{2x^3}{(1-x)^3}$ $|x| < 1$

24) $(A * B)_k = \displaystyle\sum_{j=0}^{k} A_{k-j} B_j = \sum_{s+j=k} A_s B_j$ **discrete convolution**

25) $\displaystyle\sum_{k=0}^{+\infty}(A*B)_k = \sum_{k=0}^{+\infty}A_k \sum_{k=0}^{+\infty}B_k$ if $\displaystyle\sum_{k=0}^{+\infty}|A_k| < +\infty,\;\; \sum_{k=0}^{+\infty}|B_k| < +\infty$

26)   An expression of the kind $\sum_{k=0}^{+\infty}A_k x^k$ is called

power series depending on the variable x.

For any choice of the coefficients A_k there exists R, fulfilling $0 \leq R \leq +\infty$ and called **radius of convergence** of the power series, such that the sum of the series is defined for any complex number x verifying $|x| < R$.

Appendix B

Complex numbers

The set $\mathbb{C}$ of complex numbers is defined as the set of ordered pairs (x, y) of real numbers (or, equivalently, the set of points in the Cartesian plane) endowed with the two operations of sum and product fulfilling the properties listed below.

If we denote by i the complex number $(0, 1)$ (called **imaginary unit**) then any complex number (x, y) can be represented[1] in the usual form $x + iy$. We also write $x = \operatorname{Re} z$ and $y = \operatorname{Im} z$. The real numbers x and y are called respectively **real part** and **imaginary part** of the complex number $x + iy$.

If $z_1 = x_1 + iy_1$ and $z_2 = x_2 + iy_2$ are two complex numbers, then their sum and product are defined as

$$z_1 + z_2 = (x_1 + x_2) + i(y_1 + y_2)$$
$$z_1 \cdot z_2 = (x_1 x_2 - y_1 y_2) + i(x_1 y_2 + x_2 y_1)$$

(notice that by the second definition we obtain $i^2 = -1$).

These operations verify, for any z_1, z_2, $z_3 \in \mathbb{C}$:

$$z_1 + z_2 = z_2 + z_1 \qquad\qquad z_1 z_2 = z_2 z_1$$
$$(z_1 + z_2) + z_3 = z_1 + (z_2 + z_3) \qquad (z_1 z_2) z_3 = z_1 (z_2 z_3)$$
$$z_1 + 0 = z_1 \qquad\qquad 1 \cdot z_1 = z_1$$
$$z + (-z) = 0 \qquad\qquad z = x + iy \neq 0 \quad \Rightarrow \quad \exists\, z^{-1} = \frac{x - iy}{x^2 + y^2}$$
$$(z_1 + z_2) z_3 = z_1 z_3 + z_2 z_3$$

If $z = x + iy$, then the number $\bar{z} = z - iy$ is called **conjugate** of z.

Notice that $z = \bar{z}$ if and only if z is real (that is $y = 0$).

The **modulus** of z is the number $|z| = \sqrt{x^2 + y^2}$ (arithmetic square root). Hence $|z| = 0$ if and only if $z = 0$ that is $x = y = 0$. Moreover:

$$|z_1 + z_2| \leq |z_1| + |z_2| \qquad\qquad |z_1 z_2| = |z_1| |z_2|$$

For any $z \neq 0$: $z^{-1} = \bar{z}/|z|^2$.

---

[1] $x + iy$ corresponds to $x(1, 0) + y(0, 1)$ after setting $i = (0, 1)$.
Analogously we write 0 instead of $(0, 0)$ and 1 instead of $(1, 0)$.

E. Salinelli, F. Tomarelli: *Discrete Dynamical Models.*
UNITEXT – La Matematica per il 3+2 76
DOI 10.1007/978-3-319-02291-8_B, © Springer International Publishing Switzerland 2014

The **exponential function** evaluated at the complex number $z = x + iy$ is defined by:

$$e^{x+iy} = e^x (\cos y + i \sin y)$$

and verifies $e^{z_1+z_2} = e^{z_1} e^{z_2}$ for any $z_1, z_2 \in \mathbb{C}$. In particular, $\forall \theta \in \mathbb{R}$

$$\sin \theta = \frac{e^{i\theta} - e^{-i\theta}}{2i} \qquad\qquad \cos \theta = \frac{e^{i\theta} + e^{-i\theta}}{2}.$$

For any $z = x + iy$ there exists a unique $\theta \in [0, 2\pi)$ called **argument** of z such that after setting $\rho = |z|$ we have

$$\boxed{x + iy = \rho (\cos \theta + i \sin \theta) = \rho e^{i\theta}}$$

The second and third expressions are called respectively **polar form** and **exponential form** of a complex number and are useful in product computations (whereas the Cartesian form is useful in sum computations): if $z = \rho (\cos \theta + i \sin \theta)$ and $w = r (\cos \varphi + i \sin \varphi)$, then we have **de Moivre's formulas:**[2]

$$\boxed{\begin{aligned} z\,w &= \rho r \left(\cos (\theta + \varphi) + i \sin (\theta + \varphi)\right) = \rho r\, e^{i(\theta+\varphi)} \\ z^n &= \rho^n \left(\cos (n\theta) + i \sin (n\theta)\right) = \rho^n e^{in\theta} \qquad \forall n \in \mathbb{N} \end{aligned}}$$

If n is an integer greater than zero, then any complex number $z = \rho e^{i\theta}$ different from zero (that is such that $\rho > 0$) has exactly n distinct complex n-th roots w_1, $w_2, \ldots, w_n$ (that is n numbers w_j fulfilling $w_j^n = z$ for any $j = 1, \ldots, n$):

$$w_1 = \sqrt[n]{\rho}\, e^{i\frac{\theta}{n}} \quad w_2 = \sqrt[n]{\rho}\, e^{i\frac{\theta+2\pi}{n}} \quad w_3 = \sqrt[n]{\rho}\, e^{i\frac{\theta+4\pi}{n}} \quad \ldots \quad w_n = \sqrt[n]{\rho}\, e^{i\frac{\theta+2(n-1)\pi}{n}}$$

where $\sqrt[n]{\rho}$ denotes the n-th arithmetic root of the real positive number ρ.

Definition. *A function $f : \mathbb{C} \to \mathbb{C}$ is differentiable in the complex sense at $z_0 \in \mathbb{C}$ if there exists a complex number, called **derivative** of f and denoted by $f'(z_0)$, such that*

$$\forall \varepsilon > 0 \ \ \exists \delta > 0 : \quad \left| \frac{f(z_0 + h) - f(z_0)}{h} - f'(z_0) \right| < \varepsilon \qquad \forall h \in \mathbb{C} : \ |h| < \delta.$$

The functions that are differentiable in the complex sense in the whole complex plane $\mathbb{C}$ are called **entire analytic functions**: as an automatic consequence, they have also the derivatives of any order at any point of $\mathbb{C}$ and can be expanded in convergent power series with radius of convergence equal to $+\infty$. For instance, exp, sin and cos are analytic functions and they fulfill:

$$\exp(z) = \sum_{k=0}^{+\infty} \frac{z^k}{k!} \qquad \sin(z) = \sum_{k=0}^{+\infty} (-1)^k \frac{z^{2k+1}}{(2k+1)!} \qquad \cos(z) = \sum_{k=0}^{+\infty} (-1)^k \frac{z^{2k}}{(2k)!}.$$

---

[2] Abraham de Moivre, 1667-1754.

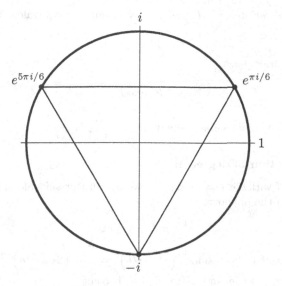

Fig. B.1 Cubic roots of i

Algebraic equations

Fundamental Theorem of Algebra. *Every polynomial P in one variable, with degree n greater than or equal to 1, has exactly n complex roots (counted with their multiplicity), that is there are $z_1, \ldots, z_n \in \mathbb{C}$ not necessarily distinct and a constant $c \in \mathbb{C} \setminus \{0\}$ such that*

$$P(z) = c(z - z_1)(z - z_2) \cdots (z - z_n). \qquad \square$$

In the sequel $a, b, c, \ldots$ are given complex numbers s.t. $a \neq 0$, x denotes the unknown.
$\sqrt{\cdot}$ always denotes the (nonnegative) arithmetic square root of a nonnegative real number.

Algebraic equation of first degree: $ax + b = 0$
Solution: $x = -b/a$

Algebraic equation of degree 2: $ax^2 + bx + c = 0$

Solutions: $x = \dfrac{-b \pm \sqrt{b^2 - 4ac}}{2a}$ if $b^2 - 4ac \geq 0$

$x = \dfrac{-b \pm i\sqrt{4ac - b^2}}{2a}$ if $b^2 - 4ac < 0$

$x = \dfrac{-b \pm \sqrt{\rho}\, e^{i\theta/2}}{2a}$ if $b^2 - 4ac = \rho e^{i\theta} \in \mathbb{C} \setminus \mathbb{R}$ with $\rho > 0$, $\theta \in [0, 2\pi)$

Algebraic equation of degree 3: $ax^3 + bx^2 + cx + d = 0$

By the change of variable $x = t - b/(3a)$ we obtain the equivalent equation in the unknown t:

$$t^3 + pt + q = 0$$

whose solutions are

$$t_k = y_k - p/(3y_k) \qquad k = 1, 2, 3$$

where y_k are the complex cubic roots of $-\dfrac{q}{2} + \sqrt{\dfrac{q^2}{4} + \dfrac{p^3}{27}}$.

Algebraic equation of degree 4: $a_4\, x^4 + a_3\, x^3 + a_2\, x^2 + a_1\, x + a_0 = 0$

By the change of variable $x = z - a_3/4a_4$ we obtain (for suitable a, b, c) the equivalent equation in the unknown z:

$$z^4 + az^2 + b = cz . \tag{B.1}$$

Adding $(2\sqrt{b} - a)z^2$ to both sides: $\left(z^2 + \sqrt{b}\right)^2 = cz + \left(2\sqrt{b} - a\right)z^2$ and this equation, after setting $\sqrt{b} = q$ and $2\sqrt{b} - a = p$, becomes

$$(z^2 + q)^2 = cz + pz^2 . \tag{B.2}$$

Adding $2tz^2 + t^2 + 2tq$ to both sides of (B.2):

$$(z^2 + q + t)^2 = \sqrt{(p + 2t)^2}\, z^2 + cz + \sqrt{(t^2 + 2tq)^2} . \tag{B.3}$$

If we can choose t in such a way that the right-hand side of (B.3) is the square of a binomial in z, then (B.3) is reduced to 2 equations of second degree in z; to this aim, it is enough to choose t such that $c = 2\sqrt{p + 2t}\,\sqrt{t^2 + 2tq}$, that is to solve the equation of third degree in t:

$$2t^3 + (p + 4q)\, t^2 + 2pq\, t - c^2/4 , \tag{B.4}$$

that is named *Ferrari's cubic resolvent* and, after its solution, it provides also the solution of (B.1).

The algebraic equations of degree 5 or higher, except in very special cases, cannot be solved in terms of radicals.

Appendix C

Basic probability

If A, B, A_j denote events whose probability of happening is defined and $P(\cdot) \in [0, 1]$ denotes the probability, then:

1) $\quad 0 \le P(A) \le 1$

2) $\quad P(A \cup B) = P(A) + P(B) - P(A \cap B)$

3) $\quad P(A \cup B) = P(A) + P(B)$ if $A \cap B = \emptyset$

4) $\quad P\left(\bigcup\limits_{j=1}^{\infty} A_j\right) = \sum\limits_{j=1}^{\infty} P(A_j)$ if $A_k \cap A_h = \emptyset$, $\forall k \ne h$

5) $\quad P(A \cap B) = P(A) \cdot P(B)$ if and only if A and B are mutually independent

6) $\quad P(A) \le P(B)$ if $A \subseteq B$

7) **Inclusion–exclusion principle:**

$$P\left(\bigcup\limits_{j=1}^{n} A_j\right) =$$

$$= \sum_{j=1}^{n} P(A_j) - \sum_{1 \le i < j \le n} P(A_i \cap A_j) + \sum_{1 \le i < j < k \le n} P(A_i \cap A_j \cap A_k) - \cdots$$

$$\cdots - (-1)^n P(A_1 \cap A_2 \cap \cdots \cap A_n)$$

8) **Conditional probability:** $\quad P(A|H) = \dfrac{P(A \cap H)}{P(H)} \qquad$ if $P(H) \ne 0$

9) **Bayes' law:**

$$P(H_i|B) = \dfrac{P(B|H_i)\,P(H_i)}{\sum\limits_{j=1}^{n} P(B|H_j)\,P(H_j)}$$

where $H_i \cap H_j = \emptyset$ if $i \ne j$, $\bigcup\limits_{j=1}^{n} H_j$ contains all events, $P(H_j) \ne 0$ for any j.

E. Salinelli, F. Tomarelli: *Discrete Dynamical Models.*
UNITEXT – La Matematica per il 3+2 76
DOI 10.1007/978-3-319-02291-8_C, © Springer International Publishing Switzerland 2014

Appendix D

Linear Algebra

An n-dimensional **vector** $\mathbf{v}$ is an ordered n-tuple of real (or complex) numbers $v_1, v_2, \ldots, v_n$ that we represent as a column:

$$\mathbf{v} = \begin{bmatrix} v_1 \\ v_2 \\ \vdots \\ v_n \end{bmatrix}.$$

The **sum** of two n dimensional vectors $\mathbf{v}$ and $\mathbf{w}$ is the vector

$$\mathbf{v} + \mathbf{w} = \begin{bmatrix} v_1 + w_1 \\ v_2 + w_2 \\ \vdots \\ v_n + w_n \end{bmatrix}.$$

The **product** of an n dimensional vector $\mathbf{v}$ times a real (or complex) number α is the vector

$$\alpha\mathbf{v} = \begin{bmatrix} \alpha v_1 \\ \alpha v_2 \\ \vdots \\ \alpha v_n \end{bmatrix}.$$

The set of n dimensional vectors, endowed with such operations is denoted by $\mathbb{R}^n$ (or $\mathbb{C}^n$).

The **norm** of a vector $\mathbf{v}$ is the real number $\|\mathbf{v}\| = \left(|v_1|^2 + |v_2|^2 + \cdots + |v_n|^2\right)^{1/2}$.

The **scalar product** of two vectors of the same dimension $\mathbf{v}$ and $\mathbf{w}$ is the number

$$\langle \mathbf{v}, \mathbf{w} \rangle = v_1\overline{w}_1 + v_2\overline{w}_2 + \cdots + v_n\overline{w}_n.$$

Hence $\|\mathbf{v}\|^2 = \langle \mathbf{v}, \mathbf{v} \rangle \geq 0$, $\forall \mathbf{v}$, and the **Cauchy-Schwarz inequality** holds:

$$|\langle \mathbf{v}, \mathbf{w} \rangle| \leq \|\mathbf{v}\|\,\|\mathbf{w}\| \qquad \forall \mathbf{v}, \mathbf{w}.$$

Two vectors $\mathbf{v}$ and $\mathbf{w}$ are called **orthogonal** if and only if $\langle \mathbf{v}, \mathbf{w} \rangle = 0$.

E. Salinelli, F. Tomarelli: *Discrete Dynamical Models.*
UNITEXT – La Matematica per il 3+2 76
DOI 10.1007/978-3-319-02291-8_D, © Springer International Publishing Switzerland 2014

k vectors (n-dimensional) $\mathbf{v}^1$, $\mathbf{v}^2$, $\ldots$, $\mathbf{v}^k$ are called **linearly dependent** if there exist k constants α_1, $\ldots$, α_k not vanishing altogether such that

$$\alpha_1 \mathbf{v}^1 + \alpha_2 \mathbf{v}^2 + \cdots + \alpha_k \mathbf{v}^k = \mathbf{0} \tag{D.1}$$

where $\mathbf{0}$ denotes the n dimensional vector whose components are all zero.

If (D.1) is verified for $\alpha_1 = \alpha_2 = \cdots = \alpha_k = 0$ only, then $\mathbf{v}^1$, $\mathbf{v}^2$, $\ldots$, $\mathbf{v}^k$ are called **linearly independent**.

Pay attention to subscripts and superscripts: $\mathbf{v}^j$ is the vector with components v_i^j:

$$\mathbf{v}^j = \begin{bmatrix} v_1^j \\ v_2^j \\ \vdots \\ v_n^j \end{bmatrix}.$$

An expression of the kind $\alpha_1 \mathbf{v}^1 + \alpha_2 \mathbf{v}^2 + \cdots + \alpha_k \mathbf{v}^k$ is called **linear combination** of the vectors $\mathbf{v}^j$. Notice that any set of k (n dimensional) vectors with $k > n$ is linearly dependent.

As set of vectors in $\mathbb{R}^n$ is a **basis** of $\mathbb{R}^n$ if these vectors fulfill this two conditions:

1) they are linearly independent;
2) they span $\mathbb{R}^n$, that is any element in $\mathbb{R}^n$ is equal to a linear combination with real coefficients of these vectors.

$\mathbb{R}^n$ has infinitely many bases, but everyone is made by exactly n vectors. A bases is called **orthonormal** if it is made by mutually orthogonal vectors with unit norm. The vectors

$$\begin{bmatrix} 1 \\ 0 \\ \vdots \\ 0 \end{bmatrix}, \quad \begin{bmatrix} 0 \\ 1 \\ \vdots \\ 0 \end{bmatrix}, \quad \ldots, \quad \begin{bmatrix} 0 \\ 0 \\ \vdots \\ 1 \end{bmatrix}$$

provide an example of orthonormal bases in $\mathbb{R}^n$; this one is called the **canonical bases** of $\mathbb{R}^n$.

Analogously, a **bases** of $\mathbb{C}^n$ is a set of n linearly independent vectors whose linear combinations with complex coefficients span $\mathbb{C}^n$.

A matrix of order $n \times m$ is a table of n rows and m columns:

$$\mathbb{A} = \begin{bmatrix} a_{11} & a_{12} & \cdots & a_{1m} \\ a_{21} & a_{22} & \cdots & a_{2m} \\ \vdots & \vdots & \ddots & \vdots \\ a_{n1} & a_{n2} & \cdots & a_{nm} \end{bmatrix}$$

where a_{ij} denotes the element of $\mathbb{A}$ on the i-th row and j-th column. For short we say order n in case of square matrices of order $n \times n$, and we will write $\mathbb{A} = [a_{ij}]$.

The vectors in $\mathbb{R}^n$ are $n \times 1$ matrices (namely they are columns).

The product of a matrix $\mathbb{A}$ times a number α is the matrix $[\alpha a_{ij}]$ whose element at place i,j is $\alpha\, a_{ij}$.

The **conjugate** $\overline{\mathbb{A}}$ of a matrix is a matrix of the same order such that $\overline{\mathbb{A}} = [\overline{a}_{ij}]$.

A matrix $\mathbb{A}$ is called **upper triangular** if $i > j$ implies $a_{ij} = 0$, **lower triangular** if $i < j$ implies $a_{ij} = 0$.

The **transpose** of a matrix $\mathbb{A}$ of order $n \times m$ is the matrix $\mathbb{A}^T$ of order $m \times n$ such that $a_{ij}^T = a_{ji}$.

A matrix $\mathbb{A}$ of order $n \times m$ can be multiplied by a matrix $\mathbb{B}$ of order $m \times l$ to give a matrix $\mathbb{D}$ of order $n \times l$ by the **"rows times columns" product**:

$$\mathbb{D} = \mathbb{A}\mathbb{B}, \qquad d_{ij} = \sum_{k=1}^{m} a_{ik} b_{kj}.$$

Notice that d_{ij} is the scalar product in $\mathbb{R}^m$ of the i-th row of $\mathbb{A}$ times the j-th column of $\mathbb{B}$. In particular, also the scalar product of two vectors is an example of the product of a (column) matrix times a (row) matrix:

$$\langle \mathbf{v}, \mathbf{w} \rangle = \overline{\mathbf{w}}^T \mathbf{v} \qquad \overline{\mathbf{w}}^T = [\overline{w}_1, \overline{w}_2, \ldots, \overline{w}_n].$$

If $n = l$, then both products $\mathbb{A}\mathbb{B}$ and $\mathbb{B}\mathbb{A}$ are defined by the rows times columns rule, nevertheless they may be different: *matrix product is noncommutative*, for example $\mathbb{A} = \begin{bmatrix} 1 & 1 \\ 1 & 1 \end{bmatrix}$ and $\mathbb{B} = \begin{bmatrix} 0 & 1 \\ 0 & 0 \end{bmatrix}$ do not commute.

Any multiple of the identity matrix commutes with any other square matrix. The square matrices with only zero elements off the main diagonal ($a_{ij} = 0$ if $i \neq j$) are called **diagonal** (or **semi-simple**) **matrices**; notice that the left (respectively right) multiplication of $\mathbb{A}$ times a diagonal matrix, corresponds to multiplying the j-th row (respectively column) of $\mathbb{A}$ times j-th element in the diagonal.

If $\mathbb{A}$ is a square matrix of order n we define its **determinant**, as a scalar-valued function of the matrix, denoted by $\det \mathbb{A}$ and defined as follows.

If $n = 2$, $\mathbb{A} = \begin{bmatrix} a & b \\ c & d \end{bmatrix}$ then $\det \mathbb{A} = ad - bc$; this value is the *oriented area of the parallelogram* associated to the vectors $\begin{bmatrix} a \\ c \end{bmatrix}$ and $\begin{bmatrix} b \\ d \end{bmatrix}$.

If $n = 3$, $\mathbb{A} = \begin{bmatrix} a_{11} & a_{12} & a_{13} \\ a_{21} & a_{22} & a_{31} \\ a_{31} & a_{32} & a_{33} \end{bmatrix}$, then the determinant of $\mathbb{A}$ is

$$\det \mathbb{A} = a_{11}a_{22}a_{33} + a_{12}a_{23}a_{31} + a_{13}a_{21}a_{32} - a_{13}a_{22}a_{31} - a_{23}a_{32}a_{11} - a_{33}a_{12}a_{21},$$

that is the *oriented volume of the parallelepiped* associated to the vectors

$$\begin{bmatrix} a_{11} \\ a_{21} \\ a_{31} \end{bmatrix} \begin{bmatrix} a_{12} \\ a_{22} \\ a_{32} \end{bmatrix} \begin{bmatrix} a_{13} \\ a_{23} \\ a_{33} \end{bmatrix}.$$

About the general case, if the square matrix $\mathbb{A}$ has order n then the definition of determinant is given in terms of a recursive formula based on the knowledge of the determinants of order $(n-1)$: given i, $1 \leq i \leq n$, then

$$\det \mathbb{A} = (-1)^{i+1} a_{i1} \det \mathbb{A}_{i1} + \cdots + (-1)^{i+n} a_{in} \det \mathbb{A}_{in} \qquad (\text{D.2})$$

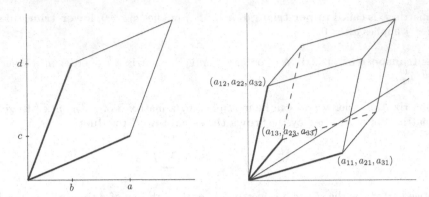

Fig. D.1 Geometrical meaning of the determinant for $n = 2, 3$

where $\mathbb{A}_{ij}$ denotes the $(n - 1) \times (n - 1)$ matrix obtained by eliminating the i-th row and j-th column in the matrix $\mathbb{A}$. Notice that this expression (called Laplace expansion with respect to the i-th row) does not depend on the choice of the row i.

Determinant properties ($\mathbb{A}$ and $\mathbb{B}$ matrices of order n)

- $\det \mathbb{A}^T = \det \mathbb{A}$;
- $\det(\mathbb{A}\mathbb{B}) = \det \mathbb{A} \det \mathbb{B}$;
- $\det(\alpha \mathbb{A}) = \alpha^n \det \mathbb{A}$;
- if $\mathbb{A}$ is either upper or lower triangular then $\det \mathbb{A} = a_{11} a_{22} \cdots a_{nn}$.

A square matrix $\mathbb{A}$ of order n is **invertible** if there exists a matrix $\mathbb{A}^{-1}$ of order n such that $\mathbb{A}\mathbb{A}^{-1} = \mathbb{I}_n$ where $\mathbb{I}_n$ denotes the **identity matrix**:

$$
\mathbb{I}_n = \begin{bmatrix} 1 & 0 & \ldots & 0 \\ 0 & 1 & \ldots & 0 \\ \vdots & \vdots & \ddots & \vdots \\ 0 & 0 & \ldots & 1 \end{bmatrix}.
$$

If $\mathbb{A}^{-1}$ exists then it is unique and it fulfills $\mathbb{A}^{-1}\mathbb{A} = \mathbb{I}_n$ too. A necessary and sufficient condition for the existence of $\mathbb{A}^{-1}$ is that $\det \mathbb{A} \neq 0$. If $\det \mathbb{A} \neq 0$, then the element at place i, j in $\mathbb{A}^{-1}$ is $(-1)^{i+j} \det \mathbb{A}_{ji} / \det \mathbb{A}$ where $\mathbb{A}_{ij}$ is the matrix obtained by eliminating i-th row and j-th column of $\mathbb{A}$, moreover $\det(\mathbb{A}^{-1}) = (\det \mathbb{A})^{-1}$.

We emphasize that the number of operations required to evaluate the determinant with formula (D.2) grows so fast with n that computations become unfeasible, even if the order n is not particularly high. Any formulation that allows to reduce the number of these operations (by exploiting structural properties of the matrix: triangular, block–structured, and so on) is extremely useful.

Systems of linear algebraic equations

The system of n equations in n unknowns x_j

$$\begin{cases} a_{11}x_1 + a_{12}x_2 + \cdots + a_{1n}x_n = b_1 \\ a_{21}x_1 + a_{22}x_2 + \cdots + a_{2n}x_n = b_2 \\ \quad \vdots \\ a_{n1}x_1 + a_{n2}x_2 + \cdots + a_{nn}x_n = b_n \end{cases}$$

can be written in the form $\mathbb{A}\mathbf{x} = \mathbf{b}$ where $\mathbb{A} = [a_{ij}]$, $\mathbf{x} = \begin{bmatrix} x_1 \\ \vdots \\ x_n \end{bmatrix}$, $\mathbf{b} = \begin{bmatrix} b_1 \\ \vdots \\ b_n \end{bmatrix}$.

Theorem D.1 (Rouché–Capelli). *The homogeneous system* $\mathbb{A}\mathbf{x} = \mathbf{0}$ *has non-trivial solutions (that is different from the vector* $\mathbf{0}$*) if and only if* $\det \mathbb{A} = 0$.
If $\det \mathbb{A} \neq 0$, *then the system* $\mathbb{A}\mathbf{x} = \mathbf{b}$ *has a unique solution given explicitly by* **Cramer's rule**

$$x_j = \frac{\det \mathbb{D}_j}{\det \mathbb{A}}$$

where $\mathbb{D}_j$ *is the matrix obtained by substituting* $\mathbf{b}$ *to the* j-*th column of* $\mathbb{A}$.
If $\det \mathbb{A} = 0$, *then either the system* $\mathbb{A}\mathbf{x} = \mathbf{b}$ *has no solution or it has infinitely many solutions: precisely, there are infinitely many solutions if and only if*

$$\langle \mathbf{b}, \mathbf{y} \rangle = 0 \qquad \forall \mathbf{y} : \mathbb{A}^T \mathbf{y} = \mathbf{0}.$$

Moreover, if $\widetilde{\mathbf{x}}$ *is a solution of* $\mathbb{A}\mathbf{x} = \mathbf{b}$, *then the others are of the kind* $\widetilde{\mathbf{x}} + \mathbf{z}$ *where* $\mathbf{z}$ *solves* $\mathbb{A}\mathbf{z} = \mathbf{0}$.

We observe that the above theorem has very important consequences in the theory, but its application to the numerical computation of the solution $\mathbf{x}$ is limited only to small values of n, due to the already mentioned difficulty in evaluating determinants. Conversely in real **numerical analysis** one has to solve high–dimensional systems; in these cases direct methods are used, for instance **Gaussian elimination**, or iterative methods that are very efficient from the computational viewpoint (see Sect. 5.8).

Linear functions and vector spaces

If $\mathbb{A}$ is an $n \times m$ matrix, then the function that maps any $\mathbf{x} \in \mathbb{R}^m$ in the vector $\mathbb{A}\mathbf{x} \in \mathbb{R}^n$ is a **linear function**, that is

$$\mathbb{A}(\alpha \mathbf{x} + \beta \mathbf{y}) = \alpha \mathbb{A}\mathbf{x} + \beta \mathbb{A}\mathbf{y} \qquad \forall \alpha, \beta \in \mathbb{R}, \ \forall \mathbf{x}, \mathbf{y} \in \mathbb{R}^m,$$

and the map $\mathbf{x} \mapsto \mathbb{A}\mathbf{x}$ itself is denoted by $\mathbb{A}$.
The **kernel** of $\mathbb{A}$ is the set

$$\ker \mathbb{A} = \{ \mathbf{x} \in \mathbb{R}^m : \ \mathbb{A}\mathbf{x} = \mathbf{0} \in \mathbb{R}^n \}.$$

The **image** of $\mathbb{A}$ is the set

$$\operatorname{im}\mathbb{A} = \{\mathbf{z} \in \mathbb{R}^n : \ \exists \mathbf{x} \in \mathbb{R}^m, \ \mathbb{A}\mathbf{x} = \mathbf{z}\}.$$

A subset V of $\mathbb{R}^n$ is a real **vector space** if it is closed with respect to the operations of sum and product times a real number, that is:

$$\forall \mathbf{v}, \mathbf{w} \in V, \forall \alpha \in \mathbb{R} \qquad\qquad \mathbf{v} + \mathbf{w} \in V, \quad \alpha\mathbf{v} \in V.$$

A subset $W \subset \mathbb{C}^n$ is a **(complex) vector space** if it is closed with respect to the operations of sum and product times a complex number, that is:

$$\forall \mathbf{v}, \mathbf{w} \in W, \forall \alpha \in \mathbb{C} \qquad\qquad \mathbf{v} + \mathbf{w} \in W, \quad \alpha\mathbf{v} \in W.$$

In general, any set that is closed with respect to the operations of sum and product times a scalar is called vector space.

A **bases of a vector space** is a set of linearly independent vectors that span the whole space. If this set is finite, then the number of its elements is called **dimension of the vector space**. $\mathbb{R}^n$ is a vector space with finite dimension (equal to n). The set of sequences of real numbers is an example vector space with infinite dimension.

Theorem D.2. *If* $\mathbb{A} : \mathbb{R}^m \to \mathbb{R}^n$ *is a linear function, then* $\ker \mathbb{A}$ *and* $\operatorname{Im} \mathbb{A}$ *are vector spaces and*

$$\dim(\operatorname{Im} \mathbb{A}) + \dim(\ker \mathbb{A}) = m.$$

Moreover, if $n = m$, *then*

$$\ker \mathbb{A} = \{\mathbf{0}\} \quad \Leftrightarrow \quad \operatorname{im} \mathbb{A} = \mathbb{R}^n \quad \Leftrightarrow \quad \mathbb{A} \text{ is bijective} \quad \Leftrightarrow \quad \det \mathbb{A} \neq 0.$$

If $\mathbf{u}^1, \mathbf{u}^2, \ldots, \mathbf{u}^n$ are n linearly independent vectors in $\mathbb{R}^n$ then they form a bases of $\mathbb{R}^n$ and the square matrix $\mathbb{U} = \left[u_j^i\right]^{-1}$ of order n (inverse of the matrix whose columns are the vectors $\mathbf{u}^i$) is called **matrix of place of basis chance:**
if $\mathbf{x}$ is a vector and $\mathbb{A}$ is a linear function whose components are respectively x_j and a_{ij} referring to the canonical bases, then the components, respectively y_i and b_{ij}, of $\mathbf{y}$ and $\mathbb{B}$ corresponding to $\mathbf{x}$ and $\mathbb{A}$ referring to the bases $\mathbf{u}^1, \mathbf{u}^2, \ldots, \mathbf{u}^n$, are given respectively by

$$\mathbf{y} = \mathbb{U}\mathbf{x} \qquad\qquad \mathbb{B} = \mathbb{U}\mathbb{A}\mathbb{U}^{-1}.$$

We call **sum** of two subspaces U and V of $\mathbb{R}^n$ the set W of vectors $\mathbf{w}$ that are sum of an element $\mathbf{u}$ of U and an element $\mathbf{v}$ of V: $\mathbf{w} = \mathbf{u} + \mathbf{v}$. This "set-theoretical" sum is a vector space too. When the representation of any element of W is unique, then we say that W is the **direct sum** of U and V and write $W = U \oplus V$. It can be proved that $W = U \oplus V$ is a direct sum of two subspaces U and V of $\mathbb{R}^n$ if and only if $U \cap V = \{\mathbf{0}\}$.

Eigenvalues and eigenvectors

Given an $n \times n$ matrix $\mathbb{A}$ and a real or complex number λ, any *non trivial* vector $\mathbf{v} \in \mathbb{C}^n$ (that is $\mathbf{v} \neq \mathbf{0}$) that solves the system

$$\mathbb{A}\mathbf{v} = \lambda\mathbf{v}$$

is called **eigenvector** of A. If $\mathbf{v}$ is an eigenvector then the corresponding (unique) value λ that solves $A\mathbf{v} = \lambda\mathbf{v}$ is called **eigenvalue** associated to the eigenvector $\mathbf{v}$. If $\mathbf{v}$ is an eigenvector associated to the eigenvalue λ, then any nontrivial multiple of $\mathbf{v}$ is an eigenvector too. Either a unique (up to its multiples) eigenvector (in this case the eigenvalue is called **simple eigenvalue**) or several linearly independent eigenvectors (**multiple eigenvalue**) may correspond to a single eigenvalue. In any case the eigenvectors associated to a given eigenvalue λ together with $\mathbf{0}$ form a vector space, called **eigenspace** of λ; the dimension of this eigenspace is called **geometric multiplicity** of the eigenvalue λ.

The set of eigenvalues of A coincides with the set of roots of $\det(A - \lambda\mathbb{I}_n)$. We notice that $\det(A - \lambda\mathbb{I}_n)$ is a polynomial in λ of degree n. This polynomial is called **characteristic polynomial** of the matrix A. Then the eigenvalues are exactly n in number, provided they are counted with their **algebraic multiplicity**.

Eigenvectors associated to distinct eigenvalues are linearly independent, and if in addition $A = \overline{A}^T$ then they are orthogonal too.

If there exists a bases of $\mathbb{R}^n$ formed by n eigenvectors $\mathbf{v}^1$, $\mathbf{v}^2$, ..., $\mathbf{v}^n$ of A and $\lambda_1, \lambda_2, \ldots, \lambda_n$ are the corresponding eigenvalues (not necessarily different from one another), then the matrix representing A in this bases is **diagonal**:

$$
\mathbb{B} = \begin{bmatrix}
\lambda_1 & 0 & 0 & \ldots & 0 \\
0 & \lambda_2 & 0 & \ldots & 0 \\
0 & 0 & \lambda_3 & \ldots & 0 \\
\vdots & \vdots & \vdots & \ddots & \vdots \\
0 & 0 & 0 & \ldots & \lambda_n
\end{bmatrix}
$$

where $\mathbb{B} = U A U^{-1}$ and $U = \left[\mathbf{u}^1, \mathbf{u}^2, \ldots, \mathbf{u}^n\right]^{-1}$. In this case, the matrix A is called **diagonalizable**.

Notice that A and $\mathbb{B}$ have the same characteristic polynomial, in fact:

$$\det(\mathbb{B} - \lambda\mathbb{I}_n) = \det\left(U(A - \lambda\mathbb{I}_n)U^{-1}\right) = \det U \det(A - \lambda\mathbb{I}_n)(\det U)^{-1} = \det(A - \lambda\mathbb{I}_n).$$

If the square matrix A of order n has exactly n distinct eigenvalues with algebraic multiplicity 1 then A is diagonalizable: we list the main related statements.

Theorem D.3 (Spectral Theorem). *Assume* $A: \mathbb{R}^n \to \mathbb{R}^n$ *is **symmetric** (that is such that* $A = A^T$*); then there exists an orthonormal bases of* $\mathbb{R}^n$ *formed by eigenvectors of* A.

Assume $A : \mathbb{C}^n \to \mathbb{C}^n$ *is **Hermitian** (that is* $A = \overline{A}^T$*); then there exists an orthonormal bases of* $\mathbb{C}^n$ *formed by eigenvectors of* A.

Theorem D.4. *If* $A : \mathbb{R}^n \to \mathbb{R}^n$ *is **positive definite**, that is*

$$\exists \alpha > 0: \quad \langle A\mathbf{x}, \mathbf{x} \rangle \geq \alpha \|\mathbf{x}\|^2 \qquad \forall \mathbf{x} \in \mathbb{R}^n$$

then all the eigenvalues λ *of* $\mathrm{sym}A = (A + A^T)/2$ *are real numbers s.t.* $\lambda \geq \alpha$, *in particular they are strictly positive.*

Theorem D.5 (Cayley–Hamilton Theorem).. *If* $\mathcal{P}(\lambda) = \det(A - \lambda\mathbb{I}_n)$, *then* $\mathcal{P}(A) = \mathbb{O}$ *(square matrix of order* n *with all elements equal to 0).*

Jordan canonical form or **Jordan normal form** of a matrix.

Diagonalizing a matrix by a change of bases is not always possible (for instance $\begin{bmatrix} 1 & 0 \\ 1 & 1 \end{bmatrix}$ and $\begin{bmatrix} 1 & 1 \\ 0 & 1 \end{bmatrix}$ are not diagonalizable). Nevertheless it is always possible to change the bases in such a way that the matrix becomes triangular. Precisely any matrix can be put in Jordan canonical form: that is triangular with possibly non zero elements only on the main diagonal (a_{ii}) and on the first subdiagonal (a_{ij} con $i = j + 1$).

We already know that if a square matrix has only simple eigenvalues then it is diagonalizable. The the matrices that might[1] be not diagonalizable are the ones that have at least an eigenvalue with non simple algebraic multiplicity. The ones that are actually not diagonalizable have at least an eigenvalue with algebraic multiplicity strictly greater than its geometric multiplicity.

Let $\mathbb{A}$ be an $n \times n$ matrix. In the sequel: V denotes $\mathbb{R}^n$ if the terms a_{ij}, λ_j are real $\forall i, j = 1, 2, \ldots, n$; V denotes $\mathbb{C}^n$ if complex coefficients and/or complex eigenvalues are present.

If λ is an eigenvalue of $\mathbb{A}$ then the eigenspace associated to $\mathbb{A}$ is $\ker(\mathbb{A} - \lambda\mathbb{I}_n)$. If this eigenspace has a dimension strictly smaller than the algebraic multiplicity of λ then $\mathbb{A}$ is not diagonalizable.

We define the **generalized eigenspace** associated to the eigenvalue λ:

$$\bigcup_{k \in \mathbb{N}} \ker(\mathbb{A} - \lambda\mathbb{I}_n)^k.$$

Theorem D.6. *For any eigenvalue λ of $\mathbb{A}$ the corresponding generalized eigenspace has dimension equal to the algebraic multiplicity of λ.*

Every element of V can be represented (in a unique way) as a linear combination of elements of the generalized eigenspaces corresponding to all eigenvalues.

There exists a bases of V (the union of the bases of all generalized eigenspaces) in which the matrix has everywhere null entries except for: the main diagonal, containing the eigenvalues ordered lexicographically[2], and the first subdiagonal which may contain either 1s or 0s. Moreover in this bases there is also a square-blocks structure along the main diagonal, and each block has dimension equal to the dimension of a generalized eigenspace.

Every generalized eigenspace can exhibit an additional substructure in such a way that the corresponding sub-blocks have only 1's on the first subdiagonal. These sub-blocks are called **Jordan blocks**. A Jordan block $\mathbb{B}$ of dimension k is characterized by the presence of an eigenvalue λ on the main diagonal and by the property that the matrix $\mathbb{T} = \mathbb{B} - \lambda\mathbb{I}_k$ verifies $\mathbb{T}^{k-1} \neq \mathbb{O}_k = \mathbb{T}^k$.

In addition to ordering according to the eigenvalues, the Jordan sub-blocks of a block related to one single eigenvalue can be listed in increasing order of dimension (see Fig. D.2).

Notice that the Jordan form can have complex terms on the diagonal even if $\mathbb{A}$ is real. In this case the complex eigenvalues (if any) appear in complex conjugate pairs of the same multiplicity, hence the generalized eigenvectors form a bases in $\mathbb{C}^n$ because $\mathbb{A}$ has order n.

---

[1] $\mathbb{I}_2$ is obviously diagonal though it has the eigenvalue 1 with algebraic and geometric multiplicity equal to 2.

[2] That is $\lambda_1 \prec \lambda_2$ if either $\mathrm{Re}\,\lambda_1 < \mathrm{Re}\,\lambda_2$ or $\mathrm{Re}\,(\lambda_1) = \mathrm{Re}\,\lambda_2$ and $\mathrm{Im}\,\lambda_1 < \mathrm{Im}\,\lambda_2$.

$$\left[\begin{array}{cccccc} \begin{bmatrix} \lambda_1 & 0 \\ 1 & \lambda_1 \end{bmatrix} & & & & & \\ & [\lambda_2] & & & & \\ & & \begin{bmatrix} \lambda_2 & 0 \\ 1 & \lambda_2 \end{bmatrix} & & & \\ & & & \begin{bmatrix} \lambda_2 & 0 \\ 1 & \lambda_2 \end{bmatrix} & & \\ & & & & \begin{bmatrix} \lambda_2 & 0 & 0 \\ 1 & \lambda_2 & 0 \\ 0 & 1 & \lambda_2 \end{bmatrix} & \\ & & & & & \begin{bmatrix} \lambda_3 & 0 & 0 & 0 \\ 1 & \lambda_3 & 0 & 0 \\ 0 & 1 & \lambda_3 & 0 \\ 0 & 0 & 1 & \lambda_3 \end{bmatrix} \end{array}\right]$$

Fig. D.2 Matrix in Jordan form: $\lambda_1 \prec \lambda_2 \prec \lambda_3$ (null elements are dropped)

However, if $\mathbb{A}$ is real, but has complex eigenvalues, we can obtain a real canonical form, through a minor bases change: we collect the pairs of conjugated eigenvalues and eigenvectors and we notice that if $\mathbb{A}\mathbf{v} = \lambda\mathbf{v}$, $\mathbb{A}\bar{\mathbf{v}} = \bar{\lambda}\bar{\mathbf{v}}$ then $\mathbf{z} = \frac{1}{2i}(\mathbf{v}-\bar{\mathbf{v}})$ and $\mathbf{w} = \frac{1}{2}(\mathbf{v}+\bar{\mathbf{v}})$ are the real vectors (though they are not eigenvectors) that we can substitute (respecting the order) to $\mathbf{v}$ and $\bar{\mathbf{v}}$ to achieve a bases of $\mathbb{R}^n$ in which the matrix has nontrivial elements only in certain blocks of the kind pointed to by the arrows in Fig. D.3.

Notice that by the identity $(a + ib)(x + iy) = (ax - by) + i(bx + ay)$ we get that the multiplication of complex numbers of the kind $z = (x + iy)$ times the complex number $\lambda = (a + ib)$ corresponds, in real coordinates $(x, y)^T$, to a rotation followed by a dilation or a contraction, as described by the multiplication times the matrix:

$$A = \begin{bmatrix} a & -b \\ b & a \end{bmatrix} = \rho \begin{bmatrix} \cos\theta & -\sin\theta \\ \sin\theta & \cos\theta \end{bmatrix} , \quad \text{where} \quad \rho = \sqrt{a^2 + b^2} , \quad \theta = \arctan(b/a) .$$

$$\begin{bmatrix} \lambda & 0 \\ 0 & \bar{\lambda} \end{bmatrix} \longmapsto \begin{bmatrix} a & -b \\ b & a \end{bmatrix} \qquad \begin{bmatrix} \lambda & 0 & 0 & 0 \\ 1 & \lambda & 0 & 0 \\ 0 & 0 & \bar{\lambda} & 0 \\ 0 & 0 & 1 & \bar{\lambda} \end{bmatrix} \longmapsto \begin{bmatrix} a & -b & 0 & 0 \\ b & a & 0 & 0 \\ 1 & 0 & a & -b \\ 0 & 1 & b & a \end{bmatrix}$$

Fig. D.3 Real Jordan blocks associated to simple and double complex eigenvalues $\lambda = a \pm ib$

By the Jordan canonical form we deduce that any matrix $\mathbb{A}$, in a suitable bases of $\mathbb{C}^n$, is represented by the sum of a semisimple (e.g. diagonal) matrix $\mathbb{S}$ and of a nilpotent **matrix** $\mathbb{T}$ (that is such that $\mathbb{T}^k = \mathbb{O}$ for a suitable integer $k \le n$): $\mathbb{A} = \mathbb{S} + \mathbb{T}$.

$$\mathbb{T} = \begin{bmatrix} 0 & 0 & 0 & 0 \\ 1 & 0 & 0 & 0 \\ 0 & 1 & 0 & 0 \\ 0 & 0 & 1 & 0 \end{bmatrix} \quad \mathbb{T}^2 = \begin{bmatrix} 0 & 0 & 0 & 0 \\ 0 & 0 & 0 & 0 \\ 1 & 0 & 0 & 0 \\ 0 & 1 & 0 & 0 \end{bmatrix} \quad \mathbb{T}^3 = \begin{bmatrix} 0 & 0 & 0 & 0 \\ 0 & 0 & 0 & 0 \\ 0 & 0 & 0 & 0 \\ 1 & 0 & 0 & 0 \end{bmatrix} \quad \mathbb{T}^4 = \mathbb{O}$$

Fig. D.4 $\mathbb{T}$ is an example of nilpotent matrix of order 4

Referring to the Jordan decomposition $\mathbb{A} = \mathbb{S} + \mathbb{T}$, the diagonal or semisimple matrix $\mathbb{S}$ commutes (in the product) with the nilpotent matrix $\mathbb{T}$ since actually the multiplication acts only between Jordan blocks, that are the sum of a multiple of the identity matrix and a nilpotent matrix.

Appendix E

Topology

Topology is a geometric theory that studies the invariant properties with respect to bijective transformations that are continuous and have continuous inverse. It provides a formal approach to the description of phenomena that naively remind of deformations of an elastic body with no splitting, glueing, cuts nor cracks.

To define a topology on a given set E means to specify a collection τ of subsets of E such that:

1) E and $\emptyset$ are elements of τ;
2) any union of elements of τ is an element of τ;
3) any finite intersection of elements of τ is an element of τ.

The elements of τ are called open sets of the topology τ.
The pair (E, τ) is called **topological space** and τ is its **topology**.

Example E.1. In the set $E = \mathbb{R}$ we consider the Euclidean topology:

$$\tau = \{A \subset \mathbb{R} : \ \forall x \in A \ \exists \varepsilon > 0 \ \text{s.t.} \ (x - \varepsilon, x + \varepsilon) \subset A\}.$$

Example E.2. In the set $E = \mathbb{R}^n$ we consider the Euclidean topology:

$$\tau = \{\Omega \subset \mathbb{R}^n : \ \forall \mathbf{x} \in \Omega \ \exists r > 0 \ \text{s.t.} \ \{\mathbf{y} : \ \|\mathbf{y} - \mathbf{x}\| < r\} \subset \Omega\}.$$

Example E.3. In the set $E = \mathbb{C}$ we consider the Euclidean topology:

$$\tau = \{\Omega \subset \mathbb{C} : \ \forall z \in \Omega \ \exists r > 0 \ \text{s.t.} \ \{w : \ |w - z| < r\} \subset \Omega\}.$$

It is easy to verify that in all the above examples the three properties that characterize a topology do hold.

Definition E.4. *Given two topological spaces $(X, \tau), (Y, \rho)$ and a function $f : X \to Y$, we say that f is continuous if $f^{-1}(A) \in \tau \ \forall A \in \rho$, that is if the preimage of any open set is an open set.*

Theorem E.5. *If $X = \mathbb{R}^n$, $Y = \mathbb{R}^m$, and τ, ρ are the corresponding Euclidean topologies, then the previous definition is equivalent to the usual metric definition of*

E. Salinelli, F. Tomarelli: *Discrete Dynamical Models.*
UNITEXT – La Matematica per il 3+2 76
DOI 10.1007/978-3-319-02291-8_E, © Springer International Publishing Switzerland 2014

continuity for $f : \mathbb{R}^n \to \mathbb{R}^m$, that we recall here:

$$\forall \mathbf{x}_0 \in X \ \forall \varepsilon > 0 \ \exists \delta > 0 \quad \text{such that} \quad \|\mathbf{x} - \mathbf{x}_0\| < \delta \ \Rightarrow \ \|f(\mathbf{x}) - f(\mathbf{x}_0)\| < \varepsilon .$$

Theorem E.6. *The composition of continuous functions is a continuous function.*

Given a topological space (E, τ), we define the collection σ of closed sets in (E, τ), as follows:

$$\sigma = \{C \subset E : \exists A \in \tau \ \text{s.t.} \ C = E \backslash A\}.$$

So a closed set is the complementary set of an open set.

It is easy to deduce the properties of closed sets by the properties of open sets:

1) E and $\emptyset$ are elements of σ;

2) any intersection of elements of σ is an element of σ;

3) any finite union of elements of σ is an element of σ.

Definition E.7. *The closure, denoted by $\overline{D}$, of a subset D of a topological space E is the smallest closed set that contains D, that is*

$$\overline{D} = \cap \{C : \ C \ \text{closed}, \ D \subset C\}.$$

Definition E.8. *The boundary of a subset D of a topological space E is*

$$\partial D = \overline{D} \cap \overline{E \backslash D}.$$

Definition E.9. *The interior $\overset{\circ}{D}$ of a subset D of a topological space is*

$$\overset{\circ}{D} = D \backslash \partial D.$$

Definition E.10. *Given two subsets A and B of a topological space, we say that A is dense in B if $B \subset \overline{A}$.*

Example E.11. In the case of the topological space $\mathbb{R}$ with the Euclidean topology and $A, B \subset \mathbb{R}$, with B open set, then A is dense in B if for any interval $(a, b) \subset B$ there exists at least an element $x \in A \cap (a, b)$.

Example E.12. $\mathbb{Q}$ is dense in $\mathbb{R}$.

Definition E.13. *An open set U contained in a topological space is called connected if no pair of nonempty disjoint open sets A and B exist whose union is U.*

Appendix F

Fractal dimension

We briefly mention some definitions related to the notions of "measure" and "dimension" for subsets of an Euclidean space. The aim is to obtain a real-valued function defined on sets that provides the length when evaluated on segments and the cardinality when evaluated on a set made of finitely many points; but we also require that this function coherently "adjusts" also on intermediate cases (for instance, on the Cantor set) and in general allows to measure the subsets of $\mathbb{R}^n$ and assign to them a dimension consistent with the notion of area and volume of objects in elementary geometry.

We define the Hausdorff dimension and the box-counting dimension and explain some relations between the two definitions.

For any subset F of $\mathbb{R}^n$ and any nonnegative real number s we define the s **dimensional Hausdorff measure** of F as

$$\mathcal{H}^s(F) = \lim_{\delta \to 0^+} \mathcal{H}^s_\delta(F) = \sup_{\delta > 0} \mathcal{H}^s_\delta(F)$$

where, for $0 < \delta < +\infty$,

$$\mathcal{H}^s_\delta(F) = \frac{\omega_s}{2^s} \inf \left\{ \sum_{j=1}^{+\infty} (\operatorname{diam} U_j)^s : \quad F \subset \bigcup_{j=1}^{\infty} U_j, \quad \operatorname{diam} U_j \leq \delta \right\}$$

and $\operatorname{diam} U_j = \sup \{|x - y| : x, y \in U_j\}$.

For any s the constant $\omega_s = \pi^{s/2} \left(\int_0^\infty e^{-x} x^{s/2} dx \right)^{-1}$ is positive and finite, moreover ω_s normalizes the measure in such a way that we obtain the expected values for segments, disks, balls and cubes in elementary geometry:

$$\omega_0 = \mathcal{H}^0(\{0\}) = 1, \qquad \omega_1 = \mathcal{H}^1([-1,1]) = 2,$$

$$\omega_2 = \mathcal{H}^2(\{(x,y) \in \mathbb{R}^2 : x^2 + y^2 \leq 1\}) = \pi,$$

$$\omega_3 = \mathcal{H}^3(\{(x,y,z) \in \mathbb{R}^3 : x^2 + y^2 + z^2 \leq 1\}) = \frac{4}{3}\pi.$$

E. Salinelli, F. Tomarelli: *Discrete Dynamical Models.*
UNITEXT – La Matematica per il 3+2 76
DOI 10.1007/978-3-319-02291-8_F, © Springer International Publishing Switzerland 2014

$\mathcal{H}^s$ is a positive measure[1], that is a function defined on sets with values in $[0, +\infty]$ that verifies some reasonable qualitative properties when measuring any subsets of $\mathbb{R}^n$, explicitly the ones listed below:

1) $\quad F = \bigcup\limits_{j=1}^{\infty} E_j \quad \Rightarrow \quad \mathcal{H}^s(F) \leq \sum\limits_{j=1}^{+\infty} \mathcal{H}^s(E_j)$ $\qquad\qquad$ **countable subadditivity**

2) $\quad \mathcal{H}^s(F + \tau) = \mathcal{H}^s(F) \quad \forall F \subset \mathbb{R}^n, \, \forall \tau \in \mathbb{R}^n$ $\qquad$ **translation invariance**

3) $\quad \mathcal{H}^s(\alpha F) = \alpha^s \, \mathcal{H}^s(F) \qquad \forall F \subset \mathbb{R}^n, \quad \forall \alpha \in \mathbb{R}$ $\qquad$ s **homogeneity of** $\mathcal{H}^s$

4) $\quad \mathcal{H}^s(\emptyset) = 0, \quad \mathcal{H}^0$ is the counting measure, $\quad \mathcal{H}^s$ vanishes identically if $s > n$

5) $\quad$ if $0 \leq s < t$ then $\mathcal{H}^t(F) > 0 \quad \Rightarrow \quad \mathcal{H}^s(F) = +\infty$.

If the sets E_j are pairwise disjoint ($j \neq k \Rightarrow E_j \cap E_k = \emptyset$) Borel sets[2], then the equality holds in 1): **countable additivity** for $\mathcal{H}^s$ restricted to Borel sets.

The **Hausdorff dimension** of a set $F \subset \mathbb{R}^n$ is defined by

$$\dim_{\mathcal{H}}(F) = \inf \{s \geq 0 : \mathcal{H}^s(F) = 0\} = \sup \{s \geq 0 : \mathcal{H}^s(F) = +\infty\}.$$

If $s = \dim_{\mathcal{H}}(F)$ then there are three possibilities:

$$\mathcal{H}^s(F) = 0, \qquad \mathcal{H}^s(F) = +\infty, \qquad 0 < \mathcal{H}^s(F) < +\infty.$$

The Hausdorff dimension of the Cantor set $\mathcal{C}$ is

$$\dim_{\mathcal{H}}(\mathcal{C}) = \frac{\ln 2}{\ln 3} = 0.6309297536\ldots$$

as can be inferred by this heuristic argument:

$$\mathcal{C} = \mathcal{C}_S \cup \mathcal{C}_D$$

where the union is disjoint:

$$\mathcal{C}_S = \mathcal{C} \cap [0, 1/3] \qquad\qquad \mathcal{C}_D = \mathcal{C} \cap [2/3, 1]$$

$$\forall s: \quad \mathcal{H}^s(\mathcal{C}) = \mathcal{H}^s(\mathcal{C}_S) + \mathcal{H}^s(\mathcal{C}_D) = \left(\frac{1}{3}\right)^s \mathcal{H}^s(\mathcal{C}) + \left(\frac{1}{3}\right)^s \mathcal{H}^s(\mathcal{C}) \qquad \text{by 3)}.$$

If we have $0 < \mathcal{H}^s(\mathcal{C}) < +\infty$ at the critical value $s = \dim \mathcal{C}$ (the property is true, but we do not prove it here), then

$$1 = 2\left(\frac{1}{3}\right)^s \quad \Rightarrow \quad s = \frac{\ln 2}{\ln 3}.$$

By $0 < \dim_{\mathcal{H}}(\mathcal{C}) < 1$ we deduce that $\mathcal{H}^1(\mathcal{C}) = 0$ and $\mathcal{H}^0(\mathcal{C}) = +\infty$.

---

[1] Precisely we should call it outer measure, but we skip this technical point.
[2] Countable unions or intersections of open and/or closed sets.

A method for devising sets with non trivial (that is non integer) Hausdorff dimension, and called **fractal sets** for this reason, is given by the procedure described below, which is due to J.E. Hutchinson.

Let $\mathcal{M} = \{M_1, \dots M_N\}$ be a finite set of similarities of $\mathbb{R}^n$ that is to say maps of the kind $M_j(\mathbf{x}) = \mathbf{p}_j + \rho_j \mathbb{M} \mathbf{x}$ where $\mathbb{M}$ is an **orthogonal matrix**[3], $0 < \rho_j < 1$. We assume also that there exists a bounded open subset U of $\mathbb{R}^n$ such that:

$$M_j(U) \subset U \qquad \text{and} \qquad M_i(U) \cap M_j(U) = \emptyset \quad \text{if } i \neq j. \qquad \text{(F.1)}$$

Theorem F.1. *If $\mathcal{M}$ verifies (F.1), then there exists a unique closed and bounded set K such that $K = M_1(K) \cup M_2(K) \cup \cdots \cup M_N(K)$.*
This set K is a self-similar[4] fractal set whose dimension $\dim_{\mathcal{H}}(K)$ is the unique real solution s of the equation

$$\sum_{j=1}^{N} \rho_j^s = 1.$$

For instance, the Cantor set is the only invariant compact set with respect to the family $\mathcal{M}$ made of these two similarities: $M_1(x) = x/3$, $M_2(x) = 1 + (x-1)/3$.

There are many other definitions of fractal dimension. They do not verify all the nice properties of the Hausdorff dimension, though their definition is more elementary and they can be evaluated more easily.

An important example is the **box-counting dimension** that first quantifies the notion of dimension at the scale $\delta > 0$, neglecting smaller scales, then performs a limit as δ tends to 0_+.

Definition F.2. *Assume that $F \subseteq \mathbb{R}^n$ and $\delta > 0$. Let $N_\delta(F)$ be the smallest number of sets with diameter less than or equal to δ that are needed for covering F. Then the box-counting dimension of F, denoted by $\dim_B(F)$, is defined by*

$$\dim_B(F) = \lim_{\delta \to 0^+} \frac{\ln N_\delta(F)}{-\ln \delta} \qquad \text{(F.2)}$$

if this limit exists.

The underlying idea in the definition is as follows: if $N_\delta(F)$ has some power-type homogeneity property with respect to δ, that is $N_\delta(F)\delta^s \to m$, where $m \in (0, +\infty)$ plays the role of the s dimensional measure of F, then $N_\delta \sim m\delta^{-s}$, $\ln N_\delta \sim \ln m - s \ln \delta$; dividing and passing to the limit as $\delta \to 0^+$, necessarily $s = \lim\limits_{\delta \to 0^+} \dfrac{\ln N_\delta(F)}{-\ln \delta}$.
Formula (F.2) is extremely useful for empirical estimates: if one plots in logarithmic scale the values of δ and N_δ obtained by many measurements, then s is the slope with reversed sign of the regression line associated to these values.

---

[3] $\mathbb{Q}$ is orthogonal if and only if itt transforms orthonormal bases in orthonormal bases (equivalently, if and only if the associated matrix in any bases, still denoted by $\mathbb{Q}$, is invertible and verifies $\mathbb{Q}^T = \mathbb{Q}^{-1}$).

[4] Self-similar means that it is the union of parts each one obtained from the set itself by a rigid motion and a scale reduction (it is an idealization, that is suggested by natural geometric structures that are present in ferns or in broccoli).

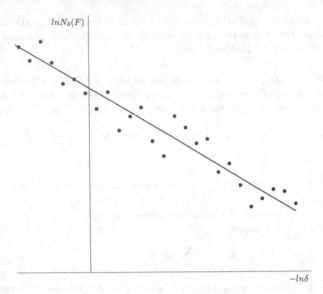

Fig. F.1 Pairs of empirical values $(\delta, N_\delta\,(F))$ represented in logarithmic scale

A first inconvenience is the fact that the limit in (F.2) might not exist (whereas the Hausdorff dimension is always defined). Nevertheless, when $\dim_B(F)$ exists then it is also a real number in $[0, n]$, and achieves the usual values on points, lines, curves and regular surfaces. Moreover $\dim_B$ verifies the natural property of monotonicity: $F \subset E \Rightarrow \dim_B(F) \le \dim_B(E)$. Finally, in the definition of $N_\delta(F)$ it is enough to use only cubes or balls with diameter δ to cover F instead of all sets with diameter less than or equal to δ as required when evaluating $\mathcal{H}_\delta^s$.

Example F.3. If F is a curve in $\mathbb{R}^2$, then $N_\delta(F)$ can be chosen as the least number of "steps" of length less than or equal to δ that are needed to "to sweep the whole curve".

Example F.4. If $F = \mathbb{Q} \cap [0,1]$, then $\dim_\mathcal{H}(F) = 0 \ne 1 = \dim_B(F)$.
This follows by the equality, true for any F whose box-counting dimension $\dim_B(F)$ is defined:

$$\dim_B(F) = \dim_B\left(\overline{F}\right)$$

where $\overline{F}$ denotes the closure of the set F.

The last example (box-counting dimension equal to 1 for a countable union of points, each one with null box-counting dimension) proves that an important property relative to countable unions fails for the box-counting dimension, contrary to the Hausdorff measure:

$$\dim_{\mathcal{H}} \left(\bigcup_{j=1}^{+\infty} F_j \right) = s \qquad \text{if} \quad \dim_{\mathcal{H}} (F_j) = s \ \ \forall j$$

$$\dim_{\mathcal{B}} \left(\bigcup_{j=1}^{+\infty} F_j \right) \geq s \qquad \text{if} \quad \dim_{\mathcal{B}} (F_j) = s \ \ \forall j$$

The strict inequality may hold in the last relation (anyway the equality is guaranteed if there are finitely many F_j).

Due to this and other reasons the less elementary notion of Hausdorff dimension is preferred. However the following result holds.

Theorem F.5. *If K is a Hutchinson self-similar fractal set (that is to say: it is the only invariant compact set associated to a collection of similarities fulfilling (F.1)), then $\dim_{\mathcal{B}} (K)$ exists and*

$$\dim_{\mathcal{H}} (K) = \dim_{\mathcal{B}} (K) = s$$

where s is the unique solution of the equation $\displaystyle\sum_{j=1}^{N} \rho_j^s = 1$ and $0 < \mathcal{H}^s (K) < +\infty$.

Theorems F.1 and F.5 allow to easily compute the dimension of some self-similar fractal sets, as shown by the subsequent examples of sets whose dimension is not an integer number.

Example F.6. The Cantor set $\mathcal{C}$ verifies $\dim_{\mathcal{H}} (\mathcal{C}) = \dim_{\mathcal{B}} (\mathcal{C}) = \ln 2 / \ln 3$.

Example F.7. The **Koch curve** K (obtained from the segment $[0,1]$ by replacing the segment $[1/3, 2/3]$ with the two sides of an equilateral triangle with side length $1/3$ and iterating the procedure) verifies $\dim_{\mathcal{H}} (K) = \dim_{\mathcal{B}} (K) = \ln 4 / \ln 3$.

Example F.8. The triangular **Sierpinski gasket** G (obtained from an equilateral triangle, with its boundary and the interior, by subdividing it into four identical

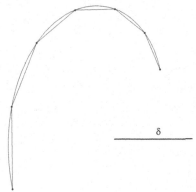

Fig. F.2. Rectification of a curve by steps of length less than or equal to δ

Fig. F.3. Koch snowflake (or Koch island, or Koch star), whose boundary is built as the Koch curve, and the first three iterations of the Sierpinski gasket

equilateral triangles, removing the central one, and iterating the procedure on each triangle left) verifies

$$\dim_{\mathcal{H}}(G) = \dim_{\mathcal{B}}(G) = \ln 3/\ln 2.$$

Example F.9. The t-middle Cantor set $\mathcal{C}_t$, $0 < t < 1$ (see Example 4.38), verifies

$$\dim_{\mathcal{H}}(\mathcal{C}_t) = \dim_{\mathcal{B}}(\mathcal{C}_t) = \frac{\ln 2}{\ln 2 - \ln(1 - t)}.$$

Appendix G

Tables of $\mathcal{Z}$-transforms

Table Z.1

X_k	$\mathcal{Z}\{X\}(z) = x(z) = \displaystyle\sum_{k=0}^{+\infty} \frac{X_k}{z^k}$	
1	$\dfrac{z}{z-1}$	
a^k	$\dfrac{z}{z-a}$	$a>1$
a^k	$\dfrac{z}{z-a}$	$0<a<1$
a^k	$\dfrac{z}{z-a}$	$-1<a<0$
$(-1)^k$	$\dfrac{z}{z+1}$	
k	$\dfrac{z}{(z-1)^2}$	
k^2	$\dfrac{z(z+1)}{(z-1)^3}$	
k^3	$\dfrac{-11z^3+4z^2+z}{(z-1)^4}$	
$\sin(ak)$	$\dfrac{z\sin(a)}{z^2-2\cos(a)z+1}$	
$\cos(ak)$	$\dfrac{z^2-z\cos(a)}{z^2-2\cos(a)z+1}$	
$\sinh(ak)$	$\dfrac{z\sinh(a)}{z^2-2\cosh(a)z+1}$	
$\cosh(ak)$	$\dfrac{z^2-z\cosh(a)}{z^2-2\cosh(a)z+1}$	

E. Salinelli, F. Tomarelli: *Discrete Dynamical Models.*
UNITEXT – La Matematica per il 3+2 76
DOI 10.1007/978-3-319-02291-8_G, © Springer International Publishing Switzerland 2014

Table Z.2.

$X = \{X_k\} \quad k = 0, 1, 2, \ldots$	$\mathcal{Z}\{X\}(z) = x(z) = \displaystyle\sum_{k=0}^{\infty} X_k z^{-k}$	Convergence domain
$\begin{cases} K^0 = \{K_k^0\}_k \quad \text{(pulse at } k=0) \\ \text{where} \\ K_0^0 = 1,\ K_k^0 = 0 \quad \forall k > 0 \end{cases}$	1	$\mathbb{C}$
$\begin{cases} K^N = \{K_k^N\}_k \quad \text{(pulse at } k=N) \\ \text{where} \\ K_N^N = 1,\ K_k^N = 0 \quad k \neq N \end{cases}$	z^{-N}	$z \neq 0$
$U = \{U_k\},\ U_k = 1 \ \forall k$	$\dfrac{z}{z-1}$	$|z| > 1$
k	$\dfrac{z}{(z-1)^2}$	$|z| > 1$
k^2	$\dfrac{z(z+1)}{(z-1)^3}$	$|z| > 1$
$e^{-\alpha k} \qquad \alpha \in \mathbb{C}$	$\dfrac{z}{z - e^{-\alpha}}$	$|z| > e^{-\mathrm{Re}\,\alpha}$
$k e^{-\alpha k} \qquad \alpha \in \mathbb{C}$	$\dfrac{z e^{-\alpha}}{(z - e^{-\alpha})^2}$	$|z| > e^{-\mathrm{Re}\,\alpha}$
$\sin(\omega_0 k) \qquad \omega_0 \in \mathbb{R}$	$\dfrac{z \sin \omega_0}{z^2 - 2z \cos \omega_0 + 1}$	$|z| > 1$
$\cos(\omega_0 k) \qquad \omega_0 \in \mathbb{R}$	$\dfrac{z[z - \cos \omega_0]}{z^2 - 2z \cos \omega_0 + 1}$	$|z| > 1$
$e^{-ak} \sin(\omega_0 k) \qquad a, \omega_0 \in \mathbb{R}$	$\dfrac{z e^{-a} \sin \omega_0}{z^2 - 2z e^{-a} \cos \omega_0 + e^{-2a}}$	$|z| > e^{-a}$
$e^{-ak} \cos(\omega_0 k) \qquad a, \omega_0 \in \mathbb{R}$	$\dfrac{z e^{-a}[z e^{a} - \cos \omega_0]}{z^2 - 2z e^{-a} \cos \omega_0 + e^{-2a}}$	$|z| > e^{-a}$
$\alpha^k \qquad \alpha \in \mathbb{C}$	$\dfrac{z}{z - \alpha}$	$|z| > |\alpha|$
$k \alpha^k \qquad \alpha \in \mathbb{C}$	$\dfrac{\alpha z}{(z - \alpha)^2}$	$|z| > |\alpha|$

Sequence	Transform	ROC
$\sinh(\omega_0 k)$ $\qquad \omega_0 \in \mathbb{R}$	$\dfrac{z\sinh(\omega_0)}{z^2 - 2z\cosh(\omega_0) + 1}$	$|z| > e^{|\omega_0|}$
$\cosh(\omega_0 k)$ $\qquad \omega_0 \in \mathbb{R}$	$\dfrac{z[z - \cosh(\omega_0)]}{z^2 - 2z\cosh(\omega_0) + 1}$	$|z| > e^{|\omega_0|}$
$\begin{cases} X_0 = 0 \\ X_k = \dfrac{1}{k}, \ k \geq 1 \end{cases}$	$\ln\left(\dfrac{1}{z-1}\right)$	$|z| > 1$
$\begin{cases} X_0 = 0 \\ X_k = \dfrac{(-1)^{k+1}}{k}, \ k \geq 1 \end{cases}$	$\ln\left(\dfrac{z+1}{z}\right)$	$|z| > 1$
$\dfrac{\alpha^k}{k!} \qquad \alpha \in \mathbb{C}$	$e^{\alpha/z}$	$|z| > 0$
$X_0 = 0, \ X_k = \alpha^{k-1}, \ k \geq 1 \quad \alpha \in \mathbb{C}, \ X = \tau_1\{\alpha^k\}$	$\dfrac{1}{z - \alpha}$	$|z| > |\alpha|$
$X_0 = X_1 = 0, \ X_k = (k-1)\alpha^{k-2} \quad k \geq 2, \ \alpha \in \mathbb{C}$	$\dfrac{1}{(z-\alpha)^2}$	$|z| > |\alpha|$
$X_0 = 0, \ X_k = \dfrac{\alpha^{k-1} - \beta^{k-1}}{\alpha - \beta} \quad k \geq 1, \ \alpha, \beta \in \mathbb{C}, \ \alpha \neq \beta$	$\dfrac{1}{(z-\alpha)(z-\beta)} = \dfrac{1}{\alpha-\beta}\left(\dfrac{1}{z-\alpha} - \dfrac{1}{z-\beta}\right)$	$|z| > max(|\alpha|, |\beta|)$
$X_0 = 0, \ X_k = \dfrac{\alpha^{k-1}}{(\alpha-\beta)(\alpha-\gamma)} + \dfrac{\beta^{k-1}}{(\beta-\alpha)(\beta-\gamma)} + \dfrac{\gamma^{k-1}}{(\gamma-\alpha)(\gamma-\beta)}$ $\alpha, \beta, \gamma \in \mathbb{C}, \ \alpha \neq \beta \neq \gamma$	$\dfrac{1}{(z-\alpha)(z-\beta)(z-\gamma)} = \dfrac{1}{(\alpha-\beta)(\alpha-\gamma)}\dfrac{1}{z-\alpha} + \dfrac{1}{(\beta-\alpha)(\beta-\gamma)}\dfrac{1}{z-\beta} + \dfrac{1}{(\gamma-\alpha)(\gamma-\beta)}\dfrac{1}{z-\gamma}$	$|z| > max(|\alpha|, |\beta|, |\gamma|)$
$X_k = \displaystyle\sum_{h=0}^{k-3}(h+1)\,\alpha^h\,\beta^{k-h-3} \quad \alpha, \beta \in \mathbb{C}, \ \alpha \neq \beta$	$\dfrac{1}{(z-\alpha)^2(z-\beta)}$	$|z| > max(|\alpha|, |\beta|)$
$X_k = \dfrac{1}{n!}\,k(k+1)(k+2)\ldots(k+n-1), \quad n \geq 1$	$\dfrac{z^n}{(1-z)^{n+1}}$	$|z| > 1$

Table Z.3.

$Y = \{Y_k\}$ $k = 0, 1, 2, \ldots$	$\mathcal{Z}\{Y\}(z) = y(z)$
$X = \{X_k\}$ $k = 0, 1, 2, \ldots$	$\mathcal{Z}\{X\}(z) = x(z)$
$Y_k := X_{k+1} = \tau_{-1} X$	$y(z) = z\,x(z) - z\,X_0$
$Y_k := X_{k+2} = \tau_{-2} X$	$y(z) = z^2 x(z) - z^2 X_0 - z\,X_1$
$Y_k := X_{k+3} = \tau_{-3} X$	$y(z) = z^3 x(z) - z^3 X_0 - z^2 X_1 - z\,X_2$
$Y_k := X_{k+n} = \tau_{-n} X$ $n \geq 0$	$y(z) = z^n x(z) - z^n X_0 - \ldots - z\,X_{n-1}$
$\begin{cases} Y_k := X_{k-n} = \tau_n X, \\ Y_k = 0 \quad k < n, \ Y_k = X_{k-n} \quad k \geq n \geq 0 \end{cases}$	$y(z) = z^{-n} x(z)$
$\begin{cases} X_k = U_{k-n} \\ X_0 = X_1 = \ldots = X_{n-1} = 0, \quad X_k = 1 \text{ se } k \geq m \end{cases}$	$\dfrac{z^{n-1}}{z-1}$
$k\,X_k$	$-z\,x'(z)$
$k^n X_k, \quad n \geq 0$	$\left(-z\dfrac{d}{dz}\right)^n x(z)$
$\dfrac{1}{k} X_k$	$-\displaystyle\int_\infty^z \dfrac{x(w)}{w}\,dw$
$\dfrac{1}{k+n} X_k \quad n \geq 0$	$-z^n \displaystyle\int_\infty^z \dfrac{x(w)}{w^{n+1}}\,dw$
$a^k X_k$	$x(z/a)$
$\displaystyle\sum_{h=0}^{k} X_h$	$\dfrac{z\,x(z)}{z-1}$
$(X * Y)_k = \displaystyle\sum_{h=0}^{k} X_h Y_{k-h}$	$x(z)\,y(z)$

Table Z.4.

Initial Value Theorem: $\displaystyle\lim_{|z|\to\infty} x(z) = X_0$.

Final Value Theorem:

if $\displaystyle\lim_{k\to\infty} X_k = X_\infty$ exists and X_∞ is finite, then there exists

$$\lim_{z\to 1,\, |z|>1} \frac{z-1}{z} x(z) = X_\infty .$$

Inverse Z transform:

$$X_k = \frac{1}{2\pi i} \oint_{C_R} z^{k-1} x(z)\,dz , \qquad \forall k \geq 0,$$

where C_R is a circle centered at 0 with radius R, swept once counterclockwise and such that $z \mapsto x(z)$ is an analytic function on $\{z \in C : |z| > R/2\}$.

Appendix H

Algorithms and hints for numerical experiments

Many numerical computations and plots in the text were made using the software Mathematica (Wolfram Research Inc.). We provide some hints useful in the study of discrete dynamical systems with this software.

Iterated function evaluation

An easy way to evaluate iterated functions consists in exploiting the command **Nest**, whose sintax is:

$$\text{Nest} [\; function, \quad initial\ value, \quad number\ of\ iterations \;]$$

For instance, a program for computing the four initial iteration of the cosine function starting from $X_0 = \pi$ is as follows:

$$\begin{aligned} &\text{Clear} [f, x] \\ &\quad f [x] = \text{Cos} [x] ; \\ &\quad\quad \text{Nest} [f, \text{Pi}, 4] \\ &\quad\quad\quad \text{N} [\%, 3] \end{aligned}$$

The first command is strongly recommended to avoid confusion with possible values of f and x already present in the memory because they were computed in a previous evaluation; the second command defines the function cosine dependent on the variable x (the semicolon prevents from printing any output); the third command defines the iteration $\cos(\cos(\cos(\cos(\pi)))) = \cos^4(\pi)$; the fourth command compute the numerical value of the symbolic value in the previous command, with three decimal digits, that is $0,654$.

If also the numerical values of the previous iterations are wanted, then we have to substitute the third command with

$$\text{NestList} [f, \; \text{Pi}, \; 4]$$

that produces $\{\pi, -1, \cos 1, \cos(\cos 1), \cos(\cos(\cos 1))\}$ that is $\{X_0, X_1, X_2, X_3, X_4\}$; as a consequence the fourth command evaluates the numerical values $X_0, dots, X_4$

E. Salinelli, F. Tomarelli: *Discrete Dynamical Models*.
UNITEXT – La Matematica per il 3+2 76
DOI 10.1007/978-3-319-02291-8_H, © Springer International Publishing Switzerland 2014

with three decimal digits, that is

$$\{3, 14 , -1 , 0, 54 , 0, 858 , 0, 654\}.$$

Equilibria of discrete dynamical systems

The DDS $\{\mathbb{R}, \cos\}$ has a unique equilibrium α that is stable, attractive and is the solution of $\{\alpha \in \mathbb{R} : \alpha = \cos\alpha\}$. This equilibrium can be evaluated in several ways:

1) *Newton's method*: the function $g(x) = \cos x - x$ is continuous and concave in $[0, \pi/2]$, positive at 0 and negative at $\pi/2$. Then, starting from $X_0 = \pi/2$, the iterations of $N_g(x) = (x \sin x + \cos x) / (\sin x + 1)$ are decreasing and converging to α. The commands for evaluating the fifth and sixth iteration of N_g are:

$$\text{Clear} [Ng , x]$$
$$Ng [x\_] = (x * \text{Sin} [x] + \text{Cos} [x]) / (\text{Sin} [x] + 1);$$
$$N [\text{Nest} [Ng , \text{Pi}/2 , 5] , 3]$$
$$N [\text{Nest} [Ng , \text{Pi}/2 , 6] , 3]$$

Output: 0.739 , 0.739.

The suffix in $x\_$ requests to perform the evaluation of the function N_g whenever it is possible: the request is implicit for elementary functions as was the case of the cosine function. Instead the command

$$Ng [x] = (x * \text{Sin} [x] + \text{Cos} [x]) / (\text{Sin} [x] + 1)$$

would have produced $N_g [N_g [N_g [N_g [N_g [1, 57]]]]]$.

2) *Iterations of g*: $g^k (X_0)$ converges (by oscillating) to α, $\forall X_0 \in \mathbb{R}$. We choose $X_0 = \pi/2$ and compute the 10th and 20th iteration:

$$\text{Clear} [f , x]$$
$$f [x] = \text{Cos}[x]$$
$$N [\text{Nest} [f , \text{Pi}/2 , 10] , 3]$$
$$N [\text{Nest} [f , \text{Pi}/2 , 20] , 3]$$

Output: 0.75   0.739

3) Using the *equation solver* "FindRoot":

$$\text{FindRoot} [\{\text{Cos} [x] == x\} , \{x , 0\}]$$

Output: 0.739085.

Fibonacci numbers

I) The Fibonacci numbers are defined by the initialization $F_0 = 0$, $F_1 = 1$ and the (two-step) recursive law $F_{k+1} = F_k + F_{k-1}$. Therefore the list of commands

$$\text{Clear } [f, x]$$
$$f[0] = 0;$$
$$f[1] = 1;$$
$$f[x\_] := f[x-1] + f[x-2]$$

generates the function $f[k] = F_k$. Even if these commands lead to the evaluation of several Fibonacci numbers, for instance $F_{20} = 6765$, already F_{50} poses hard problems in terms of time computation, since the number of operations that are required is very high: it is a poorly efficient algorithm.

II) Alternatively, we can exploit the explicit formula for F_k as the nearest integer to $\left((1+\sqrt{5})/2\right)^k /\sqrt{5}$, obtaining the approximate value of F_k by

$$N[0.5 * ((1 + \text{Sqrt} [5]) * 0.5)\,\char"5E 50] \ .$$

The function N performs an approximated numerical computation. The approximation could be improved. Nevertheless it is better to use an algebraically exact procedure with some expedient to avoid the difficulties met by the first algorithm.

III) An efficient algorithm to evaluate F_k based on the recursive definition must take into account the previous values $(F_h,\ h < k)$ of the sequence, computed before, and tabulate them in order to avoid the fast growth of the number of operations needed (if one repeats the computations already performed):

$$\text{Clear } [f, x]$$
$$f[0] = 0;$$
$$f[1] = 1;$$
$$f[x\_] := f[x] = f[x-1] + f[x-2]$$
$$f[50]$$

The penultimate command defines a function f that "is reminiscent" of all the values that are computed. The last one swiftly computes the exact value of the 50th Fibonacci's number. The effectiveness of the method is such that it can quickly provide computations like the one of F_{100}:

$$F_{50} = 12\,586\,269\,025$$
$$F_{100} = 354\,224\,848\,179\,261\,915\,075$$

IV) An even faster procedure consists in using a function that is inbuilt in the software Mathematica 4.1 and produces the k-th Fibonacci number:

$$\text{Fibonacci } [k]$$

Cobwebs of discrete dynamical systems

In the sequel we list the commands of a program (a Notebook in the software Mathematica) similar to the ones used to plot cobwebs of figures in the text. In the example we examine the logistic map h_a with parameter $a = 3.5$ and we evaluate 20 iterations starting from the initial datum $X_0 = 0.1$.

To adapt to different maps (that transform the interval $[0, 1]$ in itself), initial data and/or numbers of iterations, it is sufficient to perform the obvious changes in the rows 3, 5, 6 and 7.

We separate blocks of commands using comments (* comment *) that do not affect the syntax.

(* **Choice of the dynamical system and start** *)

```
Clear [h, x] ;
  h [x_] := 3.5 * x * (1 - x)
     id [x_] := x
        StartingValue = .1 ;
           FirstIt = 0 ;
              LastIt = 20 ;
xmin = 0;
xmax = 1;
```

(* **Evaluation of iterated map** *)

```
k = 0;
  y = N [StartingValue] ;
     DataTable = {{y, 0}, {y, h [y]}, {h [y], h [y]}};

While[k < LastIt, y = h [y] ;
           AppendTo[DataTable, {y, y}] ;
              AppendTo[DataTable, {y, h [y]}] ;
                 k = k + 1] ;
AppendTo[DataTable, {h [y], h [y]}] ;
```

(* **Plot the cobweb** *)

```
Cobweb = ListPlot [DataTable, PlotJoined - > True,
          PlotRange - > {{xmin, xmax}, {xmin, xmax}},
             Ticks - > None, AspectRatio - > 1 ,
                PlotStyle - > RGBColor[1, 0, 0],
                   DisplayFunction - > Identity] ;
```

(* **Plot the graphs of h_a (logistic map) and of the identity** *)

```
Graphh = Plot[h [x], {x, xmin, xmax},
       PlotRange - > {xmin, xmax}, AspectRatio - > 1 , Ticks - > None,
          PlotStyle - > {{Thickness [0.015] , RGBColor [0, 0, 1]}},
             DisplayFunction - > Identity] ;

Graphid = Plot[id [x], {x, xmin, xmax},
        PlotRange - > {xmin, xmax}, AspectRatio - > 1 , Ticks - > None,
           DisplayFunction - > Identity] ;
```

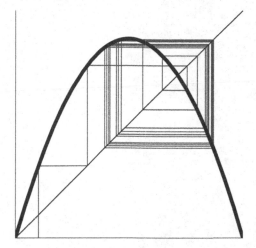

Fig. H.1 A cobweb of the logistic map h_a with $a = 3.5$

(* **Show the outcome** *)

Show[Cobweb, Graphh, Graphid, DisplayFunction $->$ $DisplayFunction].

References

1. Alseda, L., Llibre, J., Misiurewicz, M.: Combinatorial Dynamics and Entropy in Dimension One. Advanced Series in Nonlinear Dynamics, Vol. 5. World Scientific, New York (2001)
2. Block, L.S., Coppel, W.A.: Dynamics in one dimension. Lecture Notes in Mathematics, Vol. 1513. Springer-Verlag, Berlin Heidelberg (1992)
3. Boyarski, A.: Laws of Chaos. Invariant measures and dinamical systems in one dimension. Birkäuser, Boston (1997)
4. Bianchini, M., Gori, M., Scarselli, F.: Inside Page-Rank. ACM Transactions on Internet Technology, Vol. 5(1), 92–128 (2005)
5. Brin, S., Page, L.: The anatomy of a large-scale hypertextual web searching engine. Computer networks and ISDN Systems 33, 107–117 (1998)
6. Devaney, R.L.: An Introduction to Chaotic Dynamical Systems. Westview Press, Boulder, CO (2003)
7. Ekeland, I.: The Broken Dice, and Other Mathematical Tales of Chance. University of Chicago Press. Chicago, IL (1993)
8. Gantmacher, F.R.: The Theory of matrices, Vol. 1–2. American Mathematical Society, Providence RI (2001)
9. Gazzola, F., Tomarelli, F., Zanotti, M.: Analytic Functions, Integral Transforms, Differential Equations, Esculapio, Bologna (2013)
10. Graham, R.L., Knuth, D., Patashnik, O.: Concrete Mathematics. 2nd ed. Addison-Wesley. Reading, MA (1994)
11. Haveliwala, T.H., Kamvar, S.D.: The second eigenvalue of the Google matrix, technical report. Stanford University (2003)
12. Hirsch, M.W., Smale, S., Devaney, R.L.: Differential Equations. Dynamical Systems and an Introduction to Chaos. Elsevier Academic Press, New York (2004)
13. Jain, A.: Fundamentals of Digital Image Processing. Prentice-Hall, Englewood Cliff (1989)
14. Leslie, P.H.: On the use of matrices in certain population mathematics. Biometrika **33**(3), 183–212 (1945)
15. Oldham, K., Myland, J., Spanier, J.: An Atlas of functions. 2nd ed. Springer-Verlag, New York (2000)
16. Peitgen, H.O., Jürgens, H., Saupe, D.: Fractals for the Classroom: Strategic Activities, Vol. III. Springer-Verlag, New York (1999)

E. Salinelli, F. Tomarelli: *Discrete Dynamical Models.*
UNITEXT – La Matematica per il 3+2 76
DOI 10.1007/978-3-319-02291-8, © Springer International Publishing Switzerland 2014

17. Quarteroni, A., Sacco, R., Saleri, F.: Numerical Mathematics. 3th ed. Texts in Applied Mathematics, Vol. 37. Springer-Verlag, Berlin Heidelberg (2008)
18. Ruelle, D.: Chance and Chaos. Princeton University Press, Princeton (1991)
19. Seneta, E.: Non-Negative Matrices and Markov chains. Springer Series in Statistics. Springer-Verlag, New York (2006)
20. Sirovich, L.: Introduction to Applied Mathematics. Springer-Verlag, New York (1988)
21. Wolfram, S.: The Mathematica Book. 4th ed., Wolfram Media. Cambridge University Press, Champaign (1999)

Index

k-th iterate, 86

algorithm
PageRank $-$, 278
allele, 181
dominant $-$, 182
recessive $-$, 182
amortization, 20
attraction basin, 102
attractor, 102

Bayes' law, 351
bifurcation
period-doubling $-$, 137
pitchfork $-$, 137
saddle-node $-$, 137
transcritical $-$, 137
bifurcation diagram, 134
bifurcation value, 135
bond
fixed-reverse, 5
reverse-floater, 82
zero-coupon, 21
boundary, 364
box-counting dimension, 367

Casorati determinant, 64
Cauchy-Schwarz inequality, 353
causal filter, 78
causal system, 78
characteristic equation, 33
characteristic polynomial, 33
chromosomal inheritance, 181
circular convolution, 80

closure, 364
cobweb, 94
Collatz-Wielandt formula, 204
communicating class, 274
closed $-$, 274
computational complexity, 75
continued fraction, 71
contraction mapping, 99
convolution
circular $-$, 80
discrete $-$, 54
Cramer's rule, 357
cycle, 89, 267
locally asymptotically stable $-$, 114
repulsive $-$, 115

dangling page, 281
DDS
scalar-valued discrete dynamical system, 85
de Moivre's formulas, 348
DFT, 74
difference equation
linear $-$ with constant coefficients
of order n, homogeneous, 32
of order n, non-homogeneous, 39, 47
of the first order, homogeneous, 25
of the first order, non-homogeneous, 26
linear $-$ with non-constant coefficients
of order n, homogeneous, 64
of the first order, homogeneous, 61

E. Salinelli, F. Tomarelli: *Discrete Dynamical Models.*
UNITEXT – La Matematica per il 3+2 76
DOI 10.1007/978-3-319-02291-8, © Springer International Publishing Switzerland 2014

of the first order, non-homogeneous,
  61
of order n, 2
  in normal form, 2
discrete convolution, 54, 344
discrete dynamical system
  − of order n, autonomous, 175
  affine − , 87
  chaotic − , 148
  complex − , 164
  linear − , 87
  nonlinear − , 87
  of the first order, autonomous, 86
  vector-valued − , 212
  vector-valued linear −
    homogeneous, 191
  vector-valued linear positive − , 199
  vector-valued non-homogeneous
    linear − , 208
discrete Fourier transform, 74
discretization of a differential equation,
  13, 14

eigenvalue, 180
  dominant − , 199
eigenvector, 180
  dominant − , 199
equation
  − of three moments, 17
  Airy − , 12
  Bessel − , 22
  Black-Scholes − , 22
  d'Alembert − , 22
  heat − , 14
  Hermite − , 22
  Laguerre − , 22
  Laplace − , 22
  Legendre − , 22
  Riccati − , 301
equilibrium, 40, 88
  asimptotically stable − , 101
  asymptotically stable −
    from above, 106
  attractive, 41
    from above − , 106
  attractive − of a linear vector-valued
    DDS, 191
  attractive − , 40
  globallly attractive − , 101
  hyperbolic − , 106

locally attractive − , 101, 164
neutral − , 106
of a linear equation, 26
of a linear vector-valued DDS, 191
repulsive − , 104
repulsive − from above, 106
semistable
  from above − , 106
stable − , 40, 41, 100, 164
superattractive − , 106
unstable − , 101, 164
unstable − from above, 106

Fast Fourier Transform algorithm, 76
Feigenbaum constant, 143
FFT, 76
Fibonacci numbers, 9, 377
flops, 77
formula
  Collatz-Wielandt − , 204
Fourier coefficients, 72
Fourier series, 72
fractal set, 151, 367
function
  continuous − , 363
  linear rational, 67
  of transfer, 54
  topologically transitive − , 148

gambler's ruin, 12, 81, 227
general solution
  of a difference equation, 3
genetics, 181
Google, 278
graph, 266
  directed − , 266
  strongly connected − , 267

half-life, 6
Hardy-Weinberg law, 182
Hausdorff dimension, 366
Hausdorff measure, 365
heterozygous, 181
homeomorphism, 104
homozygous, 181
Hutchinson self-similar fractal, 367

induction principle, 2
initial value, 87

interest
  compound − , 4
  simple − , 4
interior, 364

Jordan canonical form, 360
Jordan normal form, 360

Koch curve, 369

leverage effect, 307
logistic growth
  − with continuous time, 160
  − with discrete time, 126
logistic map, 126
loop, 267

Möbius transformation, 67
map
  logistic − , 126
  tent − , 146
mapping
  contraction − , 99
Markov chain, 231
  absorbing − , 238
  irreducible − , 275
  reducible − , 275
  regular − , 237
matrix
  adjacency − , 268
  basis change − , 358
  Casorati − , 64
  cyclic − , 264
  elementary Frobenius form of a − , 190
  irreducible − , 259
  Jordan canonical form of a − , 360
  nilpotent − , 196, 361
  period of an irreducible − , 271
  permutation − , 254, 257
  positive − , 199
  primitive − , 264
  reducible − , 259
  rotation − , 195
  semi-simple − , 355
  sparse − , 215
  strictly positive − , 199
  transition − , 232
  weakly positive − , 199
mean lifetime, 6

method
  backward Euler − , 14, 15, 160, 218, 331
  Euler − , 14, 16, 160, 217, 330
  Gauss-Seidel − , 222
  Jacobi − , 222
  LU decomposition − , 220
  Newton's − , 160, 165
  Ruffini-Horner − , 75
  SOR − , 221
minimal polynomial, 262
model
  cobweb − , 5
  Leslie − , 11
  logistic growth − , 8
  Lotka-Volterra − , 9, 213
  Malthus − , 8
  of radioactive decay, 6
mortgage, 5
moving average, 56
multiplier
  − of the equilibrium, 177
  − of the orbit, 177

Newton's method, 160, 165
node, 266
  aperiodic − , 271
  connected − , 267
  period of a − , 270
  strongly connected − , 267
norm
  of a matrix, 213
notations, xv

orbit, 87
  periodic − of a linear vector-valued DDS, 191
  asimptotycally stable − , 103
  eventually periodic − , 89
  eventually stationary − , 89
  period of a periodic − , 89
  periodic − , 89
  stable − , 103
  stationary − , 88
orthonormal bases, 354

partial sums, 63
path, 267
period of a node, 270

period-doubling, 137
 − cascade, 139
phase plane, 196
phase portrait, 87
phase-doubling, 146
present value, 82
principle
  discrete maximum − , 291
  induction − , 2
  mutation − , 185
  selection − , 183
probability distribution
  initial − , 234
  invariant − , 235
pulse signal, 51

query, 278

recursive relationship, 2
repeller, 104
rolling mean, 56
Ruffini-Horner method, 75

Schwarzian derivative, 133
sequence, 1
  periodic − , 71
set
  t-middle Cantor − , 155
  Cantor − , 151
  Cantor-like − , 154
  closed − , 364
  connected open − , 364
  dense − , 364
  fractal − , 151, 367
  interior of a − , 364
  invariant − , 102
  Julia − , 175
  locally attractive − , 102
  open − , 363
  repulsive − , 104
  totally disconnected − , 152
Sharkovskii's ordering, 129
Sierpinski gasket, 156, 369
signal
  Heaviside − , 50
  Kronecker − , 51

linear − , 51
simplex, 234
solution
  of a difference equation, 3
state
  absorbing − , 238
  accessible − , 274
  aperiodic − , 276
  recurrent − , 276
  transient − , 276
states
  communicating − , 274
strongly connected component, 268
  period of a − , 271
sum
  of a geometric progression, 49
  of an arithmetic progression, 48
symbols
  list of − , xv
system
  ill-conditioned, 223

Theorem
  Cayley–Hamilton − , 359
  contraction mapping − , 99
  existence and uniqueness −
  for a DDS in normal form, 3
  Fatou − , 141
  Frobenius − , 263
  Fundamental − of Algebra, 349
  Li–Yorke − , 155
  Markov–Kakutani − , 250
  Perron–Frobenius − , 199
  Rouché–Capelli − , 357
  Schur − , 45
  Sharkovskii − , 130
  spectral − , 359
topological conjugacy, 157
topology, 363
  Euclidean − , 363
transition probability, 231
trapezoidal rule, 72

Vandermonde determinant, 36

Zeta transform, 50

Collana Unitext – La Matematica per il 3+2

Series Editors:
A. Quarteroni (Editor-in-Chief)
L. Ambrosio
P. Biscari
C. Ciliberto
G. van der Geer
G. Rinaldi
W.J. Runggaldier

Editor at Springer:
F. Bonadei
francesca.bonadei@springer.com

As of 2004, the books published in the series have been given a volume number. Titles in grey indicate editions out of print.
As of 2011, the series also publishes books in English.

A. Bernasconi, B. Codenotti
Introduzione alla complessità computazionale
1998, X+260 pp, ISBN 88-470-0020-3

A. Bernasconi, B. Codenotti, G. Resta
Metodi matematici in complessità computazionale
1999, X+364 pp, ISBN 88-470-0060-2

E. Salinelli, F. Tomarelli
Modelli dinamici discreti
2002, XII+354 pp, ISBN 88-470-0187-0

S. Bosch
Algebra
2003, VIII+380 pp, ISBN 88-470-0221-4

S. Graffi, M. Degli Esposti
Fisica matematica discreta
2003, X+248 pp, ISBN 88-470-0212-5

S. Margarita, E. Salinelli
MultiMath – Matematica Multimediale per l'Università
2004, XX+270 pp, ISBN 88-470-0228-1

A. Quarteroni, R. Sacco, F.Saleri
Matematica numerica (2a Ed.)
2000, XIV+448 pp, ISBN 88-470-0077-7
2002, 2004 ristampa riveduta e corretta
(1a edizione 1998, ISBN 88-470-0010-6)

13. A. Quarteroni, F. Saleri
    Introduzione al Calcolo Scientifico (2a Ed.)
    2004, X+262 pp, ISBN 88-470-0256-7
    (1a edizione 2002, ISBN 88-470-0149-8)

14. S. Salsa
    Equazioni a derivate parziali - Metodi, modelli e applicazioni
    2004, XII+426 pp, ISBN 88-470-0259-1

15. G. Riccardi
    Calcolo differenziale ed integrale
    2004, XII+314 pp, ISBN 88-470-0285-0

16. M. Impedovo
    Matematica generale con il calcolatore
    2005, X+526 pp, ISBN 88-470-0258-3

17. L. Formaggia, F. Saleri, A. Veneziani
    Applicazioni ed esercizi di modellistica numerica
    per problemi differenziali
    2005, VIII+396 pp, ISBN 88-470-0257-5

18. S. Salsa, G. Verzini
    Equazioni a derivate parziali – Complementi ed esercizi
    2005, VIII+406 pp, ISBN 88-470-0260-5
    2007, ristampa con modifiche

19. C. Canuto, A. Tabacco
    Analisi Matematica I (2a Ed.)
    2005, XII+448 pp, ISBN 88-470-0337-7
    (1a edizione, 2003, XII+376 pp, ISBN 88-470-0220-6)

20. F. Biagini, M. Campanino
    Elementi di Probabilità e Statistica
    2006, XII+236 pp, ISBN 88-470-0330-X

21. S. Leonesi, C. Toffalori
    Numeri e Crittografia
    2006, VIII+178 pp, ISBN 88-470-0331-8

22. A. Quarteroni, F. Saleri
    Introduzione al Calcolo Scientifico (3a Ed.)
    2006, X+306 pp, ISBN 88-470-0480-2

23. S. Leonesi, C. Toffalori
    Un invito all'Algebra
    2006, XVII+432 pp, ISBN 88-470-0313-X

24. W.M. Baldoni, C. Ciliberto, G.M. Piacentini Cattaneo
    Aritmetica, Crittografia e Codici
    2006, XVI+518 pp, ISBN 88-470-0455-1

25. A. Quarteroni
    Modellistica numerica per problemi differenziali (3a Ed.)
    2006, XIV+452 pp, ISBN 88-470-0493-4
    (1a edizione 2000, ISBN 88-470-0108-0)
    (2a edizione 2003, ISBN 88-470-0203-6)

26. M. Abate, F. Tovena
    Curve e superfici
    2006, XIV+394 pp, ISBN 88-470-0535-3

27. L. Giuzzi
    Codici correttori
    2006, XVI+402 pp, ISBN 88-470-0539-6

28. L. Robbiano
    Algebra lineare
    2007, XVI+210 pp, ISBN 88-470-0446-2

29. E. Rosazza Gianin, C. Sgarra
    Esercizi di finanza matematica
    2007, X+184 pp,ISBN 978-88-470-0610-2

30. A. Machì
Gruppi – Una introduzione a idee e metodi della Teoria dei Gruppi
2007, XII+350 pp, ISBN 978-88-470-0622-5
2010, ristampa con modifiche

31 Y. Biollay, A. Chaabouni, J. Stubbe
Matematica si parte!
A cura di A. Quarteroni
2007, XII+196 pp, ISBN 978-88-470-0675-1

32. M. Manetti
Topologia
2008, XII+298 pp, ISBN 978-88-470-0756-7

33. A. Pascucci
Calcolo stocastico per la finanza
2008, XVI+518 pp, ISBN 978-88-470-0600-3

34. A. Quarteroni, R. Sacco, F. Saleri
Matematica numerica (3a Ed.)
2008, XVI+510 pp, ISBN 978-88-470-0782-6

35. P. Cannarsa, T. D'Aprile
Introduzione alla teoria della misura e all'analisi funzionale
2008, XII+268 pp, ISBN 978-88-470-0701-7

36. A. Quarteroni, F. Saleri
Calcolo scientifico (4a Ed.)
2008, XIV+358 pp, ISBN 978-88-470-0837-3

37. C. Canuto, A. Tabacco
Analisi Matematica I (3a Ed.)
2008, XIV+452 pp, ISBN 978-88-470-0871-3

38. S. Gabelli
Teoria delle Equazioni e Teoria di Galois
2008, XVI+410 pp, ISBN 978-88-470-0618-8

39. A. Quarteroni
Modellistica numerica per problemi differenziali (4a Ed.)
2008, XVI+560 pp, ISBN 978-88-470-0841-0

40. C. Canuto, A. Tabacco
Analisi Matematica II
2008, XVI+536 pp, ISBN 978-88-470-0873-1
2010, ristampa con modifiche

41. E. Salinelli, F. Tomarelli
Modelli Dinamici Discreti (2a Ed.)
2009, XIV+382 pp, ISBN 978-88-470-1075-8

42. S. Salsa, F.M.G. Vegni, A. Zaretti, P. Zunino
Invito alle equazioni a derivate parziali
2009, XIV+440 pp, ISBN 978-88-470-1179-3

43. S. Dulli, S. Furini, E. Peron
Data mining
2009, XIV+178 pp, ISBN 978-88-470-1162-5

44. A. Pascucci, W.J. Runggaldier
Finanza Matematica
2009, X+264 pp, ISBN 978-88-470-1441-1

45. S. Salsa
Equazioni a derivate parziali – Metodi, modelli e applicazioni (2a Ed.)
2010, XVI+614 pp, ISBN 978 88-470-1645-3

46. C. D'Angelo, A. Quarteroni
Matematica Numerica – Esercizi, Laboratori e Progetti
2010, VIII+374 pp, ISBN 978-88-470-1639-2

47. V. Moretti
Teoria Spettrale e Meccanica Quantistica – Operatori in spazi di Hilbert
2010, XVI+704 pp, ISBN 978-88-470-1610-1

48. C. Parenti, A. Parmeggiani
Algebra lineare ed equazioni differenziali ordinarie
2010, VIII+208 pp, ISBN 978-88-470-1787-0

49. B. Korte, J. Vygen
Ottimizzazione Combinatoria. Teoria e Algoritmi
2010, XVI+662 pp, ISBN 978-88-470-1522-7

50. D. Mundici
Logica: Metodo Breve
2011, XII+126 pp, ISBN 978-88-470-1883-9

51. E. Fortuna, R. Frigerio, R. Pardini
Geometria proiettiva. Problemi risolti e richiami di teoria
2011, VIII+274 pp, ISBN 978-88-470-1746-7

52. C. Presilla
    Elementi di Analisi Complessa. Funzioni di una variabile
    2011, XII+324 pp, ISBN 978-88-470-1829-7

53. L. Grippo, M. Sciandrone
    Metodi di ottimizzazione non vincolata
    2011, XIV+614 pp, ISBN 978-88-470-1793-1

54. M. Abate, F. Tovena
    Geometria Differenziale
    2011, XIV+466 pp, ISBN 978-88-470-1919-5

55. M. Abate, F. Tovena
    Curves and Surfaces
    2011, XIV+390 pp, ISBN 978-88-470-1940-9

56. A. Ambrosetti
    Appunti sulle equazioni differenziali ordinarie
    2011, X+114 pp, ISBN 978-88-470-2393-2

57. L. Formaggia, F. Saleri, A. Veneziani
    Solving Numerical PDEs: Problems, Applications, Exercises
    2011, X+434 pp, ISBN 978-88-470-2411-3

58. A. Machì
    Groups. An Introduction to Ideas and Methods of the Theory of Groups
    2011, XIV+372 pp, ISBN 978-88-470-2420-5

59. A. Pascucci, W.J. Runggaldier
    Financial Mathematics. Theory and Problems for Multi-period Models
    2011, X+288 pp, ISBN 978-88-470-2537-0

60. D. Mundici
    Logic: a Brief Course
    2012, XII+124 pp, ISBN 978-88-470-2360-4

61. A. Machì
    Algebra for Symbolic Computation
    2012, VIII+174 pp, ISBN 978-88-470-2396-3

62. A. Quarteroni, F. Saleri, P. Gervasio
    Calcolo Scientifico (5a ed.)
    2012, XVIII+450 pp, ISBN 978-88-470-2744-2

63. A. Quarteroni
Modellistica Numerica per Problemi Differenziali (5a ed.)
2012, XVIII+628 pp, ISBN 978-88-470-2747-3

64. V. Moretti
Spectral Theory and Quantum Mechanics
With an Introduction to the Algebraic Formulation
2013, XVI+728 pp, ISBN 978-88-470-2834-0

65. S. Salsa, F.M.G. Vegni, A. Zaretti, P. Zunino
A Primer on PDEs. Models, Methods, Simulations
2013, XIV+482 pp, ISBN 978-88-470-2861-6

66. V.I. Arnold
Real Algebraic Geometry
2013, X+110 pp, ISBN 978-3-642–36242-2

67. F. Caravenna, P. Dai Pra
Probabilità. Un'introduzione attraverso modelli e applicazioni
2013, X+396 pp, ISBN 978-88-470-2594-3

68. A. de Luca, F. D'Alessandro
Teoria degli Automi Finiti
2013, XII+316 pp, ISBN 978-88-470-5473-8

69. P. Biscari, T. Ruggeri, G. Saccomandi, M. Vianello
Meccanica Razionale
2013, XII+352 pp, ISBN 978-88-470-5696-3

70. E. Rosazza Gianin, C. Sgarra
Mathematical Finance: Theory Review and Exercises. From Binomial
Model to Risk Measures
2013, X+278pp, ISBN 978-3-319-01356-5

71. E. Salinelli, F. Tomarelli
Modelli Dinamici Discreti (3a Ed.)
2014, XVI+394pp, ISBN 978-88-470-5503-2

72. C. Presilla
Elementi di Analisi Complessa. Funzioni di una variabile (2a Ed.)
2014, XII+360pp, ISBN 978-88-470-5500-1

73. S. Ahmad, A. Ambrosetti
A Textbook on Ordinary Differential Equations
2014, XIV+324pp, ISBN 978-3-319-02128-7

74. A. Bermúdez, D. Gómez, P. Salgado
   Mathematical Models and Numerical Simulation in Electromagnetism
   2014, XVIII+430pp, ISBN 978-3-319-02948-1

75. A. Quarteroni
   Matematica Numerica. Esercizi, Laboratori e Progetti (2a Ed.)
   2013, XVIII+406pp, ISBN 978-88-470-5540-7

76. E. Salinelli, F. Tomarelli
   Discrete Dynamical Models
   2014, XVI+386pp, ISBN 978-3-319-02290-1

The online version of the books published in this series is available at
SpringerLink.
For further information, please visit the following link:
http://www.springer.com/series/5418